VOLUME TWO HUNDRED AND EIGHT

PROGRESS IN
MOLECULAR BIOLOGY AND TRANSLATIONAL SCIENCE

CRISPR-Cas-Based Genome Editing for Treating Human Diseases-Part A

VOLUME TWO HUNDRED AND EIGHT

PROGRESS IN MOLECULAR BIOLOGY AND TRANSLATIONAL SCIENCE

CRISPR-Cas-Based Genome Editing for Treating Human Diseases-Part A

Edited by

VIJAI SINGH
Department of Biosciences, School of Science,
Indrashil University, Rajpur, Mehsana, India

ELSEVIER

ACADEMIC PRESS
An imprint of Elsevier

Academic Press is an imprint of Elsevier
125 London Wall, London, EC2Y 5AS, United Kingdom
50 Hampshire Street, 5th Floor, Cambridge, MA 02139, United States
525 B Street, Suite 1650, San Diego, CA 92101, United States

First edition 2024

ISBN: 978-0-443-31588-6
ISSN: 1877-1173

For information on all Academic Press publications
visit our website at https://www.elsevier.com/books-and-journals

Publisher: Zoe Kruze
Acquisitions Editor: Leticia M. Lima
Editorial Project Manager: Sneha Apar
Production Project Manager: James Selvam
Cover Designer: Gopalakrishnan Venkatraman

Typeset by MPS Limited, India

Contents

Contributors

Farah Ahmady
Fiona Elsey Cancer Research Institute; Federation University, VIC, Australia

Namra Ali
Department of Microbiology, Gargi College, University of Delhi, New Delhi, India

Rabia Amir
Atta-ur-Rahman School of Applied Biosciences, National University of Sciences and Technology, Islamabad, Pakistan

Monisa Anwer
Department of Biotechnology, Faculty of Engineering and Technology Rama University, Mandhana, Kanpur, Uttar Pradesh, India

Mustafeez Mujtaba Babar
Shifa College of Pharmaceutical Sciences, Shifa Tameer-e-Millat University, Islamabad, Pakistan; Advanced Drug Delivery and Regenerative Biomaterials Lab, Stanford University School of Medicine, Stanford University, Palo Alto, CA, United States

Abhisheka Bansal
School of Life Sciences, Jawaharlal Nehru University, New Delhi, India

Himashree Choudhury
School of Life Sciences, Jawaharlal Nehru University, New Delhi, India

Phumuzile Dube
Synthetic Nanobiotechnology and Biomachines, Synthetic Biology and Precision Medicine Centre, Future production Chemicals Cluster, Council for Scientific and Industrial Research, Pretoria, South Africa

Hunaiza Fatima
Shifa College of Pharmaceutical Sciences, Shifa Tameer-e-Millat University; Atta-ur-Rahman School of Applied Biosciences, National University of Sciences and Technology, Islamabad, Pakistan

Rania Fatima
Shifa College of Pharmaceutical Sciences, Shifa Tameer-e-Millat University, Islamabad, Pakistan

Utkarsh Gangwar
School of Life Sciences, Jawaharlal Nehru University, New Delhi, India

Alvina Gul
Atta-ur-Rahman School of Applied Biosciences, National University of Sciences and Technology, Islamabad, Pakistan

Huzaifa Ibrahim
Shifa College of Pharmaceutical Sciences, Shifa Tameer-e-Millat University, Islamabad, Pakistan

Mahnoor Ilyas
Shifa College of Pharmaceutical Sciences, Shifa Tameer-e-Millat University; Atta-ur-Rahman School of Applied Biosciences, National University of Sciences and Technology, Islamabad, Pakistan

Juveriya Israr
Institute of Biosciences and Technology, Shri Ramswaroop Memorial University, Lucknow, Barabanki, Uttar Pradesh, India

Aparna Jayachandran
Fiona Elsey Cancer Research Institute; Federation University, VIC, Australia

George Kannourakis
Fiona Elsey Cancer Research Institute; Federation University, VIC, Australia

Naina Kataria
Department of Biochemistry, Sri Venkateswara College, University of Delhi, New Delhi, India

Vanelle Larissa Kenmogne
Synthetic Nanobiotechnology and Biomachines, Synthetic Biology and Precision Medicine Centre, Future production Chemicals Cluster, Council for Scientific and Industrial Research, Pretoria; Department of Surgery, University of the Witwatersrand, Johannesburg, South Africa

Ajay Kumar
Department of Biotechnology, Faculty of Engineering and Technology, Rama University, Mandhana, Kanpur, Uttar Pradesh, India

Haiyan Li
RMIT University, VIC, Australia

Rodney B. Luwor
Fiona Elsey Cancer Research Institute; Federation University; The University of Melbourne, Parkville, VIC, Australia; Huagene Institute, Kecheng Science and Technology Park, Pukou, Nanjing, P.R. China

Indra Mani
Department of Microbiology, Gargi College, University of Delhi, New Delhi, India

Anshu Mathuria
Department of Biochemistry, Sri Venkateswara College, University of Delhi, New Delhi, India

Bernice Monchusi
Synthetic Nanobiotechnology and Biomachines, Synthetic Biology and Precision Medicine Centre, Future production Chemicals Cluster, Council for Scientific and Industrial Research, Pretoria, South Africa

Karan Murjani
Department of Biosciences, School of Science, Indrashil University, Rajpur, Mehsana, Gujarat, India

Arpita Poddar
Fiona Elsey Cancer Research Institute; Federation University; RMIT University, VIC, Australia

Shakil Ahmed Polash
RMIT University, VIC, Australia

Prashanth Prithviraj
Fiona Elsey Cancer Research Institute; Federation University, VIC, Australia

Hajra Ali Raja
Shifa College of Pharmaceutical Sciences, Shifa Tameer-e-Millat University; Health Services Academy, Ministry of Health, Islamabad, Pakistan

Jayakumar Rajadas
Advanced Drug Delivery and Regenerative Biomaterials Lab, Stanford University School of Medicine, Stanford University, Palo Alto, CA, United States

Suresh Ramakrishna
Hanyang University, Seoul, South Korea

Qasim Shah
Shifa College of Pharmaceutical Sciences, Shifa Tameer-e-Millat University, Islamabad, Pakistan

Risha Shameem
School of Life Sciences, Jawaharlal Nehru University, New Delhi, India

Ravi Shukla
RMIT University, VIC, Australia

Avinash Singh
Department of Biotechnology, Axis Institute of Higher Education, Kanpur, Uttar Pradesh, India

Vijai Singh
Department of Biosciences, School of Science, Indrashil University, Rajpur, Mehsana, Gujarat, India

Yashi Singh
Department of Biosciences & Biomedical Engineering, Indian Institute of Technology, Indore, India

Dharmisha Solanki
Department of Biosciences, School of Science, Indrashil University, Rajpur, Mehsana, Gujarat, India

Mutsa Monica Takundwa
Synthetic Nanobiotechnology and Biomachines, Synthetic Biology and Precision Medicine Centre, Future production Chemicals Cluster, Council for Scientific and Industrial Research, Pretoria, South Africa

Deepak Balaji Thimiri Govinda Raj
Synthetic Nanobiotechnology and Biomachines, Synthetic Biology and Precision Medicine Centre, Future production Chemicals Cluster, Council for Scientific and Industrial Research, Pretoria, South Africa

Renu Tripathi
Department of Biosciences, School of Science, Indrashil University, Rajpur, Mehsana, Gujarat, India

Chaitali Vora
Department of Biosciences and Biomedical Engineering, Indian Institute of Technology, Indore, India

Preface

CRISPR (Clustered Regularly Interspaced Short Palindromic Repeats) and Cas (CRISPR-associated) proteins are form defense system present in bacteria that can prevent infection from bacteriophage or invading plasmid. CRISPR-Cas systems are quite simple, rapid, sensitive and cost effective tool for genome editing of wide range of organisms. The global CRISPR-Cas technology market size was USD 2.57 billion in 2022 and expected to grow with compound annual growth rate (CAGR) of 17.15% from 2023 to 2030. The expression of single guide RNA (sgRNA) and Cas9 within the organism to forms sgRNA-Cas9 complex that binds on the desired DNA sequences in the presence of its PAM sequences. It creates a double-strand break (DSB) that is being repaired either by a homology-directed repair (HDR) pathway which follow gene correction/insertion via homologous recombination or by a non-homologous end-joining (NHEJ) pathway by creating gene deletion or addition.

If two mutations (RuvC (D10A) and HNH (H840A)) were made in Cas9 in its active sites then it lost their catalytic activity but it retains binding ability. This Cas9 is known as dead Cas9 or dCas9 that is being used as CRISPR interference (CRISPRi) for repurposing applications. CRISPRi is used for transcriptional interference by binding on promoter sequence or silencing of transcriptional elongation, used for epigenetic modification, gene activation by fusing dCas9 with transcriptional activators, dCas9 fused with fluorescent protein for genomic loci imaging and many. CRISPR-Cas13 was used for detection of RNA or DNA sequences for development of diagnostic assay. Currently, CRISPR-Cas systems have shown tremendous potential for model and non-model organisms including treating infectious and non-infectious human diseases.

This volume offers different topics from basic to advanced in CRISPR-Cas technology for genome editing of bacteria, manipulation of gut microbiome, removal of human virus and protozoa. It also covers advances in CRISPR-Cas systems for cell and gene therapy, epigenetic modification for gene regulation, genome editing for autoimmune disease, and cancer. This volume offers several aspects of the CRISPR-Cas sytems that can help the basic understanding of students, researchers, clinicians, entrepreneurs, and stakeholders to perform their research with great interest.

Vijai Singh

CHAPTER ONE

An overview and potential of CRISPR-Cas systems for genome editing

Karan Murjani, Renu Tripathi, and Vijai Singh*
Department of Biosciences, School of Science, Indrashil University, Rajpur, Mehsana, Gujarat, India
*Corresponding author. e-mail address: vijaisingh15@gmail.com; vijai.singh@indrashiluniversity.edu.in

Contents

Abstract

Genome editing involves altering of the DNA in organisms including bacteria, plants, and animals using molecular scissors that helps in treatment and diagnosis of various

Progress in Molecular Biology and Translational Science, Volume 208
ISSN 1877-1173, https://doi.org/10.1016/bs.pmbts.2024.07.009

diseases. Genome editing technology is exponentially growing and have been developed for enabling precise genomic alterations and the addition, removal, and correction of genes. These modifications begin with the creation of double-stranded breaks (DSBs) that is generated by nucleases and can be joined through homology-directed repair (HDR) or non-homologous end-joining (NHEJ). NHEJ is quick but increases mutation chances due to deletions and insertions of nucleotides at the break site, while HDR uses homologous templates for precise repair and targeted DNA specific to the gene or sequence. Other methods such as zinc-finger protein is a transcription factor that binds with DNA and binds specific to that sequence, which uniquely recognise 3-base pairs of DNA. TALENs consists of two domains: TALE domain, a transcription activator and FokI that is a restriction endonuclease that cuts the DNA at specific sites. CRISPR-Cas systems are clustered regularly interspersed short palindromic repeats present in various bacterial species. These sequences activate RNA-guided DNA cleavage, aiding in the development of an adaptive immune defence against foreign DNA. CRISPR-Cas9 is widely used for genome editing, regulation, diagnostic and many.

Abbreviations

Cas CRISPR-associated protein
CRISPRi CRISPR interference
CRISPRs Clustered regularly interspersed short palindromic repeats
DSBs Double-strand breaks
HDR Homology-directed repair
NHEJ Non-homologous end-joining
TALENs Transcription activator-like effector nucleases
tracrRNA Trans-activating CRISPR RNA
ZFNs Zinc finger nucleases

1. Introduction

Genome editing is a revolutionary technology for editing of the DNA of organisms like bacteria, plants, and animals. Editing of DNA enables researchers to change the trait of an organism, like skin colour, eye colour, and the diseased trait which can help in treatment and diagnostic of various diseases. To achieve these, researchers use various methods, and these methods function as scissors, molecular scissors (as like scissors cut the paper) cuts the DNA at specific region. With the help of these molecular scissors researchers can add or delete any region of the DNA.

There is one common way researchers use genome editing that is to diagnose the various diseases that affects the large number of humans. And to do this, they need a model organism, in which they can edit the genome of these model organisms, as many of the human genes are of same as *Drosophila*. *Drosophila* contains genes that are 75% similar to humans disease. By targeting a single or multiple genes in *Drosophila*, the researchers can predict how targeting these genes can affect the human health.

From last few years, genome editing methods has developed well, this helps to understand the importance of single gene in diseased organism. Genome editing may be performed *in vitro* or *in vivo* by administering the editing system directly at the site, effectively enabling the addition, removal, and correction of genes, along with other precise genomic alterations.[1,2] Targeted DNA modifications start with the creation of double-stranded breaks (DSBs) which is induced by nucleases, and then at the process of repair mechanism this can lead to the insertions or deletions. Moreover, at the time of repair mechanism some exogenous DNA also integrates, researchers can use this mechanism to insert or delete the genes. One of the two main repair mechanism for nuclease-induced DNA DSBs which exists in almost all cell types and organisms are, non-homologous end-joining (NHEJ) and homology-directed repair (HDR). These processes lead to either targeted gene integration or gene disruptions, respectively.[3] NHEJ directly repairs the DSBs without need of homologous template. It is quick repair system that ligates the DSBs, but the chances of mutations are also more because of deletions and insertions at break site. HDR uses homologous template to repair the DSBs and are more accurate than NHEJ. HDR results in precise repair and are used to target the DNA specific to the gene or sequence. There are various methods to create DSBs with help of nucleases and some of the methods are described in this chapter.

2. Zinc finger nucleases (ZFNs)

The zinc finger protein was first identified in 1985 as a component of transcription factor IIIa in *Xenopus oocytes*. It binds with DNA and has a site-specific binding capacity.[4] The zinc-finger domain's functional specificity is comprised of multiple Cys2His2 zinc fingers (ZFs) that are generated by their zinc-finger domains interacting with homologous DNA sequences in a highly conserved manner.[3] Each zinc fingers (ZFs) uniquely recognises a 3-base pair of DNA, along with ZFs it consists of *Fok*I type II restriction endonuclease. The recognition of the target sequence and how specific ZFNs binds which depends on three factors:

(i) the sequence of an amino acid in each ZFs, (ii) the number of ZFs and (iii) interaction between nuclease domain.[3] To cleave the DNA, *Fok*I must be dimerised to achieve this two ZFs are required to bind with target site in specific position.[5] After cleavage is done by ZFNs, the repair mechanism of the cell starts and it can correct these DSBs by the NHEJ or HDR repair system.

3. Transcription activator-like effector nucleases (TALENs)

Another type of genome editing method in which DSBs can be achieved by nucleases are TALENs which has higher efficiency than ZFNs. TALENs consists of two domains, first is TALE domain which is a transcription activator like effector protein derived from *Xanthomonas* bacteria. Furthermore, DNA is cleaved at a specific site by the restriction endonuclease *Fok*I. TALEs are naturally occurring proteins that consist of a series of repetitive DNA-binding motifs, typically 10–30 repeats, which bind to specific sites on the DNA.[6] The repeat-variable di-residue (RVD) pair, which is conferred by two neighbouring amino acids on one of the four DNA base pairs, determines the length of each repetition, which ranges from 33 to 35 amino acids.[7] Thus, there is a direct connection between each base pair and tandem repeats. The DNA binding sites are designated by RVDs, where ND binds to C nucleotides, HN to A or G nucleotides, NH to G nucleotides, and NP to all nucleotides.[8] In order to break DNA at a particular site, the FokI is dimerised. After cleavage, the DNA is repaired by NHEJ or HDR. Thus, like zinc fingers, TALEN modules are made in pairs and have the purpose of bind DNA target site that are opposite to each other, with a suitable distance of 12–30 bp between the two binding sites.[9] On the other hand, unlike zinc finger proteins, which target specific site of DNA, the linkage between repeats that make up long arrays of TALEs does not need to be redesigned.[3] One other benefit of TALENs over ZFNs is that site selection appears to be less restricted; theoretically, every base pair of a random DNA sequence can have many potential TALEN pairings.[10] A major disadvantage of TALEN is that their size is quite large as compared to ZFN. The size of the cDNA that encodes TALEN is around 3 kb and ZFN is around 1 kb, thus it is more difficult to express the TALENs.[7]

4. CRISPR-Cas systems

Clustered regularly interspersed short palindromic repeats (CRISPRs), were first identified in *E. coli* in early 1987 and later they were also found in several other bacterial species.[11] The purpose of short repeat sequences was unknown for long time. Following that, numerous studies described that these sequences are like phage sequences. However, later experiments showed that these sequences activated RNA-guided DNA cleavage, which helps bacteria and archaea to develop an adaptive immune defence against foreign DNA.[12–14] CRISPR-Cas systems often fall into two classes based on the structural variety of the Cas genes and their organisation.[12] Six CRISPR-Cas types and at least 29 subtypes have been described so far, and the field is growing continuously. Furthermore, Class 1 CRISPR-Cas systems to function a complex of several proteins must cooperate with one another is called a multiprotein effector complexes, while Class 2 CRISPR-Cas systems are made up of a single protein that can carry out all required tasks on its own called a single effector protein.[15,16] The frequently utilised CRISPR system subtype is type II CRISPR-Cas9. It targets particular DNA sequences using a single Cas protein from *Streptococcus pyogenes* (SpCas9), making it useful tool for genome editing.[17] A single-stranded guide RNA (sgRNA) and Cas9 endonuclease are the two main components of CRISPR-Cas systems. The DNA target sites are particularly complementary to the unique 20 base pair (bp) sequence that made up of sgRNA. This sequence must be succeeded by a short upstream DNA segment, termed the "protospacer adjacent motif" (PAM), which is necessary for the Cas9 compatibility. Usually, the PAM sequence is usually "NGG" or "NAG," where N codes for any nucleotide.[18,19] After CRISPR-Cas9 activity, NEHJ or HDR modifies the sequences by insertion or deletion.

By modifying the 20-bp guide RNA protospacer, which is achieved by subcloning this nucleotide sequence into the gRNA plasmid backbone, CRISPR-Cas9 can be readily customised for use with any genomic sequence. Compared to ZFNs and TALENs, which necessitate the recoding of proteins using long DNA segments (500–1500 bp), respectively, targeting new target sites is not as challenging.[7] The protein component of Cas9 is unmodified. When it comes to produce a wide range of vectors to target several places, CRISPR-Cas9 is more user-friendly than ZFNs and TALENs.[20] As depicted in Fig. 1 give the idea about how these three different genome editing techniques work.

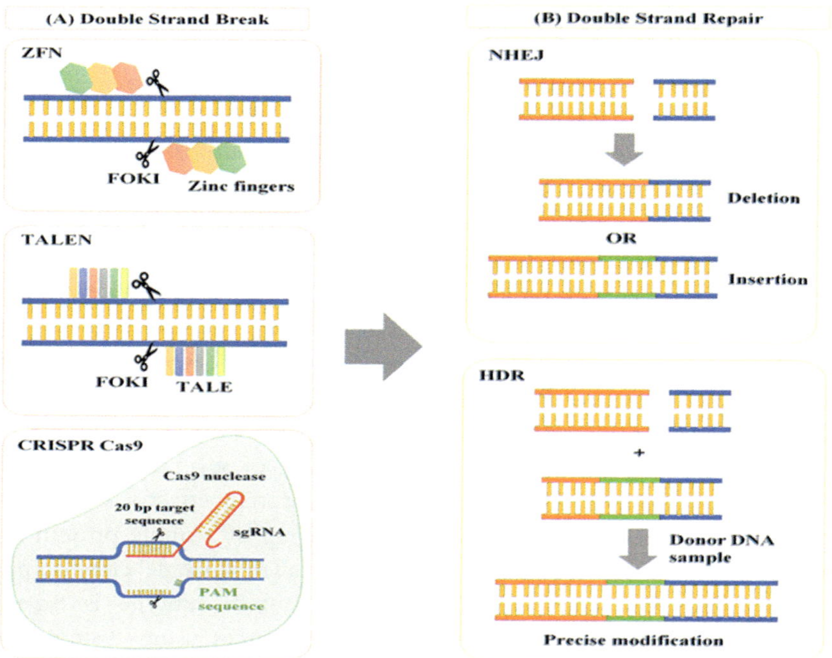

Fig. 1 (A) Shows the different methods such as ZFNs, TALENs and CRISPR-Cas9 by which DSB can be generated (B) Shows the repair mechanism using NHEJ and HDR.

5. Evolution of CRISPR-Cas system

CRISPR, was first identified in the DNA sequences of *Escherichia coli* and subsequently described by Ishino et al. in 1987 from Osaka University (Japan).[11] Initial study identified repeated genomic structures seen in bacteria and archaea, which led the way for the creation of the powerful CRISPR-Cas9 system for genome editing. At the early stages the biological function was still not known, therefore scientists have proposed a various ways to study this system. They have decided to genotype different strains of bacteria, first on *Mycobacterium tuberculosis* and followed by *Streptococcus pyogenes*.[21] A few years later, the genome of the halophilic Archaea *Haloferax mediterranei* revealed identical, repeating structures, consisting of 14 nearly completely preserved sequences of 30 bp, repeated at regular intervals.[22] The discovery of CRISPR-associated (Cas) genes, a class of genes exclusive to prokaryotes carrying CRISPR and always located next to CRISPRs, this marked a significant advancement in comprehending the role of CRISPR. Prokaryotes carry these specific sequences which seemed to be immune to infection. According to these correlated study, spacer sequences serve as a "memory of past genetic aggressions" that plays an important role in prokaryotic defence against foreign DNA invasion.[14] Protospacer sequences are the sequences that were found only a few nucleotides apart from short sequence motifs.[13,23] Subsequently, these designs were later identified as protospacer adjacent motifs (PAMs).[24] The human pathogen *Streptococcus pyogenes* and the Class II, Type-II CRISPR-Cas system in *Streptococcus thermophilus* provided the crucial data that identified CRISPR-Cas9 as a tool for genome editing.[25] This system has four Cas genes: three (cas1, cas2, and csn2) are crucial for spacer acquisition, and the fourth (cas9; formerly known as cas5 and csn1) is necessary for interference.[26] In order to further identify the elements required for immunity, the *S. thermophilus* CRISPR-Cas system was introduced into *Escherichia coli* to give heterologous protection against phage infection and horizontal transfer of plasmid. This experimental model was used to determine which system components needed to be inactivated to provide protection. The investigation clearly demonstrated that the two protein nuclease domains, HNH and RuvC, were both required for this outcome. Cas9 protein by itself was sufficient for the CRISPR-encoded interference step.[27] Charpentier and colleagues found an area 210 bp upstream that was home to an abundant RNA species and an active CRISPR locus in *S. pyogenes*, as evidenced by the development of pre-crRNA and mature crRNA molecules.[28] The transcript is known as trans-activating CRISPR RNA (tracrRNA), included a 25-nucleotide length that nearly complementary to the repeat sections of the CRISPR locus (1-nt mismatch), indicating base pairing with pre-crRNA.[28] Charpentier and colleagues suggested that the Cas9 protein serves as a molecular anchor to facilitate base pairing between pre-crRNA and tracrRNA, hence facilitating the recognition and cleavage of the resultant molecule by the host RNase III protein.[28] According to research done by, Charpentier and Doudna reported that any double-stranded DNA sequence can be cleaved by the Cas9 endonuclease when it is programmed with guide RNA produced as a single transcript.[12] Due to the discovery of these two noble laureates, the CRISPR system is now widely used as potent and adaptable tool for genome editing.

6. Mechanism of CRISPR-Cas systems

When a bacteriophage attacks on its bacteria, that is it never faced a bacteriophage in entire life cycle, is not aware of any CRISPR defence mechanism. Therefore, when a bacteriophage infects the bacteria, the Cas proteins come into the picture. The Cas proteins carry the DNA of phage and insert into the CRISPR loci. CRISPR loci are transcribed into crRNA (CRISPR RNA) which then forms essential proteins. This protein recognises and cuts the foreign DNA which is complementary to that sequence. Three different stages are involved in the fundamental mechanism of CRISPR: adaptation (or acquisition), expression and maturation (or biogenesis) and interference (or targeting) (Fig. 2).[29]

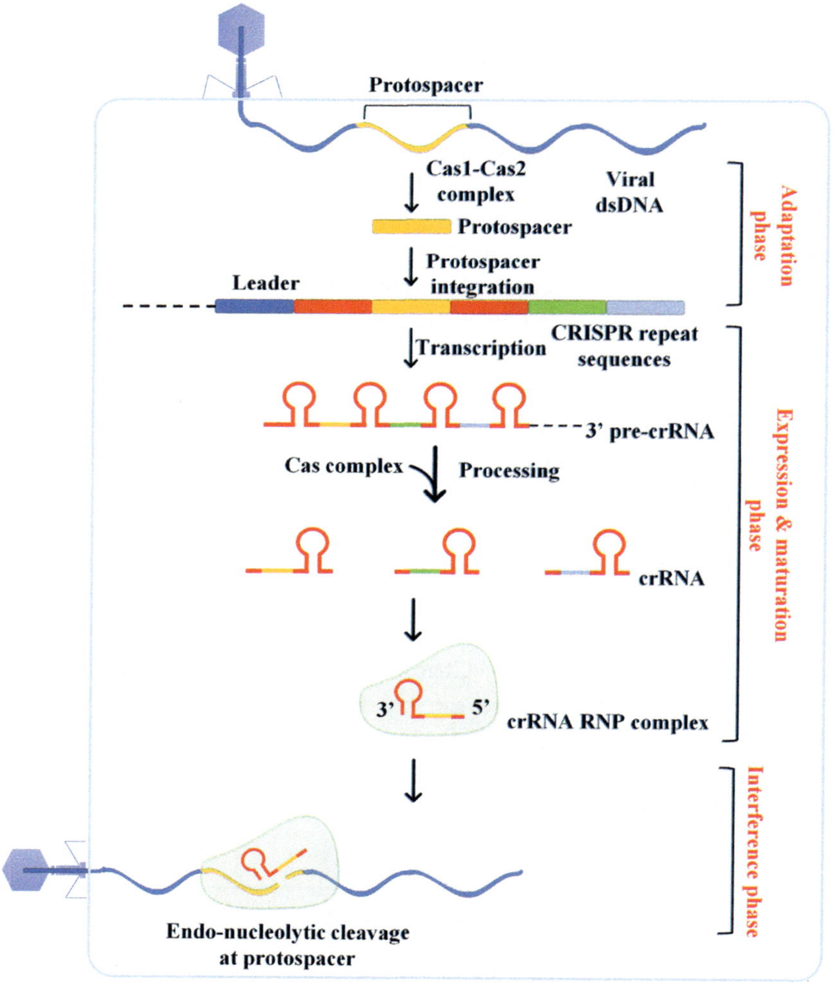

Fig. 2 Schematic representation of different phases of CRISPR mechanism.

Although the synthesis of crRNA and targeting vary across different classes of CRISPR but the adaption process is nearly identical for all of them.[30]

6.1 Adaptation phase

The first step in the process of developing a defence mechanism against any foreign DNA or viral DNA or plasmid, is adaptation phase. This step in the CRISPR locus entails taking out a piece of invading DNA called a "protospacer" and putting it in between two nearby repeats. This can be done with the help of Cas1-Cas2 complex, which is made up of two Cas1 dimers and one Cas2 dimer.[29] A memory of the invasive genetic elements is created when the invasive DNAs are directionally integrated as new CRISPR spacers into a CRISPR array that is divided by repetitive sequences.

6.2 Expression and maturation phase

As the protospacer sequences are introduced into the CRISPR loci, CRISPR transcription starts in the expression and maturation phase, forming the pre crRNA. After that, this pre-crRNA matures into crRNA. Spacer sequences are present in every crRNA and are connected to partial repeat sequences. These crRNAs combine to create a complex known as ribonucleoprotein (RNP) with the aid of Cas proteins.[29]

6.3 Interference phase

In the interference phase, the RNPs recognises the invaded foreign DNA and binds with its complementary sequence. Once the RNPs binds with its complementary sequence with the help of crRNA, the cleavage of foreign DNA occurs.[29] Thus, the bacteria are protected by CRISPR mechanism from viral infection.

7. Types of CRISPR-Cas systems

The CRISPR-Cas systems are mainly divided into two main classes, class 1, and class 2. Based on the protein subunits, these two classifications are further divided into numerous types. Class 1 is divided on the basis of multiprotein subunits: type I, type III, and type IV. Class 2 is divided on the basis of single protein subunit: type II, type V, and type VI (Fig. 3).

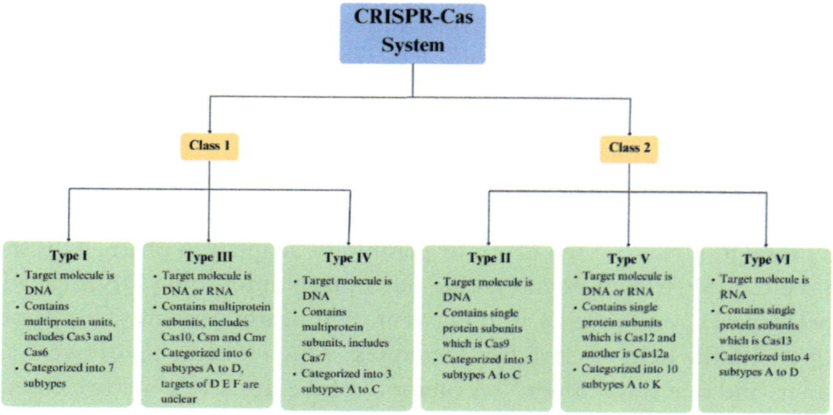

Fig. 3 Schematic representation of different types of CRISPR Cas systems.

7.1 Type I CRISPR-Cas systems

The majority of *Cas* genes are encoded by one or more operons and are found in type I systems. Six proteins are present in them, consists of the primary enzyme in the interference phase, Cas3, which possesses both helicase and nuclease functions.[31] In the type I system, Cas endoribonuclease, Cas6, is responsible for pre-cRNA cleavage. A ribonucleoprotein complex is formed by mature cRNA and cascade. At the time of interference stage which involves the recruitment of Cas nuclease Cas3 to cleave the invasive target DNA.[32]

7.2 Type II CRISPR-Cas systems

A distinguishing component, such as the Cas9 protein, which comes in three subtypes II-A, II-B, and II-C defines type II CRISPR-Cas systems. This system consists of Cas9 protein, crRNA, trans-activating crRNA (tracrRNA). Nowadays, researchers use the Cas9 protein which consists of the recognition lobe (REC) and the nuclease lobe (NUC) and has 1368 amino acids. Two highly conserved nuclease domains, one called HNH and the other RuvC, are present in the NUC domain.[31] Furthermore, the single strand that the former cleaves is complementary to the crRNA sequence, whereas the single strand that the later cleaves is non-complementary to the protospacer sequence. The two of them act simultaneously to create a blunt end at a certain point in the target sequence.[33] The PAM region of Cas9, whose sequence is 5′-NGG-3′, is located at the 3′ end of the target sequence. A hairpin RNA called tracrRNA is produced by transcription of a repeat sequence. The tracrRNA is the one that leads to the activation of RNase III, which promotes the maturation of pre-crRNA. tracrRNA, precursor crRNA (pre-crRNA), and Cas9 protein subsequently come together to form a complex.[28] The cleavage process is initiated by the combination of mature crRNA, tracrRNA, and cas9. The sgRNA, a fusion of crRNA and tracrRNA that facilitates the efficient identification of certain sequences that regulates the activity of Cas9, making the genome editing process simple.[12]

7.3 Type III CRISPR-Cas systems

It consists of Cas10 protein with RNase activity and a cascade makes up the type III system does not dependent on PAM sequences. The function of cascade indicates the type I system. The Cas10 protein is crucial for the development of crRNA and the breakdown of foreign DNA that invades the cells.[31] There are four subtypes of type III systems, designated A–D. The mRNA is the interference target of type III-A, whereas DNA is the interference target of type III-B, which is identical to that of type I and II CRISPR-Cas systems. However, it is unclear what C and D are trying to interfere with them.[31]

7.4 Type IV CRISPR-Cas systems

The type IV CRISPR-Cas systems are distinguished by their simple structures and are mostly found on plasmids.[34,35] They usually lack the adaptation genes found in other CRISPR variants, but they do encode a unique *cas7*-like gene (*csf2*). A full understanding of their mechanics and biological activities are complicated due to their these unique features.[36] Three subtypes of type IV systems are recognised as IV-A, IV-B, and IV-C. A new study findings suggests that, type IV CRISPR systems target and hamper mobile genomic elements by using multi-subunit ribonucleoprotein complexes.[37] For example, it has been shown that the *Pseudomonas oleovorans* type IV-A system targets double-stranded DNA in a PAM-dependent way, acting similarly to CRISPR interference (CRISPRi) but without cleaving the DNA.[37]

7.5 Type V CRISPR-Cas systems

The Cas12 effector protein is the primary characteristic of type V CRISPR systems. Based on the particular Cas12 proteins that these systems encode, they are divided into a number of subtypes, such as V-A, V-B, V-U, and others.[38] The RuvC-like nuclease domain, which is found in the Cas12 proteins and it is responsible of their endonuclease activity and it can be distinguished feature of type V systems.[39] Cas12 proteins cleaves double-stranded DNA in a staggered pattern that is advantageous for genome editing techniques involving DNA repair. It also has its own gRNA, which leads to an improved multiplexing capacity.

7.6 Type VI CRISPR-Cas systems

The primary characteristic of type VI CRISPR systems is their dependence on the Cas13 effector protein, which is essential to their ability to target RNA. With just one Cas13 protein and one CRISPR RNA (crRNA) molecule that is required for action, these systems have a somewhat simple structure.[40] The four subtypes of type VI CRISPR-Cas systems that have been identified are VI-A (Cas13a), VI-B (Cas13b), VI-C (Cas13c), and VI-D (Cas13d). These subtypes vary in size and sequence, but they all have two HEPN (higher eukaryotes and prokaryotes nucleotide-binding) domains that are involved in their nucleolytic activity.[40] A long pre-crRNA is first transcribed into mature crRNAs by the Cas13 protein, which is the first step in the Type VI system's mechanism. As the crRNA-Cas13 complex searches for complementary single-stranded RNA (ssRNA) sequences, the protospacer flanking sequence (PFS) causes conformational modifications in Cas13, activates the HEPN domains which results in nonspecific RNA cleavage.[41]

8. Applications of CRISPR-Cas systems

CRISPR systems have developed as vital research genome editing technology that have made significant advances possible in a wide range of scientific domains. Following are the some application of CRISPR-Cas systems discussed here.

8.1 CRISPR-Cas based genome editing

CRISPR-Cas systems can precisely edit the genome of bacteria, plants, and animals. Understanding the roles that genes play in biological processes and the mechanisms behind the disease are essential. To learn more about the roles and interactions of particular genes, researchers may modify or knockout these genes. CRISPR-Cas systems can be used to modify the genome of model organisms like *Drosophila*, Zebra fish, and mice that contains genes that are similar to humans. Thus, it is used for treatment and diagnosis of various diseases by editing the genes of the organism. A conditional mutagenesis method mediated by CRISPR-Cas9 has been created. To effectively and precisely inhibit gene expression in a controlled manner, they have linked the expression of Cas9, which is tissue-specific and driven by a Gal4/upstream activating site (UAS), with a variety of widely expressed sgRNAs.[42] This study demonstrates how several sgRNAs can be used to target a single gene, leading to the high degree of mutations that are subsequently created in the targeted tissue.[43]

8.2 CRISPR-Cas assisted metabolic pathways engineering

CRISPR-Cas systems are essential tool for genome engineering. Researchers can improve the synthesis of medicines, biofuels, and other important substances by altering the metabolic pathways in micro-organism.[44] This is the sustainable way of developing medicines and biofuels because there will be less usage of conventional resources. Zhan et al. used CRISPR method to increase the production of lycopene in *Corynebacterium glutamicum* by overexpressing the genes and knockout some genes.[45]

8.3 CRISPR-Cas systems in gene therapy

Diseases such as cystic fibrosis, duchenne muscular dystrophy, huntington's disease, and haemophilia are caused due to mutation in their corresponding genes. To correct these mutations CRISPR-Cas systems have been used to correct the mutation for restoring their gene function. Mutation in the factor IX gene, F9, results in an inadequate amount of factor IX protein, which causes haemophilia B, a blood clotting X-linked genetic disorder. Gene therapy may also be used for treatment of haemophilia.[43]

8.4 CRISPR-Cas systems for removal human viruses

The ability of CRISPR-Cas systems to specifically target and inactivate viral genomes, including those of hepatitis B and HIV, are explored. This method offers an innovative approach to antiviral treatments by attempting to eradicate the virus from afflicted cells.[46] HBsAg of HBV, which is an envelope protein of HBV has been targeted by CRISPR-Cas9 in cell culture and *in vivo* systems. This has been verified by quantitative polymerase chain reaction (qPCR) and quantitative enzyme-linked immunosorbent assay (ELISA). Following CRISPR-Cas9 system, the total amount of HBsAg secreted into the cell culture and in the mouse, serum was decreased.[43]

8.5 CRISPR-Cas systems in creating resistance crops

Crop characteristics like drought resistance, pest resistance, and nutritional value are improved by the application of CRISPR. For instance, changes are made to wheat and rice to increase the production and stress tolerance. By precisely editing the genome to introduce gain-of-function mutations or remove negative genetic factors that cause undesirable phenotypes, CRISPR–Cas technology offers a quick and efficient alternative to traditional breeding methods for creating optimum germplasms.[47]

8.6 CRISPR-Cas systems in disease diagnostics

Due to the growing pandemics, accurate and time-efficient diagnostic methods are required. Nucleic acid detection is an accurate and targeted diagnostic technique, but it is expensive and requires expertise. Using CRISPR is an interesting new advancement in molecular sensing. Within a few years, CRISPR-based diagnostics have evolved from an experimental tool to a clinically useful diagnostic technology that is portable, sensitive, quick, cheap, and fast. NASBACC (Nucleic Acid Sequence-Based Amplification-CRISPR Cleavage), one of the earliest CRISPR-based diagnostic techniques, uses Cas9 to identify double-stranded DNA that is used to identify the Zika virus at low femtomolar concentrations.[48] By using Cas13 variations, SHERLOCK (Specific High-sensitivity Enzymatic Reporter un-LOCKing) and the upgraded SHERLOCKv2 detects the single DNA or RNA molecules, such as ssRNA from infections like dengue, the Zika virus, and pneumonia.[49,50] The upgraded SHERLOCKv2 additionally enables multiplexed nucleic acid detection at the zeptomolar range (10^{-21} M), whereby gold nanoparticles coupled with antibodies are used to detect cleaved reporter molecules on a paper strip. To create DETECTR (DNA Endonuclease-Targeted CRISPR Trans Reporter), Chen et al. also employed the Cas12a variant to achieve very accurate HPV16–HPV18 differentiation.[51] Most recently, CRISPR-Cas13a has been developed by integration of recombinase polymerase amplification (RPA) coupled with a lateral flow assay for targeting *Klebsiella pneumoniae* housekeeping *rpoB* gene (for species-specific identification) and *rmpA* capsular polysaccharide hypervirulent strains.[52] Thus, CRISPR-Cas is a promising tool for disease diagnostic and it helps in future treatment of non-infectious and infectious diseases.

9. Conclusion and future challenges

Genome editing has been transformed by CRISPR-Cas technology, which has led to important breakthroughs in many areas, most notably in health. However, to be more widely used, it must overcome several challenges. An extensive transfection approach for CRISPR-Cas system must prioritise high transfection efficiency, remarkable capacity, and ease of mass manufacturing. Nevertheless, the approaches used now are still far short of achieving these objectives. One of the drawbacks of the CRISPR-Cas technology is the off-target effect, which is seen as a serious concern *in vivo* gene therapy. While the design of sgRNA has been optimised by several computer tools although its specificity cannot be completely assured. For example, CRISPR-Cas can be used to treat cancer, but in case of humans, the chances of mutations are more as compared to other model organisms, which leads to the off-target effect. And in case of cancer, the number of mutations are more as compared to the other disease. So, if one wants to treat cancer using CRISPR system, it will be very difficult because the number of targets will more, and there are greater chance to mutations in the cell thus this may become more laborious work for any researcher. Thus, more effort is required for developing a CRISPR-Cas technology to be more precise and accurate. It is hoped that the researchers would find a solution for off-targeting and delivery mechanism, so that it becomes easy to use and it might be used for treating more and more harmful diseases.

Acknowledgement

K.M. acknowledges financial support from Indrashil University, Rajpur, Mehsana for carry out research work. Authors acknowledge infrastructure and facility of Indrashil University to carry out this work.

References

1. Ghosh D, Venkataramani P, Nandi S, Bhattacharjee S. CRISPR–Cas9 a boon or bane: the bumpy road ahead to cancer therapeutics. *Cancer Cell Int.* 2019;19:1–10.
2. Gaj T, Gersbach CA, Barbas CF. ZFN, TALEN, and CRISPR/Cas-based methods for genome engineering. *Trends Biotechnol.* 2013;31:397–405.
3. Li H, Yang Y, Hong W, Huang M, Wu M, Zhao X. Applications of genome editing technology in the targeted therapy of human diseases: mechanisms, advances and prospects. *Signal Transduct Target Ther.* 2020;5:1.
4. Diakun G, Fairall L, Klug A. EXAFS study of the zinc-binding sites in the protein transcription factor IIIA. *Nature.* 1986;324:698–699.
5. Smith J, Bibikova M, Whitby FG, Reddy A, Chandrasegaran S, Carroll D. Requirements for double-strand cleavage by chimeric restriction enzymes with zinc finger DNA-recognition domains. *Nucleic acids Res.* 2000;28:3361–3369.
6. Bogdanove AJ, Voytas DF. TAL effectors: customizable proteins for DNA targeting. *Science.* 2011;333:1843–1846.

7. Gupta RM, Musunuru K. Expanding the genetic editing tool kit: ZFNs, TALENs, and CRISPR-Cas9. *J Clin Investig.* 2014;124:4154–4161.
8. Boch J, Scholze H, Schornack S, et al. Breaking the code of DNA binding specificity of TAL-type III effectors. *Science.* 2009;326:1509–1512.
9. Li T, Huang S, Jiang WZ, et al. TAL nucleases (TALNs): hybrid proteins composed of TAL effectors and *Fok*I DNA-cleavage domain. *Nucleic acids Res.* 2011;39:359–372.
10. Reyon D, Tsai SQ, Khayter C, Foden JA, Sander JD, Joung JK. FLASH assembly of TALENs for high-throughput genome editing. *Nat Biotechnol.* 2012;30:460–465.
11. Ishino Y, Shinagawa H, Makino K, Amemura M, Nakata A. Nucleotide sequence of the iap gene, responsible for alkaline phosphatase isozyme conversion in Escherichia coli, and identification of the gene product. *J Bacteriol.* 1987;169:5429–5433.
12. Jinek M, Chylinski K, Fonfara I, Hauer M, Doudna JA, Charpentier E. A programmable dual-RNA–guided DNA endonuclease in adaptive bacterial immunity. *Science.* 2012;337:816–821.
13. Bolotin A, Quinquis B, Sorokin A, Ehrlich SD. Clustered regularly interspaced short palindrome repeats (CRISPRs) have spacers of extrachromosomal origin. *Microbiology.* 2005;151:2551–2561.
14. Pourcel C, Salvignol G, Vergnaud G. CRISPR elements in Yersinia pestis acquire new repeats by preferential uptake of bacteriophage DNA, and provide additional tools for evolutionary studies. *Microbiology.* 2005;151:653–663.
15. Makarova KS, Wolf YI, Alkhnbashi OS, et al. An updated evolutionary classification of CRISPR–Cas systems. *Nat Rev Microbiol.* 2015;13:722–736.
16. Makarova KS, Haft DH, Barrangou R, et al. Evolution and classification of the CRISPR–Cas systems. *Nat Rev Microbiol.* 2011;9:467–477.
17. Jiang W, Bikard D, Cox D, Zhang F, Marraffini LA. RNA-guided editing of bacterial genomes using CRISPR-Cas systems. *Nat Biotechnol.* 2013;31:233–239.
18. Sternberg SH, Redding S, Jinek M, Greene EC, Doudna JA. DNA interrogation CRISPR RNA-guided endonuclease Cas9. *Biophys J.* 2014;106:695a.
19. Deveau H, Barrangou R, Garneau JE, et al. Phage response to CRISPR-encoded resistance in Streptococcus thermophilus. *J Bacteriol.* 2008;190:1390–1400.
20. Mali P, Yang L, Esvelt KM, et al. RNA-guided human genome engineering via Cas9. *Science.* 2013;339:823–826.
21. Gostimskaya I. CRISPR–cas9: a history of its discovery and ethical considerations of its use in genome editing. *Biochemistry (Mosc).* 2022;87:777–788.
22. Mojica FJ, Juez G, Rodriguez-Valera F. Transcription at different salinities of Haloferax mediterranei sequences adjacent to partially modified *Pst*I sites. *Mol Microbiol.* 1993;9:613–621.
23. Horvath P, Romero DA, Coûté-Monvoisin A-C, et al. Diversity, activity, and evolution of CRISPR loci in Streptococcus thermophilus. *J Bacteriol.* 2008;190:1401–1412.
24. Mojica FJ, Díez-Villaseñor C, García-Martínez J, Almendros C. Short motif sequences determine the targets of the prokaryotic CRISPR defence system. *Microbiology.* 2009;155:733–740.
25. Gustafsson C. *Scientifc Background on the Nobel Prize in Chemistry 2020: A Tool for Genome Editing.* The Royal Swedish Academy of Sciences; 2020.
26. Barrangou R, Fremaux C, Deveau H, et al. CRISPR provides acquired resistance against viruses in prokaryotes. *Science.* 2007;315:1709–1712.
27. Sapranauskas R, Gasiunas G, Fremaux C, Barrangou R, Horvath P, Siksnys V. The Streptococcus thermophilus CRISPR/Cas system provides immunity in Escherichia coli. *Nucleic acids Res.* 2011;39:9275–9282.
28. Deltcheva E, Chylinski K, Sharma CM, et al. CRISPR RNA maturation by trans-encoded small RNA and host factor RNase III. *Nature.* 2011;471:602–607.
29. Hille F, Charpentier E. CRISPR-Cas: biology, mechanisms and relevance. *Philos Trans R Soc B: Biol Sci.* 2016;371:20150496.

30. Kumar P. Biophysics and Molecular Biology: Fundamentals and Techniques. Pathfinder Publication; 2020.
31. Liu Z, Dong H, Cui Y, Cong L, Zhang D. Application of different types of CRISPR/Cas-based systems in bacteria. *Microb Cell Factories*. 2020;19:1–14.
32. Charpentier E, Richter H, van der Oost J, White MF. Biogenesis pathways of RNA guides in archaeal and bacterial CRISPR-Cas adaptive immunity. *FEMS Microbiol Rev*. 2015;39:428–441.
33. Jiang F, Doudna JA. CRISPR–Cas9 structures and mechanisms. *Annu Rev Biophys*. 2017;46:505–529.
34. Koonin EV, Makarova KS. Origins and evolution of CRISPR-Cas systems. *Philos Trans R Soc B*. 2019;374:20180087.
35. Pinilla-Redondo R, Mayo-Muñoz D, Russel J, et al. Type IV CRISPR–Cas systems are highly diverse and involved in competition between plasmids. *Nucleic Acids Res*. 2020;48:2000–2012.
36. Taylor HN, Laderman E, Armbrust M, et al. Positioning diverse type IV structures and functions within class 1 CRISPR-Cas systems. *Front Microbiol*. 2021;12:671522.
37. Guo X, Sanchez-Londono M, Gomes-Filho JV, et al. Characterization of the self-targeting Type IV CRISPR interference system in Pseudomonas oleovorans. *Nat Microbiol*. 2022;7:1870–1878.
38. Tong B, Dong H, Cui Y, Jiang P, Jin Z, Zhang D. The versatile type V CRISPR effectors and their application prospects. *Front Cell Dev Biol*. 2021;8:622103.
39. Beckett MQ, Ramachandran A, Bailey S. Type V CRISPR-Cas systems. *CRISPR: Biol Appl*. 2022:85–97.
40. Perčulija V, Lin J, Zhang B, Ouyang S. Functional features and current applications of the RNA-targeting type VI CRISPR-Cas systems. *Adv Sci*. 2021;8:2004685.
41. Huang Z, Fang J, Zhou M, Gong Z, Xiang T. CRISPR-Cas13: a new technology for the rapid detection of pathogenic microorganisms. *Front Microbiol*. 2022;13:1011399.
42. Xue Z, Wu M, Wen K, et al. CRISPR/Cas9 mediates efficient conditional mutagenesis in Drosophila. *G3: Genes Genom Genet*. 2014;4:2167–2173.
43. Singh V, Braddick D, Dhar PK. Exploring the potential of genome editing CRISPR-Cas9 technology. *Gene*. 2017;599:1–18.
44. Huang J, Zhou Y, Li J, Lu A, Liang C. CRISPR/Cas systems: delivery and application in gene therapy. *Front Bioeng Biotechnol*. 2022;10:942325.
45. Zhan Z, Chen X, Ye Z, et al. Expanding the CRISPR toolbox for engineering lycopene biosynthesis in Corynebacterium glutamicum. *Microorganisms*. 2024;12:803.
46. Morshedzadeh F, Ghanei M, Lotfi M, et al. An update on the application of CRISPR technology in clinical practice. *Mol Biotechnol*. 2024;66:179–197.
47. Zhu H, Li C, Gao C. Applications of CRISPR-Cas in agriculture and plant biotechnology. *Nat Rev Mol Cell Biol*. 2020;21:661–677.
48. Pardee K, Green AA, Takahashi MK, et al. Rapid, low-cost detection of Zika virus using programmable biomolecular components. *Cell*. 2016;165:1255–1266.
49. Kellner MJ, Koob JG, Gootenberg JS, Abudayyeh OO, Zhang F. SHERLOCK: nucleic acid detection with CRISPR nucleases. *Nat Protoc*. 2019;14:2986–3012.
50. Gootenberg JS, Abudayyeh OO, Kellner MJ, Joung J, Collins JJ, Zhang F. Multiplexed and portable nucleic acid detection platform with Cas13, Cas12a, and Csm6. *Science*. 2018;360:439–444.
51. Chen JS, Ma E, Harrington LB, et al. CRISPR-Cas12a target binding unleashes indiscriminate single-stranded DNase activity. *Science*. 2018;360:436–439.
52. Bhattacharjee G, Gohil N, Khambhati K, Gajjar D, Abusharha A, Singh V. A paper-based assay for detecting hypervirulent *Klebsiella pnuemoniae* using CRISPR-Cas13a system. *Microchem J*. 2024;203:110931.

CHAPTER TWO

Advances in CRISPR-Cas systems for human bacterial disease

Anshu Mathuria[a,1], Chaitali Vora[b,1], Namra Ali[c], and Indra Mani[c,*]
[a]Department of Biochemistry, Sri Venkateswara College, University of Delhi, New Delhi, India
[b]Department of Biosciences and Biomedical Engineering, Indian Institute of Technology, Indore, India
[c]Department of Microbiology, Gargi College, University of Delhi, New Delhi, India
*Corresponding author. e-mail address: indra.mani@gargi.du.ac.in; indramanibhu@gmail.com

Contents

Abstract

Prokaryotic adaptive immune systems called CRISPR-Cas systems have transformed genome editing by allowing for precise genetic alterations through targeted DNA cleavage. This system comprises CRISPR-associated genes and repeat-spacer arrays,

[1] *Anshu Mathuria and Chaitali Vora contributed equally the first authors.*

Progress in Molecular Biology and Translational Science, Volume 208
ISSN 1877-1173, https://doi.org/10.1016/bs.pmbts.2024.07.013

which generate RNA molecules that guide the cleavage of invading genetic material. CRISPR-Cas is classified into Class 1 (multi-subunit effectors) and Class 2 (single multi-domain effectors). Its applications span combating antimicrobial resistance (AMR), targeting antibiotic resistance genes (ARGs), resensitizing bacteria to antibiotics, and preventing horizontal gene transfer (HGT). CRISPR-Cas3, for example, effectively degrades plasmids carrying resistance genes, providing a precise method to disarm bacteria. In the context of ESKAPE pathogens, CRISPR technology can resensitize bacteria to antibiotics by targeting specific resistance genes. Furthermore, in tuberculosis (TB) research, CRISPR-based tools enhance diagnostic accuracy and facilitate precise genetic modifications for studying *Mycobacterium tuberculosis*. CRISPR-based diagnostics, leveraging Cas endonucleases' collateral cleavage activity, offer highly sensitive pathogen detection. These advancements underscore CRISPR's transformative potential in addressing AMR and enhancing infectious disease management.

1. Introduction

Research has demonstrated that the system known as CRISPR-Cas (clustered regularly interspaced short palindromic repeats/CRISPR-associated) works as an adaptive immune mechanism, allowing prokaryotes to defend themselves against foreign genetic material (mainly viruses and plasmids) by removing the foreign DNA and RNA.[1] It is composed of a group of genes known as CRISPR-associated (cas) genes that result in endonuclease-producing Cas proteins and CRISPR repeat-spacer arrays, which can be translated into trans-activating CRISPR RNA (tracrRNA) and CRISPR RNA (crRNA).[2] Cas proteins can break foreign genetic material into short fragments when they infect prokaryotes. After that, the CRISPR array incorporates these fragments as new spacers.[3] The host is protected when the same invader returns because crRNA promptly identifies and couples with the foreign DNA, causing the Cas protein to cleave certain foreign DNA sequences.[3]

A CRISPR-Cas system is often composed of a CRISPR locus and a matching cas operon. It conducts immunity in three stages: adaptation, crRNA synthesis, and interference.[4] Because CRISPR-Cas systems allow us to recognize, remove, alter, and annotate specific nucleic acid sequences in a variety of live creatures, they have completely changed the field of genetics research. This is a result of their exceptional programmability.[5]

1.1 Structure and mechanism of the CRISPR-Cas system

The CRISPR system consists of a set of CRISPR-associated (Cas) genes and a CRISPR array.[6] The CRISPR array is composed of short repetitive sequences separated by spacer sequences.[7] The CRISPR array is often preceded by an A-T-rich leader sequence.[8,9] This sequence includes spacer and repetition sequences as well as a promoter that initiates their transcription.[10] The intricacy of their effector modules distinguishes two kinds of CRISPR-Cas systems.[11] The class 1 systems, which include the type I, III, and IV systems, use a multi-subunit effector complex in conjunction with an additional Cas nuclease to change the target, whereas the class 2 type II, V, and VI systems use a single multi-domain effector.

There are three phases to the CRISPR/Cas systems' bacterial defense mechanism.[12,13] (i) During the adaptation stage, acquiring spacer sequences.[14,15] (ii) Expression stage: synthesis of crRNA and the Cas protein;[16] (iii) Interference stage: the crRNA-directed cleavage of nucleic acid targets.[17,18] A nucleic acid-protein complex that can identify sequences that resemble mature crRNA in nucleic acid and utilize endonuclease activity to break down the nucleic acid close to the recognition site is formed when mature crRNA interacts with Cas protein.[19] By precisely identifying and removing particular drug-resistance genes, the CRISPR system effectively inhibits horizontal gene transfer (HGT) and prevents the emergence of antibiotic resistance.[20]

1.2 Immune recognition and defense mechanism of CRISPR-Cas system I and II

The Cas3 protein in the Type I CRISPR-Cas system, which is a member of Class 1, has two domains: a helicase and a phosphohydrolase.[21] This method targets DNA with the help of a complex called cascade, which is made up of several proteins.[22] Cascade enlists Cas3 to aid in DNA cleavage following binding of DNA.[23] Cas1 and Cas2 are engaged in the process of adaptation of Type I CRISPR-Cas. They do this by introducing foreign DNA fragments into the CRISPR array between its leader sequence and the first repeat. As a result, fresh spacers are produced. During expression, Cas proteins transform pre-CRISPR RNA (pre-crRNA) into mature crRNA, which combines with cascade to instruct Cas3 to cut off foreign target sequences.[24]

The class 2 Type II CRISPR-Cas9 technique is widely used for gene editing and immune protection.[25] CrRNA maturation and double-stranded DNA cleavage depend on RuvC at the amino-terminal end and HNH2 in the core of the Cas9 protein's active site.[26] Concurrent with the transcription of pre-crRNA is the synthesis of a tracrRNA, which joins forces with CRISPR repeats to trigger Cas9 and the double-stranded

RNA-specific RNase III nuclease, therefore transforming pre-crRNA into mature crRNA.[27] After being processed, a complex consisting of crRNA, tracrRNA, and Cas9 identifies and attaches to the foreign genomic material of mobile genetic elements (MGE). The RuvC endonuclease domain of Cas9 cleaves the non-complimentary strand, whereas the HNH endonuclease domain cleaves the complementary protospacer strand, resulting in double-stranded DNA breaks (DSBs) in the invasive MGE.[28] In the protospacer adjacent motif (PAM) region, the CRISPR-Cas9 cleavage site is often located near to the crRNA complementary sequence at the NGG site. By precisely cleaving nucleotides that are near the target sequence, this cleavage creates a double-strand break (DSB).[5,29,30]

2. Application of CRISPR-Cas system in human bacterial diseases

CRISPR-Cas systems have demonstrated a lot of promise in various applications for treating and managing human bacterial diseases. These applications leverage the precise gene-editing capabilities of CRISPR-Cas to target and modify bacterial genomes, combat antibiotic resistance, and develop new therapeutic strategies. Here are some key applications:

2.1 Combating antibiotic resistance

2.1.1 Targeting resistance genes

It is possible to precisely target and interfere with CRISPR-Cas and disrupt antibiotic resistance genes (ARGs) on bacterial chromosomes and plasmids. This approach has been successful in resensitizing bacteria to antibiotics. For example, targeting the *mcr-1* gene in gram-negative bacteria has reversed polymyxin resistance.[31]

2.1.2 Plasmid curing

By targeting and eliminating plasmids carrying multiple resistance genes, CRISPR-Cas can reduce the spread of multidrug resistance. This has been demonstrated in pathogens like *E. coli*, where CRISPR-Cas9 systems successfully eliminated plasmids carrying resistance genes like *blaNDM-1* and *blaCTX-M-15*.[32]

2.2 Targeted bactericidal effects

2.2.1 Direct killing of pathogenic bacteria

Pathogenic bacteria's critical genes can be targeted by CRISPR-Cas systems, which will cause death of cells. This targeted approach can be used to specifically eliminate harmful bacteria without affecting the beneficial microbiota. For instance, targeting the *nuc* gene in *Staphylococcus aureus* has been shown to kill the bacteria effectively.[33]

2.2.2 Phage-mediated delivery

The specificity and effectiveness of bacterial targeting are increased when CRISPR-Cas systems are delivered via bacteriophages, which are viruses that infect bacteria. Bacteria resistant to antibiotics in mixed populations have been targeted for selective killing using phage-delivered CRISPR-Cas systems.[34,35]

2.3 Modulating the microbiome

2.3.1 Precision microbiome editing

CRISPR-Cas can be used to selectively modify the composition of the human microbiome by targeting specific bacterial strains. This approach has potential applications in treating dysbiosis-related conditions, such as inflammatory bowel disease (IBD) and obesity, by restoring a healthy balance of microbial species.[36]

2.3.2 Preventing horizontal gene transfer (HGT)

By targeting and disrupting genes involved in HGT, the transmission of genes that show resistance among bacterial populations can be stopped by CRISPR-Cas, thereby maintaining the efficacy of existing antibiotics.[37]

2.4 Developing new antimicrobials

2.4.1 Novel antibacterial agents

Systems utilizing CRISPR-Cas can be designed to develop new classes of antimicrobials that are highly specific to pathogenic bacteria. These engineered systems can be used to create bactericidal agents that target and kill bacteria through precise genetic disruption.[38]

2.4.2 Synergistic therapies

Combining CRISPR-Cas-based antimicrobials with traditional antibiotics can enhance treatment efficacy. By weakening bacterial defenses through CRISPR-mediated gene disruption, antibiotics can become more effective at lower doses.[39]

2.5 Diagnostic applications

2.5.1 Rapid and accurate diagnostics

Bacterial pathogens can be quickly and accurately identified using CRISPR-Cas systems, such as CRISPR-Cas12 and CRISPR-Cas13. These systems can identify specific DNA or RNA sequences associated with bacterial infections, providing quick diagnostic results that guide appropriate treatment strategies.[40,41]

2.5.2 Detection of resistance genes

Additionally, CRISPR-based diagnostics can identify genes for resistance in bacterial samples, allowing for the identification of resistant strains and informing treatment decisions to avoid ineffective antibiotics.[40]

2.6 Vaccines and prophylactics

2.6.1 CRISPR-Cas vaccines

Research is exploring the use of CRISPR-Cas systems to develop vaccines that target specific bacterial antigens. These vaccines could provide targeted immunity against bacterial pathogens, reducing the incidence and severity of infections.[42]

2.6.2 Preventative measures

CRISPR-Cas systems can be engineered to create "probiotic" strains that carry CRISPR constructs designed to target and eliminate pathogenic bacteria in the gut, thereby preventing infections before they occur.

All things considered, the adaptability and accuracy of CRISPR-Cas systems have enormous potential for transforming the management and treatment of bacterial illnesses and providing fresh hope in the struggle against bacterial infections and antibiotic resistance.

3. CRISPR-Cas systems to combat antimicrobial resistance

The CRISPR-Cas system has the ability to directly eliminate pathogenic bacteria since it can target genes on chromosomes and plasmids.[43] CRISPR-Cas9 specifically cleaves target genes on bacterial chromosomes; *Salmonella, S. aureus,* and *Streptococcus pneumoniae* have all shown this to be the case.[42] Park et al. selected the *nuc* gene unique to *S. aureus* as the target gene to verify that the CRISPR-Cas9 system specifically destroys the target bacteria. Further investigation showed that the primary determinant of the CRISPR-Cas9 bactericidal impact was the system's ability to reach the intended target bacterium. With the use of a phage carrier, the researchers were able to administer CRISPR-Cas9 and successfully eradicate *S. aureus*.[44] CRISPR-Cas carefully and specifically targets ARGs in complex microbial communities to eradicate dangerous bacteria while sparing other bacterial species, in contrast to standard antibiotics, which typically lack specificity.[45] Fig. 1 shows a schematic example of the CRISPR-Cas system, which functions as an antibacterial agent by neutralizing ARGs, increasing bacterial susceptibility to medications.[46]

Recent studies have demonstrated that CRISPR-Cas, which targets plasmid-borne ARGs, can successfully resensitize drug-resistant varieties of pathogenic bacteria. For instance, it has been shown that CRISPR-Cas can restore *S. aureus*' kanamycin sensitivity[33] and methicillin.[47] Yosef et al. demonstrated the system's ability to remove multiple drug resistance gene-carrying plasmids simultaneously.[32] Kim et al. targeted conserved sequences in TEM- and SHV-type extended-spectrum beta-lactamase (ESBLs) using CRISPR-Cas9, effectively combating β-lactam resistance in *E. coli*.[48]

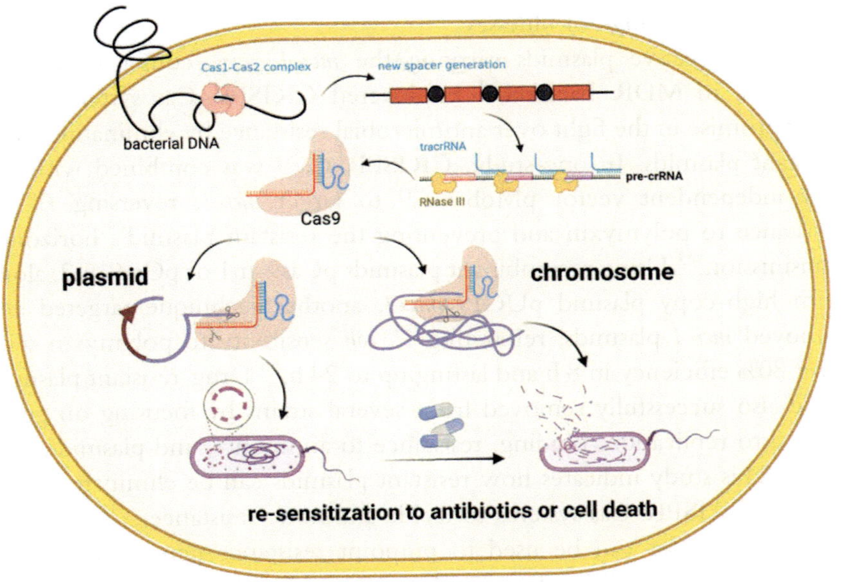

Fig. 1 The basic mechanism of use of CRISPR-Cas systems in resensitization of antibiotics or cell death. The target can be localized on the plasmid or/and the chromosome. *Adapted from Wu Y, Battalapalli D, Hakeem MJ, et al. Engineered CRISPR-Cas systems for the detection and control of antibiotic-resistant infections. J Nanobiotechnol. 2021;19(1):401. https://doi.org/10.1186/s12951-021-01132-8.*

Tagliaferri et al. showed that CRISPR-Cas9 can selectively remove high-copy plasmids from bacterial cells, leading to regained sensitivity to antibiotics like ampicillin and cephalosporins.[49] Liu et al. successfully targeted *blaNDM-1*-encoding plasmids with CRISPR-Cas9, achieving rapid and extensive clearance of target plasmids in bacterial infections, restoring sensitivity to kanamycin in mouse models and preventing the emergence of resistance mutations.[50] By focusing on critical genes or including several cleavage sites, the CRISPR-Cas system can more effectively target clinically complicated multidrug resistance (MDR) plasmids. In combating *S. aureus* resistance to Lysostaphin (Lst), Wu et al. used CRISPR-dCas9 to downregulate *tarH*, *tarO*, and *tarG* genes, enhancing susceptibility to Lst and eliminate bacteria within 24 h in vitro.[51] Wu et al. edited and removed drug-resistance genes in *Shewanella algae* with CRISPR-Cas9, resensitizing them to multiple antibiotics.[52] Similar to this, Sun et al. showed how CRISPR-Cas9 can inactivate *K. pneumoniae* resistance genes, restoring susceptibility to polymyxin and tigecycline.[53] All things considered, our findings demonstrate the efficacy of CRISPR-Cas systems as tools for increasing bacterial susceptibility to antibiotics and combating resistance to antibiotics in a range of illnesses.

Gram-negative plasmids carrying the *mcr-1* gene confer polymyxin resistance in MDR bacteria.[54] Engineered CRISPR-Cas systems show great promise in the fight over antimicrobial resistance by eliminating these resistant plasmids. In one study, CRISPR-Cas9 was combined with the host-independent vector pMob-Cas9 to target *mcr-1*, reversing *E. coli* resistance to polymyxin and preventing the resistant plasmid's horizontal transmission.[31] Using recombinant plasmids pCas9-m1 or pCas9-m2, along with high-copy plasmid pUC19-*mcr-1*, another technique targeted and removed *mcr-1* plasmids, reinstating *E. coli* sensitivity to polymyxin with over 80% efficiency in 8 h and lasting up to 24 h.[55] Drug-resistant plasmids were also successfully removed from several strains by focusing on genes related to replication, binding, resistance to antibiotics, and plasmid stability.[56] This study indicates how resistant plasmids can be eliminated using tailored CRISPR-Cas systems to tackle antibiotic resistance.

CRISPR-Cas can be used to pinpoint resistance genes on bacterial chromosomes in addition to neutralizing resistance genes on plasmids, and a diagrammatic illustration is given in Fig. 2.[46] For instance, quinolone-sensitive *E. coli's gyrA* gene changed due to targeted gene modification using CRISPR-Cas9, which changed particular amino acids and reversed quinolone resistance. This verified that quinolone resistance and *gyrA*

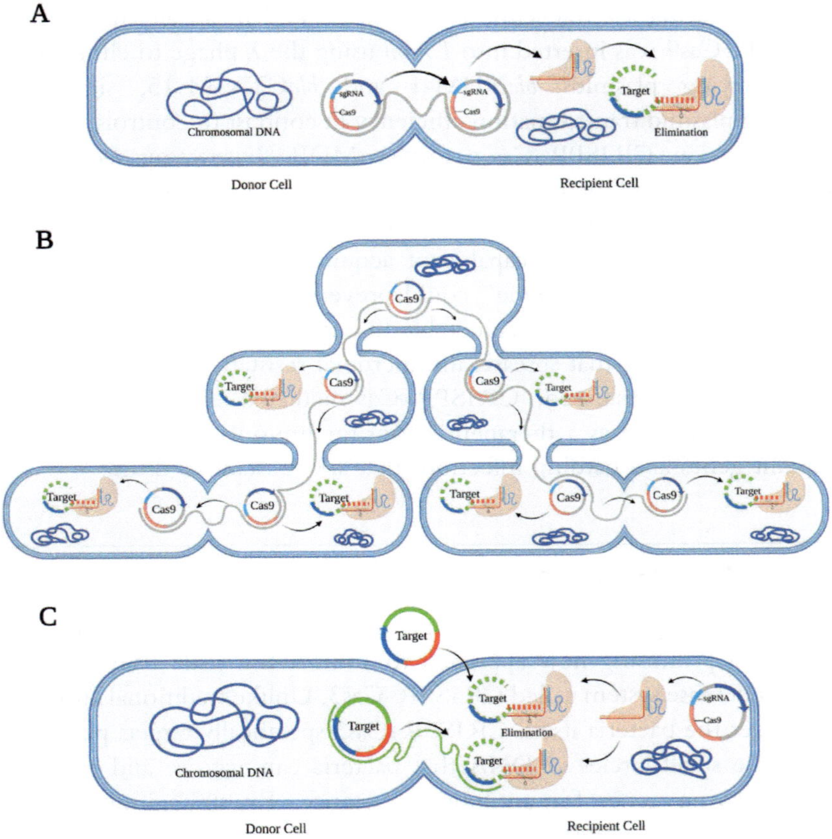

Fig. 2 Diagram showing the target genes for antibiotic resistance that are targeted by a modified CRISPR-Cas9 system. (A) The conjugative CRISPR-Cas plasmid eliminated the *mcr-1* gene-containing plasmid. (B) A host-independent conjugative plasmid was used to transfer the designed CRISPR-Cas9 to the target bacterium, where it spread throughout the microflora and affected target sequence-containing plasmids that were medication-resistant. (C) By focusing on DNA, the modified CRISPR-Cas9 system prevents drug-resistant plasmids from conjugating and transforming in bacteria. *Adapted from Wu Y, Battalapalli D, Hakeem MJ, et al. Engineered CRISPR-Cas systems for the detection and control of antibiotic-resistant infections. J Nanobiotechnology. 2021;19(1):401. https://doi.org/10.1186/s12951-021-01132-8.*

mutations are closely related.[57] Plasmids carrying antibiotic resistance genes (ARGs) are efficiently kept from transforming and conjugating thanks to the CRISPR-Cas system, so essentially stopping drug resistance in its tracks. Tetracycline resistance was not seen in one investigation where methicillin-sensitive *S. aureus* was targeted with a phagemid to remove a plasmid carrying a tetracycline resistance gene.[33] In an alternative study, CRISPR-Cas9 was inserted into *E. coli* using the λ phage to eliminate the gene-carrying plasmids *bla*NDM-1 and *bla*CTX-M-15, significantly reducing plasmid transformation efficiency in contrast to controls.[32] Studies conducted on CRISPR-Cas-negative, MDR *Enterococcus faecalis* have shown that because these bacteria lack genomic defenses, they quickly pick up drug-resistance genes.[58] Nevertheless, the introduction of CRISPR-Cas rendered *E. faecalis* incapable of acquiring ARGs, suggesting that a CRISPR-Cas-based "vaccine" could prevent the entry of drug-resistant genes into bacteria that are vulnerable to antibiotics. This approach might help halt the HGT that spreads antimicrobial resistance (AMR).

All things considered, CRISPR-Cas-mediated targeted deletion of ARGs has promise as a therapeutic tool for limiting the spread of drug-resistant genes and battling infections that are resistant to drugs.

4. CRISPR in precision antibacterials

The emergence of AMR is a major threat to world health. This study explores a promising new approach to combat this issue using a natural bacteria defense system called CRISPR-Cas3. Unlike traditional antibiotics that target the bacteria itself, CRISPR-Cas3 specifically targets plasmids.[59] These are small circles of DNA that bacteria can acquire and exchange, often carrying genes for antibiotic resistance. By strategically utilizing CRISPR-Cas3, scientists can disarm bacteria from these antibiotic resistance tools.[59]

CRISPR-Cas3 acts like a molecular scalpel, precisely cutting and degrading the plasmid DNA. This process renders the ARGs inert, essentially stripping the bacteria of their defenses against antibiotics.[10] The primary benefit of this approach is in its ability to leverage a bacterium's own CRISPR system. Scientists can identify and extract these spacer sequences to design a CRISPR-Cas3 system that targets specific plasmids. This focused strategy reduces the possibility that the bacteria would become resistant to the antibiotic itself, which is a common problem with conventional antibiotics.[60]

The effectiveness of this method in eliminating AMR plasmids in controlled laboratory experiments is high. Success in a wax moth infection model, highlighting the potential for real-world application was also depicted. Furthermore, researchers envision combining this approach with bacteriophages, viruses that specifically target and kill bacteria. By using CRISPR-Cas3 to disarm bacteria of their antibiotic resistance and then deploying phages to eliminate the weakened bacteria, could form a powerful one-two punch against AMR infections. This combination strategy holds promise for tackling even complex and highly resistant bacterial strains. All things considered, this work opens the door for a potentially revolutionary strategy in the battle against antibiotic resistance.[61]

5. CRISPR in ESKAPE pathogens

A major global threat is posed by the growth of AMR bacteria and the lack of new antimicrobial medications. The ESKAPE group, which includes *Pseudomonas aeruginosa, Acinetobacter baumannii, Staphylococcus aureus, Enterococcus spp.*, and *Klebsiella pneumoniae*, is particularly harmful because of its high toxicity, high transmissibility, and resistance to therapy. By 2050, antimicrobial resistance is predicted to cause a 1.1–3.8% decline in global GDP.[62] Similar resistance mechanisms, such as decreased drug uptake, drug target change, drug inactivation, and drug efflux pump activation, are employed by ESKAPE infections despite their genetic variations. Comprehending these pathways is essential for creating innovative remedies against these drug-resistant microbes.[63]

5.1 Drug resistance mechanisms

5.1.1 Innate resistance

Some bacteria are naturally resistant to antibiotics. For instance, *Pseudomonas aeruginosa* is often impervious to many antibiotics due to porins in its outer membrane.[64]

5.1.2 Acquired resistance

Bacteria can potentially develop antibiotic resistance by repeated exposure, genetic exchange, or mutation. *Acinetobacter baumannii* has gained extensive resistance by enzymatic modification of antibiotics, target gene alterations, changed outer membrane permeability, and enhanced efflux pumps.[64]

One effective strategy for battling antibiotic resistance is the CRISPR-Cas system. Antibiotic-resistant bacterial cells may be resensitized by this RNA-guided DNA nuclease system, which can precisely break ARGs.[65] CRISPR-Cas systems in antibiotic-resistant enterococci are often non-functional. Studies show that CRISPR-Cas systems can stop the accumulation of genes that cause antibiotic resistance. For example, hospital-acquired *E. faecalis* strains, which frequently lack CRISPR-Cas, are more resistant than those from endodontic or oral infections. Reactivating CRISPR-Cas in MDR *Enterococci* can make them sensitive to drugs once again.[10] In an effort to tackle methicillin-resistant Staphylococcus aureus (MRSA), CRISPR-Cas systems have been deployed. The Cas13a enzyme has an antibacterial effect by targeting resistance genes. Furthermore, CRISPR-dCas9 systems can inhibit the transcription of resistance genes, greatly lowering antibiotic resistance in MRSA.[59] However, CRISPR-Cas systems do not work for all *S. aureus* strains, especially for those lacking the *mecA* gene. Furthermore, *S. aureus* can be targeted by downregulating genes for wall teichoic acid production, increasing susceptibility to lysostaphin.[66]

In *K. pneumoniae*, the absence of CRISPR-Cas correlates with higher antibiotic resistance. CRISPR-Cas-positive strains are more susceptible to aminoglycosides and beta-lactams.[67] Researchers have resensitized carbapenem-resistant *K. pneumoniae* to antibiotics such as tigecycline and colistin by utilizing CRISPR-Cas9 to inactivate particular genes. CRISPR-Cas encoding plasmids have demonstrated significant efficacy in resensitizing resistant strains to meropenem and imipenem.[68] *Acinetobacter baumannii* has developed resistance through various mechanisms such as enzymatic alteration, target gene mutation, and efflux pump activation. Research has not explicitly mentioned the use of CRISPR systems in this pathogen, but these systems could potentially be applied similarly as in other ESKAPE pathogens.[69] *P. aeruginosa* has developed mechanisms to evade CRISPR-Cas systems. Certain proteins and phages can inhibit CRISPR activity, making these bacteria resistant to antibiotics. As an illustration, the AmrZ protein suppresses CRISPR-Cas immunity. Nonetheless, comprehending these systems offers valuable perspectives for formulating innovative approaches to surmount resistance.[70] Specifically targeting the carbapenemase genes of carbapenem-resistant Enterobacteriaceae (CRE), pCasCure is a unique CRISPR/Cas9-based plasmid-curing method. This system effectively cuts resistance genes in various species, demonstrating more than 94% efficiency. The removal of these genes significantly reduces the minimum inhibitory concentration (MIC) of carbapenem antibiotics, enhancing their effectiveness.[71]

The viruses known as ESKAPE pose a significant threat because of their strong resistance to antibiotics. CRISPR-Cas technology, however, presents a viable way to overcome this resistance. To address the threat that these drug-resistant bacteria represent to global health, it may be essential to continue researching and developing CRISPR-based remedies.[68]

6. CRISPR in tuberculosis

6.1 Historical and epidemiological context of tuberculosis

Tuberculosis (TB) has a long history, with evidence of the disease found in ancient Egyptian mummies and historical texts describing its impact over centuries. *Mycobacterium tuberculosis* (Mtb) is the bacterium that causes tuberculosis (TB). It is a serious worldwide health concern, especially in developing countries where it remains one of the leading causes of infectious disease-related deaths. The epidemiology of TB reflects its widespread nature, affecting millions annually and posing significant public health challenges due to its ability to spread through airborne particles. The pathophysiology of Mtb involves complex interactions with the host immune system, leading to latent infections that can reactivate, making eradication difficult. Current diagnostic procedures, such as sputum microscopy and culture, are frequently ineffective due to their limited sensitivity and lengthy turnaround times. Additionally, the rise of MDR-TB strains complicates treatment, highlighting the urgent need for innovative diagnostic and therapeutic approaches (https://globaltb.njms.rutgers.edu/abouttb/historyoftb.php).

6.2 Mechanisms and variants of CRISPR-Cas systems

In bacteria and archaea, CRISPR-Cas systems represent a novel class of adaptive immunity systems that recognize and cleave foreign genetic material. The most researched variant, CRISPR-Cas9, allows for precision genome editing by introducing DSBs at predetermined sites using a dual-RNA-guided DNA endonuclease. Beyond Cas9, other CRISPR systems like Cas12 and Cas13 expand the toolbox, offering distinct nuclease activities and targeting capabilities; Cas12, for example, possesses single-strand DNA cleavage activity, while Cas13 targets RNA substrates. These systems have catalyzed advancements in various fields, from functional genomics and genetic engineering to innovative applications in diagnostics and therapeutics, underscoring their versatility and transformative impact on modern molecular biology.[72]

6.3 Advances in CRISPR-Cas systems for tuberculosis diagnostics

Current TB diagnostic methods face significant limitations, including low sensitivity in detecting Mtb in paucibacillary conditions, prolonged turnaround times, and high costs associated with advanced molecular techniques. A novel method to tuberculosis detection has been made possible by the development of CRISPR-based diagnostic instruments, which make use of the outstanding precision and flexibility of CRISPR-Cas systems. For quick and accurate detection, CRISPR-Cas12 and Cas13 systems are made to identify and cut nucleic acid sequences unique to tuberculosis, producing fluorescence or lateral flow readouts. Case studies have demonstrated the efficacy of CRISPR diagnostics in identifying TB with superior sensitivity and specificity compared to conventional methods. These CRISPR-based assays provide practical advantages, including minimal equipment requirements, ease of use, and potential for point-of-care applications, making them suitable for use in resource-limited settings.

6.4 CRISPR in TB treatment and research

CRISPR technology has revolutionized TB research by offering precise genetic manipulation tools for studying Mtb. Traditional TB treatment regimens face significant challenges, such as prolonged treatment duration and the emergence of drug-resistant strains. CRISPR technology addresses these issues by enabling detailed genetic studies and the development of new therapeutic approaches.

Mtb has been genetically edited using CRISPR-Cas systems, namely CRISPR-Cas9 and CRISPR-Cas12a. This makes it possible for scientists to create tailored gene knockouts or introduce particular mutations in order to examine the physiology and pathogenic pathways of the bacteria thoroughly.[73,74]

For instance, CRISPR-Cas9 has been used to alter drug-resistant genes, offering insights into how Mtb avoids standard treatments and pointing to possible targets for novel therapeutics.

Moreover, CRISPR-based diagnostics, such as those using CRISPR-Cas12a, have shown promise in rapidly detecting Mtb and its drug-resistant variants from clinical samples. These diagnostic tools offer high sensitivity and specificity, significantly improving the speed and accuracy of TB detection.[73]

CRISPR also holds potential for therapeutic applications. By targeting specific genes essential for Mtb survival or virulence, CRISPR-based strategies could develop novel treatments that are more effective and have fewer side effects compared to current antibiotics. Reprogramming Mtb's endogenous CRISPR-Cas system, for example, may aid in the degradation of plasmids containing ARGs, restoring the efficacy of currently available medications.[74]

To sum up, CRISPR technology is an effective tool in the fight against tuberculosis, opening up new paths for research, diagnostics, and therapies. Its capacity to precisely modify the Mtb genome opens up new avenues for studying TB pathogenesis and overcoming medication resistance.

6.5 Significance of CRISPR in advancing TB research and treatment

CRISPR technology has revolutionized TB research and treatment, offering significant advancements in several key areas. One of the primary benefits is the ability to perform precise genetic modifications in Mtb, the pathogen responsible for TB. This allows researchers to systematically study gene function and identify virulence factors, which are crucial for understanding the pathogen's mechanisms of infection and persistence.[75]

Furthermore, CRISPR enables the development of novel diagnostic tools. Particularly in low-resource environments, CRISPR-based diagnostics, such SHERLOCK (Specific High-Sensitivity Enzymatic Reporter UnLOCKing), have demonstrated promise in accurately and highly specifically diagnosing tuberculosis. This is crucial for prompt diagnosis and treatment, both of which are necessary to stop the spread of tuberculosis.[76]

CRISPR technology holds promise for accelerating the development of new therapeutic interventions. Researchers may develop models of drug-resistant tuberculosis strains by targeting and altering certain genes, allowing them to screen novel medications and identify viable treatment regimens more efficiently. Additionally, CRISPR-based gene editing could be employed to develop phage therapy approaches, where bacteriophages are engineered to specifically target and kill drug-resistant TB bacteria.[77]

Moreover, CRISPR technology shows potential for improving the efficacy of tuberculosis vaccinations. By precisely modifying the genetic makeup of *Mycobacterium bovis* BCG (Bacillus Calmette-Guérin), the only currently available TB vaccine, scientists can potentially improve its protective efficiency and widen its use to other populations.[78]

7. CRISPR in pathogen detection

Cas endonucleases provide a strong host defensive mechanism by precisely identifying intruding nucleic acids and amplifying the signal by promiscuous collateral cleavage activity. This method guarantees a strong reaction to infections, enhancing the effectiveness of the immune system.[79]

By splitting a reporter nucleic acid collaterally, the promiscuous activity can also be employed to increase target detection. This ensures that the signal is gradually amplified even when just a little amount of the target sequence is present. The employment of Cas endonucleases with collateral cleavage in nucleic acid detection is supported by this concept. One possible way to achieve more Cas13a-based RNA detection is to label a reporter RNA with both a fluorophore and a quencher. The fluorophore is released from the quencher by the reporter RNA's collateral cleavage upon Cas13a's binding to its target, which increases the fluorescence signal. This approach could differentiate target sequences at picomolar concentrations within complicated RNA mixtures, such as endogenous transcripts within total cellular RNA.[79]

The gRNA/Cas protein combination can selectively recognize foreign fragments by constructing the guide RNA (gRNA) sequence, and it can then trigger the nucleic acid-cleaving enzyme activity of Cas protein to cleave foreign components in the system. For applications that need precise targeting, such as genome editing and disease detection, this selectivity is essential. By using reporter probes customized with fluorescence and quenching categories, trans-cleavage movement can increase the detection signal and enable real-time target identification. Because of the CRISPR/Cas system's exceptional sensitivity and specificity, it has been used extensively in clinical pathogen detection.[80] Recombinase-aided amplification (RAA) in conjunction with the CRISPR-Cas system was used to create the unique Mtb identification technique known as TB-CRISPR.[80] Recombinase (UvsX), single chain binding protein, and DNA polymerase (Klenow) are the three enzymes used in isothermal nucleic acid amplification (RAA), which efficiently amplifies DNA at 37 °C in 5–20 min. RAA is a useful technique in diagnostic applications because of its quick amplification process.

A new method for identifying *S. aureus* strains to harbor the dangerous *pvl* virulence gene combines CRISPR-Cas13a for precise targeting, RAA for efficient detection, and ERASE (Easy-Readout and Sensitive Enhanced1) strips for user-friendly visual readout.[81] This selectivity is critical for applications like disease detection and genome editing, where precise targeting is required. By using reporter probes customized with fluorescence and quenching categories, trans-cleavage movement can increase the detection signal and enable real-time target identification. Because of the CRISPR/Cas system's exceptional sensitivity and specificity, it has been used extensively in clinical pathogen detection.[81]

This technique allows for detection even in the event of a single target site mutation by employing Cas12a from the Lachnospiraceae bacterium, which possesses two unique crRNAs and targets the real-time reverse-transcription loop-mediated isothermal amplification (RT-LAMP) amplicons. Four viral RNA copies/μL is the detection limit for both the fluorescence and lateral flow readout techniques. In order to reduce the sensitivity of CRISPR-based diagnostics to viral genome alterations, this technique highlights the necessity of mixing two different crRNAs in a single reaction. This innovation provides a significant advancement in ensuring the reliability of CRISPR-based diagnostics in the face of evolving viral infection trends.[82]

8. Ethical and regulatory considerations

Using CRISPR in infectious disease management presents a complex interplay of ethical, regulatory, public health, and risk management considerations. Ethically, the use of CRISPR raises concerns about the potential for unintended consequences and the alteration of germ lines, as highlighted by the National Academies of Sciences, Engineering, and Medicine.[83] Additionally, the accessibility and affordability of CRISPR technology may exacerbate existing health disparities, raising equity concerns. The regulatory landscape surrounding CRISPR applications is multifaceted, with varying degrees of oversight and approval processes across different jurisdictions. While some countries, such as China, have taken relatively progressive stances on CRISPR research and clinical applications, others, like the European Union and the United States, have adopted more cautious approaches, emphasizing rigorous safety and ethical evaluations.[84] Public health implications of CRISPR in infectious disease control are significant, promising advancements in precision medicine and targeted interventions. However, challenges persist in ensuring equitable access to CRISPR-based treatments, as well as addressing potential biosafety and biosecurity risks. Policy considerations must balance innovation with safeguards to mitigate these risks and promote equitable distribution of benefits. Risk assessment and management strategies are essential in navigating the complexities of CRISPR technology. This involves rigorous evaluation of potential off-target effects, long-term health impacts, and ecological consequences of genetically modified organisms. Strategies such as gene drive containment measures and community engagement are crucial for responsible deployment of CRISPR in infectious disease control.[83] In conclusion, while CRISPR holds immense potential for transforming infectious disease management, its ethical, regulatory, public health, and risk management implications necessitate careful consideration and robust governance frameworks to ensure responsible and equitable implementation.

9. Conclusion and future perspectives

In conclusion, CRISPR-Cas technology represents a transformative tool in combating antibiotic resistance and managing bacterial diseases. Its precision gene-editing capabilities offer versatile applications across multiple fronts, from directly targeting antibiotic resistance genes on bacterial chromosomes and plasmids to modulating microbiomes and developing novel antimicrobials. By effectively disarming bacteria of their resistance mechanisms and preventing horizontal gene transfer, a possible remedy for the issues brought about by pathogens that are resistant to many drugs is to use CRISPR-Cas systems. Furthermore, CRISPR-based diagnostics enable rapid and accurate identification of bacterial infections and resistance profiles, guiding more informed treatment decisions. The system's potential in clinical settings is highlighted by its capacity to improve the effectiveness of currently available antibiotics through synergistic therapy. Looking ahead, continued research and development in CRISPR technology promise to refine and expand its applications, potentially revolutionizing the treatment landscape for bacterial diseases. As we confront the growing threat of antibiotic resistance globally, CRISPR-Cas stands poised as a powerful ally in our efforts to safeguard public health and advance therapeutic strategies.

In the future, CRISPR technology has the potential to completely transform how bacterial illnesses in humans are treated. CRISPR-Cas antimicrobials exhibit several potential advantages over conventional antimicrobials. There are at least 33 distinct CRISPR-Cas system subtypes.[11] Its capacity to accurately target and alter the genomes of bacteria presents hitherto unheard-of possibilities for customized medicine and more efficient therapeutic approaches. By focusing on the genes that cause resistance to antibiotics, CRISPR may be able to increase the efficacy of presently prescribed medicines. This is significant since multidrug-resistant bacteria are becoming more and more harmful in the fight against them.

Furthermore, CRISPR-based diagnostics are poised to enhance the speed and accuracy of bacterial infection detection, providing clinicians with invaluable tools for rapid diagnosis and targeted therapy. As research progresses, CRISPR's applications may extend to microbiome engineering, where it could help restore, microbial balance disrupted by diseases or treatments. In order to regulate the composition of the microbiome, CRISPR-Cas systems may be developed as "smart" antimicrobials that can differentiate between pathogenic and helpful bacteria, eliminate AMR pathogens, and stop the spread of ARGs.

References

1. Marraffini LA. CRISPR-Cas immunity in prokaryotes. *Nature*. 2015;526:55–61. https://doi.org/10.1038/nature15386
2. Koonin EV, Makarova KS. CRISPR-Cas: an adaptive immunity system in prokaryotes. *F1000 Biol Rep*. 2009;1:95.
3. Makarova KS, Haft DH, Barrangou R, Brouns SJ, Charpentier E. Evolution and classification of the CRISPR-Cas systems. *Nat Rev Microbiol*. 2011;9:467–477.
4. Nussenzweig PM, Marraffini LA. Molecular mechanisms of CRISPR-Cas immunity in bacteria. *Annu Rev Genet*. 2020;54:93–120. https://doi.org/10.1146/annurev-genet-022120-112523
5. Pickar-Oliver A, Gersbach CA. The next generation of CRISPR-Cas technologies and applications. *Nat Rev Mol Cell Biol*. 2019;20(8):490–507. https://doi.org/10.1038/s41580-019-0131-5
6. Rath D, Amlinger L, Hoekzema M, Devulapally PR, Lundgren M. Efficient programmable gene silencing by Cascade. *Nucleic Acids Res*. 2015;43(1):237–246. https://doi.org/10.1093/nar/gku1257
7. Hillary VE, Ignacimuthu S, Ceasar SA. Potential of CRISPR/Cas system in the diagnosis of COVID-19 infection. *Expert Rev Mol Diagn*. 2021;21(11):1179–1189. https://doi.org/10.1080/14737159.2021.1970535
8. Alkhnbashi OS, Shah SA, Garrett RA, et al. Characterizing leader sequences of CRISPR loci. *Bioinformatics*. 2016;32(17):i576–i585. https://doi.org/10.1093/bioinformatics/btw454
9. Hu T, Cui Y, Qu X. Characterization and comparison of CRISPR Loci in Streptococcus thermophilus. *Arch Microbiol*. 2020;202(4):695–710. https://doi.org/10.1007/s00203-019-01780-3
10. Tao S, Chen H, Li N, Liang W. The application of the CRISPR-Cas system in antibiotic resistance. *Infect Drug Resist*. 2022;15:4155–4168. https://doi.org/10.2147/IDR.S370869
11. Makarova KS, Wolf YI, Iranzo J, et al. Evolutionary classification of CRISPR–Cas systems: a burst of class 2 and derived variants. *Nat Rev Microbiol*. 2020;18:67–83. https://doi.org/10.1038/s41579-019-0299-x
12. Hille F, Charpentier E. CRISPR-Cas: biology, mechanisms and relevance. *Philos Trans R Soc Lond B Biol Sci*. 2016;371(1707) https://doi.org/10.1098/rstb.2015.0496
13. Jackson SA, McKenzie RE, Fagerlund RD, et al. CRISPR-Cas: adapting to change. *Science*. 2017;356(6333) https://doi.org/10.1126/science.aal5056
14. Nuñez JK, Kranzusch PJ, Noeske J, et al. Cas1-Cas2 complex formation mediates spacer acquisition during CRISPR-Cas adaptive immunity. *Nat Struct Mol Biol*. 2014;21(6):528–534. https://doi.org/10.1038/nsmb.2820

15. Nuñez JK, Lee AS, Engelman A, Doudna JA. Integrase-mediated spacer acquisition during CRISPR-Cas adaptive immunity. *Nature*. 2015;519(7542):193–198. https://doi.org/10.1038/nature14237
16. Makarova KS, Wolf YI, Alkhnbashi OS, et al. An updated evolutionary classification of CRISPR-Cas systems. *Nat Rev Microbiol*. 2015;13(11):722–736. https://doi.org/10.1038/nrmicro3569
17. Plagens A, Richter H, Charpentier E, Randau L. DNA and RNA interference mechanisms by CRISPR-Cas surveillance complexes. *FEMS Microbiol Rev*. 2015;39(3):442–463. https://doi.org/10.1093/femsre/fuv019
18. Charpentier E, Richter H, van der Oost J, White MF. Biogenesis pathways of RNA guides in archaeal and bacterial CRISPR-Cas adaptive immunity. *FEMS Microbiol Rev*. 2015;39(3):428–441. https://doi.org/10.1093/femsre/fuv023
19. Le Rhun A, Escalera-Maurer A, Bratovič M, Charpentier E. CRISPR-Cas in Streptococcus pyogenes. *RNA Biol*. 2019;16(4):380–389. https://doi.org/10.1080/15476286.2019.1582974
20. Teng M, Yao Y, Nair V, Luo J. Latest advances of virology research using CRISPR/Cas9-based gene-editing technology and its application to vaccine development. *Viruses*. 2021;13(5):779. https://doi.org/10.3390/v13050779
21. Makarova KS, Zhang F, Koonin EV. SnapShot: class 1 CRISPR-Cas systems. 946.e1 *Cell*. 2017;168(5):946. https://doi.org/10.1016/j.cell.2017.02.018
22. Zheng Y, Li J, Wang B, et al. Endogenous type I CRISPR-Cas: from foreign DNA defense to prokaryotic engineering. *Front Bioeng Biotechnol*. 2020;8:62. https://doi.org/10.3389/fbioe.2020.00062
23. Westra ER, van Erp PB, Künne T, et al. CRISPR immunity relies on the consecutive binding and degradation of negatively supercoiled invader DNA by Cascade and Cas3. *Mol Cell*. 2012;46(5):595–605. https://doi.org/10.1016/j.molcel.2012.03.018
24. Liu Z, Dong H, Cui Y, Cong L, Zhang D. Application of different types of CRISPR/Cas-based systems in bacteria. *Microb Cell Fact*. 2020;19(1):172. https://doi.org/10.1186/s12934-020-01431-z
25. Shmakov S, Smargon A, Scott D, et al. Diversity and evolution of class 2 CRISPR-Cas systems. *Nat Rev Microbiol*. 2017;15(3):169–182. https://doi.org/10.1038/nrmicro.2016.184
26. Mir A, Edraki A, Lee J, Sontheimer EJ. Type II-C CRISPR-Cas9 biology, mechanism, and application. *ACS Chem Biol*. 2018;13(2):357–365. https://doi.org/10.1021/acschembio.7b00855
27. Nishimasu H, Ran FA, Hsu PD, et al. Crystal structure of Cas9 in complex with guide RNA and target DNA. *Cell*. 2014;156(5):935–949. https://doi.org/10.1016/j.cell.2014.02.001
28. Komor AC, Kim YB, Packer MS, Zuris JA, Liu DR. Programmable editing of a target base in genomic DNA without double-stranded DNA cleavage. *Nature*. 2016;533(7603):420–424. https://doi.org/10.1038/nature17946
29. Fu YW, Dai XY, Wang WT, et al. Dynamics and competition of CRISPR-Cas9 ribonucleoproteins and AAV donor-mediated NHEJ, MMEJ and HDR editing. *Nucleic Acids Res*. 2021;49(2):969–985. https://doi.org/10.1093/nar/gkaa1251
30. Cencic R, Miura H, Malina A, et al. Protospacer adjacent motif (PAM)-distal sequences engage CRISPR Cas9 DNA target cleavage. *PLoS One*. 2014;9(10):e109213. https://doi.org/10.1371/journal.pone.0109213
31. Dong H, Xiang H, Mu D, Wang D, Wang T. Exploiting a conjugative CRISPR/Cas9 system to eliminate plasmid harbouring the mcr-1 gene from *Escherichia coli*. *Int J Antimicrob Agents*. 2019;53(1):1–8.
32. Yosef I, Manor M, Kiro R, Qimron U. Temperate and lytic bacteriophages programmed to sensitize and kill antibiotic-resistant bacteria. *Proc Natl Acad Sci U S A*. 2015;112(23):7267–7272.

33. Bikard D, Euler CW, Jiang W, et al. Exploiting CRISPR-Cas nucleases to produce sequence-specific antimicrobials. *Nat Biotechnol.* 2014;32(11):1146–1150.
34. Khambhati K, Bhattacharjee G, Gohil N, et al. Phage engineering and phage-assisted CRISPR-Cas delivery to combat multidrug-resistant pathogens. *Bioeng Transl Med.* 2022;8(2):e10381. https://doi.org/10.1002/btm2.10381
35. Citorik RJ, Mimee M, Lu TK. Sequence-specific antimicrobials using efficiently delivered RNA-guided nucleases. *Nat Biotechnol.* 2014;32(11):1141–1145.
36. Mimee M, Citorik RJ, Lu TK. Microbiome therapeutics—advances and challenges. *Adv Drug Deliv Rev.* 2016;105:44–54.
37. Marraffini LA, Sontheimer EJ. CRISPR interference limits horizontal gene transfer in staphylococci by targeting DNA. *Science.* 2008;322(5909):1843–1845.
38. Hsu PD, Lander ES, Zhang F. Development and applications of CRISPR-Cas9 for genome engineering. *Cell.* 2014;157(6):1262–1278.
39. Beisel CL, Barrangou R. CRISPR-Cas systems and their applications in prokaryotic and eukaryotic cells. *FEMS Microbiol Rev.* 2014;38(3):415–429.
40. Gootenberg JS, Abudayyeh OO, Kellner MJ, Joung J, Collins JJ, Zhang F. Multiplexed and portable nucleic acid detection platform with Cas13, Cas12a, and Csm6. *Science.* 2018;360(6387):439–444.
41. Chen JS, Ma E, Harrington LB, et al. CRISPR-Cas12a target binding unleashes indiscriminate single-stranded DNase activity. *Science.* 2018;360(6387):436–439.
42. Gomaa AA, Klumpe HE, Luo ML, Selle K, Barrangou R, Beisel CL. Programmable removal of bacterial strains by use of genome-targeting CRISPR-Cas systems. *mBio.* 2014;5(1):e00928-13.
43. Vercoe RB, Chang JT, Dy RL, et al. Cytotoxic chromosomal targeting by CRISPR/Cas systems can reshape bacterial genomes and expel or remodel pathogenicity islands. *PLoS Genet.* 2013;9(4):e1003454.
44. Park JY, Moon BY, Park JW, Thornton JA, Park YH, Seo KS. Genetic engineering of a temperate phage-based delivery system for CRISPR/Cas9 antimicrobials against Staphylococcus aureus. *Sci Rep.* 2017;7:44929.
45. Greene AC. CRISPR-based antibacterials: transforming bacterial defense into offense. *Trends Biotechnol.* 2018;36(2):127–130.
46. Wu Y, Battalapalli D, Hakeem MJ, et al. Engineered CRISPR-Cas systems for the detection and control of antibiotic-resistant infections. *J Nanobiotechnol.* 2021;19(1):401. https://doi.org/10.1186/s12951-021-01132-8
47. Kang YK, Kwon K, Ryu JS, Lee HN, Park C, Chung HJ. Nonviral genome editing based on a polymer-derivatized CRISPR nanocomplex for targeting bacterial pathogens and antibiotic resistance. *Bioconjug Chem.* 2017;28(4):957–967.
48. Kim JS, Cho DH, Park M, et al. CRISPR/Cas9-mediated re-sensitization of antibiotic-resistant *Escherichia coli* harboring extended-spectrum β-lactamases. *J Microbiol Biotechnol.* 2016;26(2):394–401.
49. Tagliaferri TL, Guimarães NR, Pereira MPM, et al. Exploring the potential of CRISPR-Cas9 under challenging conditions: facing high-copy plasmids and counteracting beta-lactam resistance in clinical strains of enterobacteriaceae. *Front Microbiol.* 2020;11:578.
50. Liu H, Li H, Liang Y, et al. Phage-delivered sensitisation with subsequent antibiotic treatment reveals sustained effect against antimicrobial resistant bacteria. *Theranostics.* 2020;10(14):6310–6321.
51. Wu X, Zha J, Koffas MAG, Dordick JS. Reducing Staphylococcus aureus resistance to lysostaphin using CRISPR-dCas9. *Biotechnol Bioeng.* 2019;116(12):3149–3159.
52. Wu ZY, Huang YT, Chao WC, Ho SP, Cheng JF, Liu PY. Reversal of carbapenem-resistance in Shewanella algae by CRISPR/Cas9 genome editing. *J Adv Res.* 2019;18:61–69.

53. Sun Q, Wang Y, Dong N, et al. Application of CRISPR/Cas9-based genome editing in studying the mechanism of pandrug resistance in Klebsiella pneumoniae. *Antimicrob Agents Chemother.* 2019;63(7):e00113-19.
54. Liu YY, Wang Y, Walsh TR, et al. Emergence of plasmid-mediated colistin resistance mechanism MCR-1 in animals and human beings in China: a microbiological and molecular biological study. *Lancet Infect Dis.* 2016;16(2):161–168.
55. Wan P, Cui S, Ma Z, et al. Reversal of mcr-1-mediated colistin resistance in *Escherichia coli* by CRISPR-Cas9 system. *Infect Drug Resist.* 2020;13:1171–1178.
56. Wang P, He D, Li B, et al. Eliminating mcr-1-harbouring plasmids in clinical isolates using the CRISPR/Cas9 system. *J Antimicrob Chemother.* 2019;74(9):2559–2565.
57. Qiu H, Gong J, Butaye P, et al. CRISPR/Cas9/sgRNA-mediated targeted gene modification confirms the cause-effect relationship between gyrA mutation and quinolone resistance in *Escherichia coli. FEMS Microbiol Lett.* 2018;365(13):3.
58. Price VJ, Huo W, Sharifi A, Palmer KL. CRISPR-Cas and restriction-modification act additively against conjugative antibiotic resistance plasmid transfer in Enterococcus faecalis. *mSphere.* 2016;1(3):e00064-16.
59. Mayorga-Ramos A, Zúñiga-Miranda J, Carrera-Pacheco SE, Barba-Ostria C, Guamán LP. CRISPR-Cas-based antimicrobials: design, challenges, and bacterial mechanisms of resistance. *ACS Infect Dis.* 2023;9(7):1283–1302. https://doi.org/10.1021/acsinfecdis.2c00649
60. Qian Y, Zhou D, Li M, et al. Application of CRISPR-Cas system in the diagnosis and therapy of ESKAPE infections. *Front Cell Infect Microbiol.* 2023;13:1223696. https://doi.org/10.3389/fcimb.2023.1223696
61. Buckner MMC, Ciusa ML, Piddock LJV. Strategies to combat antimicrobial resistance: anti-plasmid and plasmid curing. *FEMS Microbiol Rev.* 2018;42(6):781–804. https://doi.org/10.1093/femsre/fuy031
62. Mancuso G, Midiri A, Gerace E, Biondo C. Bacterial antibiotic resistance: the most critical pathogens. *Pathogens.* 2021;10(10):1310. https://doi.org/10.3390/pathogens10101310
63. Venkateswaran P, Vasudevan S, David H, et al. Revisiting ESKAPE pathogens: virulence, resistance, and combating strategies focusing on quorum sensing. *Front Cell Infect Microbiol.* 2023;13:1159798. https://doi.org/10.3389/fcimb.2023.1159798
64. Munita JM, Arias CA. Mechanisms of antibiotic resistance. *Microbiol Spectr.* 2016;4(2) https://doi.org/10.1128/microbiolspec.VMBF-0016-2015
65. Serajian S, Ahmadpour E, Oliveira SMR, Pereira ML, Heidarzadeh S. CRISPR-Cas technology: emerging applications in clinical microbiology and infectious diseases. *Pharmaceuticals (Basel).* 2021;14(11):1171. https://doi.org/10.3390/ph14111171
66. Mikkelsen K, Bowring JZ, Ng YK, et al. An endogenous Staphylococcus aureus CRISPR-Cas system limits phage proliferation and is efficiently excised from the genome as part of the SCC*mec* cassette. *Microbiol Spectr.* 2023;11(4):e0127723. https://doi.org/10.1128/spectrum.01277-23
67. Jwair NA, Al-Ouqaili MTS, Al-Marzooq F. Inverse association between the existence of CRISPR/Cas systems with antibiotic resistance, extended spectrum β-lactamase and carbapenemase production in multidrug, extensive drug and pandrug-resistant *Klebsiella pneumoniae. Antibiotics (Basel).* 2023;12(6):980. https://doi.org/10.3390/antibiotics12060980
68. González de Aledo M, González-Bardanca M, Blasco L, et al. CRISPR-Cas, a revolution in the treatment and study of ESKAPE infections: pre-clinical studies. *Antibiotics (Basel).* 2021;10(7):756. https://doi.org/10.3390/antibiotics10070756
69. Wang Y, Yang J, Sun X, et al. CRISPR-Cas in Acinetobacter baumannii contributes to antibiotic susceptibility by targeting endogenous *AbaI. Microbiol Spectr.* 2022;10(4):e0082922. https://doi.org/10.1128/spectrum.00829-22

70. Qin S, Xiao W, Zhou C, et al. Pseudomonas aeruginosa: pathogenesis, virulence factors, antibiotic resistance, interaction with host, technology advances and emerging therapeutics. *Signal Transduct Target Ther.* 2022;7(1):199. https://doi.org/10.1038/s41392-022-01056-1
71. Hao M, He Y, Zhang H, et al. CRISPR-Cas9-mediated carbapenemase gene and plasmid curing in carbapenem-resistant *Enterobacteriaceae. Antimicrob Agents Chemother.* 2020;64(9):e00843-20. https://doi.org/10.1128/AAC.00843-20
72. Gleerup JL, Mogensen TH. CRISPR-Cas in diagnostics and therapy of infectious diseases. *J Infect Dis.* 2022;226(11):1867–1876. https://doi.org/10.1093/infdis/jiac145
73. Xiao J, Li J, Quan S, et al. Development and preliminary assessment of a CRISPR-Cas12a-based multiplex detection of *Mycobacterium tuberculosis* complex. *Front Bioeng Biotechnol.* 2023;11:1233353. https://doi.org/10.3389/fbioe.2023.1233353
74. Zein-Eddine R, Refrégier G, Cervantes J, et al. The future of CRISPR in Mycobacterium tuberculosis infection. *J Biomed Sci.* 2023;30:34. https://doi.org/10.1186/s12929-023-00932-4
75. Rock JM, Hopkins FF, Chavez A, et al. Programmable transcriptional repression in mycobacteria using an orthogonal CRISPR interference platform. *Nat Microbiol.* 2017;2:16274. https://doi.org/10.1038/nmicrobiol.2016.274
76. Gootenberg JS, Abudayyeh OO, Lee JW, et al. Nucleic acid detection with CRISPR-Cas13a/C2c2. *Science.* 2017;356(6336):438–442. https://doi.org/10.1126/science.aam9321
77. Cobb RE, Wang Y, Zhao H. High-efficiency multiplex genome editing of Streptomyces species using an engineered CRISPR/Cas system. *ACS Synth Biol.* 2015;4(6):723–728. https://doi.org/10.1021/sb500351f
78. Shi L, Gu R, Long J, Duan G, Yang H. Application of CRISPR-cas-based technology for the identification of tuberculosis, drug discovery and vaccine development. *Mol Biol Rep.* 2024;51(1):466. https://doi.org/10.1007/s11033-024-09424-6
79. Sashital DG. Pathogen detection in the CRISPR-Cas era. *Genome Med.* 2018;10(1):32. https://doi.org/10.1186/s13073-018-0543-4
80. Zhang X, He X, Zhang Y, et al. A new method for the detection of Mycobacterium tuberculosis based on the CRISPR/Cas system. *BMC Infect Dis.* 2023;23(1):680. https://doi.org/10.1186/s12879-023-08656-4
81. Jin L, Hu X, Tian Y, et al. Detection of *Staphylococcus aureus* virulence gene *pvl* based on CRISPR strip. *Front Immunol.* 2024;15:1345532. https://doi.org/10.3389/fimmu.2024.1345532
82. Malcı K, Walls LE, Rios-Solis L. Rational design of CRISPR/Cas12a-RPA based one-pot COVID-19 detection with design of experiments. *ACS Synth Biol.* 2022;11(4):1555–1567. https://doi.org/10.1021/acssynbio.1c00617 Epub 2022 Apr 1.
83. National Academies of Sciences, Engineering, and Medicine. *Gene Drives on the Horizon: Advancing Science, Navigating Uncertainty, and Aligning Research with Public Values.* National Academies Press; 2016.
84. Mulvihill JJ, Capps B, Joly Y, Lysaght T, Zwart HAE, Chadwick R. International Human Genome Organisation (HUGO) Committee of Ethics, Law, and Society (CELS). Ethical issues of CRISPR technology and gene editing through the lens of solidarity. *Br Med Bull.* 2017;122(1):17–29. https://doi.org/10.1093/bmb/ldx002

CHAPTER THREE

CRISPR-Cas based genome editing for eradication of human viruses

Dharmisha Solanki, Karan Murjani, and Vijai Singh*
Department of Biosciences, School of Science, Indrashil University, Rajpur, Mehsana, Gujarat, India
*Corresponding author. e-mail address: vijaisingh15@gmail.com; vijai.singh@indrashiluniversity.edu.in

Contents

Abstract

Clustered regularly interspaced short palindromic repeats (CRISPR)—Cas system possess a broad range of applications for genetic modification, diagnosis and treatment of infectious as well as non-infectious disease. The CRISPR-Cas system is found in bacteria and archaea that possess the Cas protein and guide RNA (gRNA). Cas9 and gRNA forms a complex to target and cleave the desired gene, providing defense against viral infections. Human immunodeficiency virus (HIV), hepatitis B virus (HBV), herpesviruses, human papillomavirus (HPV), and severe acute respiratory syndrome

Progress in Molecular Biology and Translational Science, Volume 208
ISSN 1877-1173, https://doi.org/10.1016/bs.pmbts.2024.07.012

coronavirus-2 (SARS-CoV-2) cause major life threatening diseases which cannot cure completely by drugs. This chapter describes the present strategy of CRISPR-Cas systems for altering the genomes of viruses, mostly human ones, in order to control infections.

Abbreviations

CCR5	C-C chemokine receptor type 5
CRISPR	Clustered regularly interspaced short palindromic repeats
CXCR4	C-X-C chemokine receptor type 4
GAPDH	Glyceraldehyde-3-phosphate dehydrogenase
gRNA	guide RNA
HBV	Hepatitis B virus
HIV	Human immunodeficiency virus
HPV	Human papillomavirus
HTLV-3	Human T-lymphotropic virus 3
LAV	Lymphadenopathy associated virus
MGEs	Mobile genetic elements
SARS-CoV-2	Severe acute respiratory syndrome coronavirus 2

1. Introduction

CRISPR-Cas systems (clustered regularly interspaced short palindromic repeats and CRISPR-associated protein (Cas)) are immune systems that are acquired by the majority of bacteria and archaea. The most reliable genome editing and engineering methods are used to be RNA-guided nucleases from CRISPR-Cas systems. The first-time evidence of their presence was found in the *Escherichia coli* genome in 1987 conducting a genetics study involved in phosphate metabolism. This sequence of unique repeating DNA was later identified as a CRISPR.[1] The highly versatile and heritable CRISPR-Cas system are integrated short sequences from viruses and various other mobile genetic elements (MGEs) into the host's CRISPR locus that directs the elimination of invasive nucleic acids through transcription and processing.[2]

Most notably, the intestinal microbiota and viral infections that change host-virus associations can now be accurately altered in a range of animal species thanks to recent advancements in CRISPR-Cas genome editing technology. As scientific interest in gene editing research grows, a new area of healthcare based on CRISPR-Cas9 altering technology is moving into the clinical stage for the treatment of viral infections.[3] Many different acute and chronic illnesses are brought on by viruses, and some of these illnesses can worsen and become fatal, similar to the ongoing coronavirus disease

2019 (COVID-19) pandemic. Some of these, like the herpes simplex virus, only result in mild infections.[4] At the moment, serious viral infectious illnesses like the human papillomavirus (HPV), hepatitis B virus (HBV), and human immunodeficiency virus (HIV) are endangering both health of people and global security at risk.[5] Meanwhile, mutant strains of human viruses are capable of spreading and even switching across classes, making pandemics possible, treating viral illnesses that can be difficult.[6] Consequently, numerous antiviral approaches have been created, including antibody-based treatments and medications with genetic modifications. Though more work has to be done to combat human viruses, the CRISPR-Cas genome modification system is a groundbreaking advancement in gene therapy and biomedicine.[3]

While there are a number of strategies that have great promise for fighting dangerous viruses that infect humans, the CRISPR-Cas genome editing method stands out as a groundbreaking development for microbial genome editing, gene therapy and healthcare.[3] In this chapter, the application studies that used the CRISPR-Cas9 technology to fight a number of infectious human viruses are highlighted.

2. Exploring CRISPR-Cas systems for genome editing of human viruses

2.1 Human immunodeficiency virus

The virus known as HIV, or human immunodeficiency virus, mainly targets and assaults the body's immune system, specifically the CD4 cells, which is also known as T cells, which are vital for fighting infections. Following infection, the virus multiplies and gradually erodes the host's immune system by integrating its genetic material within the host's cells.[7] HIV is a type of single-stranded, positive-sense, enveloped RNA virus that belongs to the genus lentivirus under the family retroviridae. HIV is categorised into two types: HIV-1 and HIV-2. The majority of HIV infections globally are caused by HIV-1, formerly known as lymphadenopathy associated virus (LAV) and human T-lymphotropic virus 3 (HTLV-III). HIV-1 is a more virulent and contagious virus. Due to its restricted ability to spread, HIV-2 is primarily restricted to West Africa and is less contagious, resulting in fewer infections per exposure.[8]

In the Fig. 1 of HIV, there are 72 glycoprotein projections on each HIV virion, which are made up of gp120 and gp41. The viral receptor for CD4

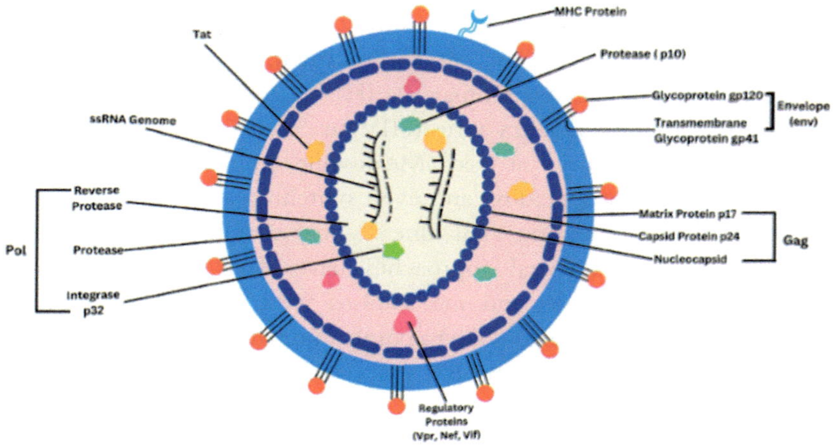

Fig. 1 Shows the structure of HIV.

on host cells is gp120, whereas gp41 is a transmembrane protein. The host-cell membrane proteins, such as MHC (major histocompatibility complex) molecules, are also present in the viral envelope. With an outer p24 protein layer and an outer p17 protein layer, the nucleocapsid is enclosed within the envelope. Two copies of ssRNA are present in the HIV genome along with protease, integrase (p32), reverse transcriptase (p64), and nucleoid proteins (p10).[8]

HIV infection can progress over time to the most advanced form, known as AIDS (acquired immunodeficiency syndrome), if treatment is not received. HIV is the virus that causes the infection, while AIDS is a condition that develops when HIV has significantly damaged the immune system. AIDS is characterized by a drastically reduced number of CD4 cells (less than 200 cells per cubic millimeter of blood) or the occurrence of certain opportunistic infections that take advantage of the weakened immune system.[9] The International Antiviral Society-USA Panel's 2020 recommendations on antiretroviral drugs for the treatment and prevention of HIV infection in adults.[10]

2.2 CRISPR-Cas systems for genome editing of HIV

HIV infections can be treated with antiretroviral therapy, but only for as long as the medications are taken. As of right now, there is no reliable vaccination or treatment for the virus. CRISPR-Cas-based gene editing systems have been considered as promising instruments in the creation of HIV gene therapeutics.[11] Researchers are looking into using CRISPR to find and remove specific parts of HIV DNA from infected cells' genetic makeup without messing with the rest of the genes. By deactivating proviral DNA, introducing therapeutic genes, and eliminating the coreceptors CCR5 and CXCR4, CRISPR-Cas systems can target HIV. Other tactics include "block and lock" virus suppression, "shock and kill" therapy, and turning on quiet restriction factors. These strategies seek to lower the latent HIV reservoir and boost HIV-resistant CD4 cells, resulting in a "functional cure" in which the virus remains dormant even in the absence of Highly active antiretroviral therapy (HAART).[12,13]

2.3 HIV-1 resistance through CRISPR-Cas9-mediated CCR5 knockout

HIV penetrates into cells through the cell receptor CD4 and the C-X-C chemokine receptor type 4 (CXCR4) or C-C chemokine receptor type 5 (CCR5) receptors. Type R5 HIV strains attach to CCR5, whereas type X4 HIV strains attach to CXCR4. R5-tropic strains predominate when an infection first arises, but as the sickness progresses, X4-tropic germs appear.[14] In order to efficiently stop, MT4CCR5 cells from expressing the CCR5 receptor is reported. Khamaikawin et al. used CRISPR/Cas9 to knock out the CCR5 gene in these cells. Once the CCR5 deletion cells were genetically modified, they were exposed to two distinct strains of HIV-1: X4-tropic HIV-1NL4-3, which targets the CXCR4 receptor, and R5-tropic HIV-1BaL, which targets the CCR5 receptor. A vector known as pLVX-C46-AcGFP1 was used to transduce the cells. This vector expresses a fusion inhibitor, C46, which stops the virus from fusing with the cell membrane and hence inhibits infection. They investigated these transduced cells with and without the CCR5 deletion. Both HIV-1 strains were introduced into the cells at two distinct multiplicities of infection (MOIs), 1 and 10, which corresponds to low and high viral exposure levels. Mock-treated MT4CCR5 cells and transduced cells using a non-antiviral vector (pLVX-AcGFP1) were utilized as controls. Using flow cytometry and 7AAD labeling to quantify cell death, the post-infection cell viability was evaluated. Similar to control cells, the CCR5 deletion cells were protected against R5-tropic HIV-1BaL but it continued to be vulnerable to X4-tropic HIV-1NL4-3. On the other hand, defense against R5-tropic HIV-1BaL and X4-tropic HIV-1NL4-3 infections were provided by the combined use of CCR5 knockout and C46 HIV-1 fusion inhibitor. This indicates that targeting both entry receptors (CCR5 and CXCR4) in addition to using a fusion inhibitor can offer strong protection against a variety of HIV-1 strains.[15]

2.4 HIV-1 therapy through targeted gene knock-in using SORTS and CRISPR-Cas9 system

The goal of surface oligopeptide knock-in for rapid target selection (SORTS) is interfered with the target gene by introducing an epitope tag into the human glycophosphatidylinositol (GPI)-anchored protein CD52 through an expression cassette.[16] To effectively transfer peptides to the plasma membrane and use them as identifiers for the fluorescence-activated cell sorting (FACS) separation of transformed cells, CD52 is the shortest human GPI-anchored protein.[12] SORTS system has been developed to fight against HIV-1. This methodology presents a promising strategy for HIV-1 therapy by utilizing the CRISPR-Cas9 system for specific gene knock-in to improve the selection and isolation of modified cells.[17]

The hemagglutinin (HA) tag has been inserted into the HIV-1 capsid protein p24 using this technique, enabling the separation of CEM cells and primary CD4+ T cells harboring inactivated HIV-1 proviruses. GAPDH (Glyceraldehyde-3-phosphate dehydrogenase) is a gene crucial for energy metabolism that was targeted GPI-construct knock-in experiment to evaluate its effectiveness. In order to stop NHEJ-mediated inactivation of the second GAPDH allele, which would cause cell death, a gRNA was engineered to target intron-2 which is close to the splice donor location. In order to ensure proper splicing and expression, a variety of 3′-UTR sequences with polyadenylation signals were examined by Zotova et at., in comparison to the splice donor site, the SV40 pA exhibits higher expression levels, and that the human β-globin's 49 bp polyA signal performed much better than the SV40 pA and soluble neuropilin-1's 17 bp polyA signal (sNRP-1). Even though poor splicing resulted in a decreased productivity of the splice donor construct, both GPI proteins were appropriately expressed on the cell surface to enable sorting and proliferation of the altered cells.[16]

This technique improves the selection and isolation of altered cells to increase the effectiveness of possible HIV-1 therapeutics by allowing the targeted insertion of therapeutic genes, such as those encoding HIV-1 inhibitors, into particular loci. The SORTS technique, when paired with CRISPR-are Cas9-mediated knock-in strategies, is a major step forward in the development of HIV-1 gene therapies. It allows for the accurate and effective selection of gene-edited cells and holds promise for effective and long-lasting treatments that may eventually result in an HIV-1 cure.[12]

3. Hepatitis B virus

As of 2022, the World Health Organization (WHO) estimates that 254 million individuals worldwide were chronically affected by hepatitis B. Furthermore, 1.2 million new infections occur each year.[18] A small DNA virus that causes hepatitis B differs from retroviruses in the aspect that it replicates via an RNA bridge and is able to be incorporated into the host genome. Hepatocytes, or liver cells, are specific target of the hepatitis B virus.[19] Cirrhosis, liver fibrosis, and hepatocellular carcinoma (HCC), a kind of liver cancer, are among the severe adverse effects that can result from a prolonged HBV infection.[20] The severity of an acute HBV infection might range from insignificant sickness to serious liver damage. Conversely, if treatment is not given, a chronic HBV infection persists for more than six months and might continue a lifetime.[21] Targeting cells with CRISPR-Cas systems are effective which destroy HBV DNA in cell cultures and animal models, as illustrated by recent investigations.[22] These encouraging findings raise the potential for CRISPR-Cas to be a viable therapy for chronic HBV infection. Current investigations seek to improve CRISPR delivery systems, improve the accuracy and efficacy of CRISPR targeting, and guarantee the safety of this strategy for possible clinical application.[23] In the fact that infected hepatocyte nuclei contain covalently closed circular DNA (cccDNA) which presents a significant treatment issue.[24] After therapy ends, viral replication resumes because this cccDNA acts as a reservoir for viral persistence.[25] To achieve complete removal of the virus from the body, it is generally believed that all cells harboring cccDNA infected cells must be destroyed.[26]

As shown in Fig. 2, DNA polymerase and other replication-related enzymes are found in the inner protein shell of the virus, also referred to as the core particle or "HBcAg," which is house of the viral DNA. The outer membrane is known as the "surface antigen," or "HBsAg," which is made up of protein and lipid. "HBV polymerase" is denoted by HBV pol and "relaxed circular DNA" by rcDNA.[27]

3.1 CRISPR-Cas systems to treatment of hepatitis B

The HBV vaccine does not treat chronic infections that already exist, it is only useful in preventing new infections. For patients who are already chronically infected with HBV, CRISPR-Cas systems may be able to target and eradicate the viral infection in affected cells.[28] Some individuals get resistant to the antiviral drugs that are already on the market. Rather

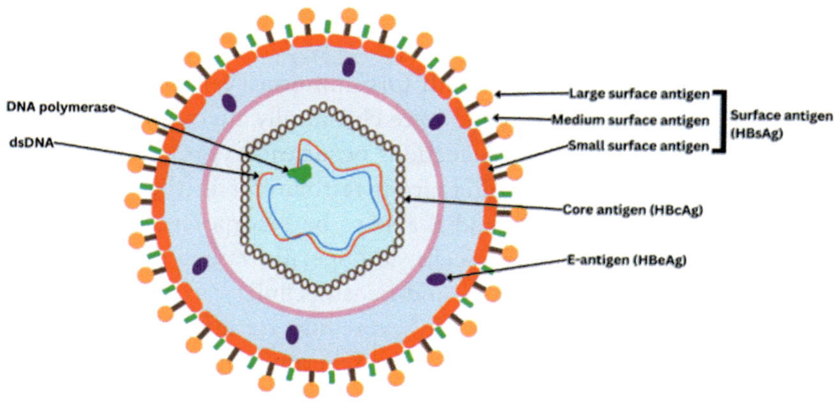

Fig. 2 shows the basic HBV structure.

from addressing the problem of drug resistance, CRISPR-Cas systems directly change the viral DNA, providing an alternative mode of action. Antiviral treatments can lower the viral load considerably, but they doesn't completely eradicate the viral infection.[29] Combining CRISPR-Cas with antiviral drugs may increase the likelihood that the viral infection can be totally removed from the patient's body. Covalently closed circular DNA, or cccDNA, is a template for the replication of viruses that the virus uses to survive in liver cells during an extended HBV infection.[30] Antiviral treatments available today can impede the growth of viruses, but they cannot destroy cccDNA. By disrupting or breaking down cccDNA, CRISPR-Cas systems may be able to treat the underlying source of persistent infection.[31]

3.2 Hepatitis B core and surface antigens

CRISPR-Cas9 ribonucleoproteins (RNPs) have the ability to edit the DNA of the hepatitis B virus (HBV), which may offer therapeutic alternatives for persistent HBV infections. Shortly after delivery, these RNPs, which are made up of the Cas9 enzyme and sgRNA causes site-specific double-strand breaks in DNA. The RNPs are broken down by endogenous nucleases and proteases in less than a day, resulting in effective on-target editing and a decrease in off-target effects.[32] Three HBV-specific sgRNAs (St3, St4, or St10) and *Staphylococcus aureus* Cas9 (StCas9) were used to construct RNPs. Recombinant covalently closed circular DNA (rcccDNA) was successfully cleaved using an in vitro cleavage assay. HBV transcription and replication were significantly reduced when these RNPs were transfected into HepG2 cells, with St10 attaining a reduction in rcccDNA levels of more than 98%. Twenty hours after transfection, it was confirmed by Bartosh et al. that StCas9 had been successfully delivered intranuclearly, and twenty hours later, Cas9 was no longer detectable. Furthermore, the levels of the hepatitis B surface antigen (HBsAg) and core antigen (HBcAg) were markedly decreased by StCas9/St10 RNPs.[26] These results highlight CRISPR-Cas9 RNPs' capacity to precisely target and damage HBV DNA, thereby lowering viral activity and providing a potentially effective treatment option for long-term HBV infections.[33]

3.3 Application of CRISPR-Cas9 RNPs for HBV elimination

By delivering CRISPR-Cas9 transiently as ribonucleoproteins (RNPs), a quicker and safer way to edit target sequences can be provided, possibly leading to the removal of HBV cccDNA from infected cells. This methodology guarantees effective on-target performance with some off-target consequences. Research employing CRISPR-Cas9 RNPs demonstrated a substantial decrease in HBV cccDNA levels. However, HBV cccDNA reassembled and viral replication continued, perhaps as a result of HBV rcDNA recycling, meaning that total eradication of the virus was not possible.[26]

3.4 Targeting HBV RNAs using CRISPR-Cas13b

The 30-nucleotide crRNA that makes CRISPR-Cas13b makes it a viable alternative for focusing on HBV RNAs. CRISPR-Cas13b showed remarkable efficiency with up to 95% and 96% in a reductions of HBeAg and HBsAg expression, respectively, as well as substantial suppression of viral replication. These outcomes surpass the previous research employing antisense oligonucleotides (ASOs) and siRNAs. ASO studies demonstrated a 68% reduction in HBsAg levels while siRNA studies demonstrated an 80% reduction in HBeAg levels. However, due to differences in experimental methods, it is difficult to directly compare the effectiveness of CRISPR-Cas13b to siRNA and ASO investigations in HepAD38 and NTCP-HepG2 cells.[34]

4. Human papillomavirus

HPV is a member of the Papillomaviridae family, which is a non-enveloped, double-stranded, circular DNA virus. The virus penetrates through damaged skin or mucosa, the epithelium and infects basal stem cells. Seven early (E) and two late (L) phase genes necessary for viral proliferation are found in its genome. The viral DNA may persist as an isolated episome for a while, before merging into the host's genome. HPV typically integrates into human DNA weak points, where strand breaks easily. HPV is responsible for malignancies of the throat, mouth, lung, and anogenital regions. Condylomata and low-grade precancerous tumors are the usual manifestations of low-risk which are subtypes of HPV 6 and 11. HPV subtypes 16 and 18 carry a high risk of causing high-grade intraepithelial lesions that leads to malignancy, particularly cervical cancer.[35]

As depicted in Fig. 3, the diameter of HPVs is approximately 50–60 nm, and they have a spherical shape. The capsid of the virus consists of proteins L1 and L2. The double-stranded, 8000 base pair viral genome is circularly coiled and linked to proteins that resemble histones. It is recognized that high-risk human papillomavirus (Hr-HPV), specifically 16 and 18 strains, are the main cause of cervical cancer. E7 interferes with the retinoblastoma (RB) protein, whereas E6 lowers the tumor suppressor protein P53. These interactions drive the carcinogenesis of cervical epithelial cells. Consequently, the E6 and E7 genes are considered to be crucial targets for preventing and treating cervical cancer.[36]

4.1 Gene knockout chain reaction targets (GKCR) in HPV18 E6 and E7

This technique elevated P53 and RB proteins, substantially knocked out HPV18 E6 and E7 oncogenes in HeLa cells, and prevented HeLa cell

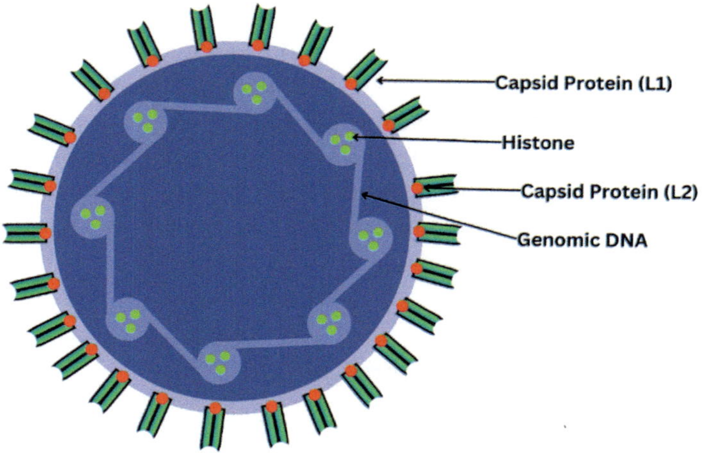

Fig. 3 Shows the structure of HPV.

motility and proliferation. GKCR offers a viable approach to treat cervical cancer by effectively disrupting multicopied HPV viral genes.[37,38] Tian et al. destroyed the HPV18 E6 and E7 genes in HeLa cervical cancer cells using the GKCR (Gene Knockout Chain Reaction) technique.[38] The control technique involved sporadically manufacturing the Cas9 protein and sgRNAs without introducing any donor DNA. On the other hand, the GKCR approach made use of the identical Cas9 and sgRNAs introduced them into the genome at the desired location, enabling ongoing expression. These strategies were merged in the combined GKCR+ control method. Because of the continuous splitting function, this method severely damaged the viral genes. The method significantly eliminated out AAVS1 alleles in HEK293T cells. The AAVS1 locus successfully integrated the GKCR donor. Additionally, it significantly eliminated the HPV18 E6 and E7 oncogenes in HeLa cells, stimulating P53 and RB proteins which prevents the cancer cells' ability to multiply and also prevents metastasis.[38]

The findings showed that, in comparison to the GKCR+ control method, the GKCR approach improved the Cas9/gRNA recordings integration rate and produced a higher indel (insertion/delation) rate than the conventional CRISPR-Cas9 strategy. This advantage over the other approaches is probably attributable to the high degree of linearized donor insertion resulting in persistent interruption of the gene. Reverse insertion of the Cas9/gRNA cassette might potentially come back a target sequence for sgRNA, leading to repetitive editing until destruction. This could explain why the forward insertion rate of the cassette was significantly higher than the reverse insertion rate.[38,39]

4.2 CRISPR-Cas13a targets HPV E6 gene to combat cancer

With an emphasis on high-risk strains like HPV16 and HPV18, the study sought to determine whether the HPV E6 gene might be specifically targeted and rendered inactive through the use of a reprogrammed CRISPR-Cas13a system. Studies have been shown that CRISPR-Cas13 can successfully address and destroy HPV RNA, which causes a increase in cell death and a decrease in cell growth in HPV-infected cells. By targeting the virus at the RNA level as opposed to altering the host DNA directly, this technique offers a novel and promising means of treating HPV cancer.[40] The CRISPR-Cas13a method was created to target the E6 gene of several HPV strains using a guide RNA-expressing vector. Keratinocytes were then exposed to the system to evaluate its effectiveness in targeting the E6 gene of several HPV strains. The effectiveness of this approach was assessed using western blot analysis, which demonstrated that E6-transformed keratinocytes had lower levels of E6 protein. Furthermore, CCK-8 tests were utilized to measure the E6-transformed cells' growth rate, which revealed a notable reduction in proliferation (14 ± 1.8% on average). Hoechst 33258 staining and ELISA for caspase-3 were used to detect apoptosis. The results showed that treated cells had a markedly higher average amount of apoptosis (80.2 ± 3.2%), whereas normal keratinocytes showed no change in this regard. These findings show that the HPV E6 gene is efficiently targeted and rendered inactive by the CRISPR-Cas13a system, hence limiting cell division and triggering apoptosis in HPV-positive cells . This strategy offers a viable substitute for the present therapeutic approaches by demonstrating CRISPR-Cas13a's potential for the treatment of cancers linked to HPV.[23]

5. Severe acute respiratory syndrome-Coronavirus-2 (SARS-CoV-2)

The coronavirus family, which includes SARS-CoV-2, the virus that caused the COVID-19 pandemic, is known to cause a variety of ailments, from the common cold to more serious conditions like severe acute respiratory syndrome (SARS) and Middle East Respiratory Syndrome (MERS).[41] After being discovered for the first time in Wuhan, China in December 2019, the virus is thought to have come from a zoonotic source likely bats and may have spread to humans through intermediate hosts like pangolins.[42] The genome of SARS-CoV-2 is a positive-sense, single-stranded RNA virus that is roughly 29.9 kilobases (kb) long. Numerous auxiliary proteins as well as various structural proteins, such as the spike, envelope, membrane, and nucleocapsid proteins, are encoded by this large RNA genome. This structure is crucial for the ability of virus to replicate and cause infection to other organisms.[43]

5.1 CRISPR-Cas based detection of SARS-CoV-2

Different methods have been employed to diagnose SARS-CoV-2, including as immunology methods, sequencing-based methods, and RT-qPCR. However, the most common techniques of diagnosing SARS-CoV-2 (RT-PCR, laboratory tests) have been limited by issues such low precision and specificity due to the preparation of samples, chemicals, equipment, and diverse types of clinical specimens.[44] Thus, more investigation is required to discover sensitive and low-cost techniques for detecting SARS-CoV-2. The

SHERLOCK technique, identifies RNA viruses like dengue and zika. An FDA-approved biosciences company's SHERLOCK-based test, which can identify several SARS-CoV-2 genes in less than an hour, was approved during the COVID-19 pandemic. Another CRISPR-based detection system, DETECTR was created by Broughton et al. which makes use of CRISPR-Cas12.[45] Initially DETECTR was developed by Chen et al. to identify human papillomavirus (HPV).[46] It takes around 30 min to identify SARS-CoV-2. In contrast to SHERLOCK, which targets ssRNA with Cas13, DETECTR recognizes dsDNA with Cas12, these two techniques provide a versatile tool for detecting DNA-based pathogens and expanding the range of CRISPR-based diagnostic applications.[47]

5.2 CRISPR-crafted SARS-CoV-2 vaccine

SARS-CoV-2, the virus responsible for COVID-19 disease, has led to a increase in the development of novel vaccine approaches to tackle this worldwide pandemic. Several vaccine platforms have been investigated, utilizing diverse technological developments to improve safety and effectiveness. CRISPR-Cas9 technology has emerged as a potent method for optimizing vaccine development and treatment approaches. Viral vectors and antigens have been engineered can easily produce potent vaccines and thanks to the CRISPR-Cas9 technique. Researchers have been able to create vaccine candidates with high immunological responses and protection against pathogens like SARS-CoV-2 due to its accuracy and adaptability. Using a canarypox viral vector system (ALVAC-VLPs), Wang et al. have used CRISPR/Cas9 to create a vaccine against SARS-CoV-2 virus-like particles (VLPs).[48] Using CRISPR/Cas9, this vaccine modifies the Canarypox virus to include SARS-CoV-2 antigens, producing VLPs that closely resemble the virus but do not spread illness. ALVAC-VLP vaccination has been shown to strongly induce T and B cell responses specific to SARS-CoV-2, protecting against lethal challenges with virus strains that have been modified for mice. These results show that the safe ALVAC-VLPs vaccine is both immunogenic and effective against SARS-CoV-2.[48]

6. Conclusion and future remarks

The biotechnology field of CRISPR-Cas9 has emerged as a promising area, enabling basic research and offering potential applications in the field of genetic engineering. Due to the widespread usage of CRISPR-Cas9 and ongoing work to enhance its editing capabilities, its full impact in society as well as in medicine is significant. The ongoing efforts to develop and implement new methods for delivering genetically modified components inside the cells may result in important pharmaceutical applications for the detection and treatment of cancer, genetic diseases, and viral infections. Through a variety of methods, CRISPR gene-editing technology may contribute to the prevention of upcoming worldwide pandemics. Emerging disease outbreaks might be tracked considerably more effectively and in real-time with CRISPR-based tools for diagnosis for quick point-of-care testing, eliminating the obstacles associated with conventional testing techniques. Before being used in therapeutic settings, CRISPR/Cas9 will need significant refinement and improvement.

Acknowledgments

Authors acknowledge the Gujarat State Biotechnology Mission (GSBTM Project ID—GSBTM/JD(R&D)/663/2023-24/02004140), Government of Gujarat, Gujarat, India for financial assistance.

References

1. Ishino Y, Krupovic M, Forterre P. History of CRISPR-Cas from encounter with a mysterious repeated sequence to genome editing technology. *J Bacteriol*. 2018;200. https://doi.org/10.1128/jb.00580-17
2. Hryhorowicz M, Lipiński D, Zeyland J, Słomski R. CRISPR/Cas9 immune system as a tool for genome engineering. *Arch Immunol Ther Exp*. 2017;65:233–240.
3. Tripathi S, Khatri P, Fatima Z, Pandey RP, Hameed S. A landscape of CRISPR/Cas technique for emerging viral disease diagnostics and therapeutics: progress and prospects. *Pathogens*. 2022;12:56.
4. Lin H, Li G, Peng X, et al. The use of CRISPR/Cas9 as a tool to study human infectious viruses. *Front Cell Infect Microbiol*. 2021;11:590989.
5. Morens DM, Fauci AS. Emerging infectious diseases: threats to human health and global stability. *PLoS Pathog*. 2013;9:e1003467.
6. Parrish CR, Holmes EC, Morens DM, et al. Cross-species virus transmission and the emergence of new epidemic diseases. *Microbiol Mol Biol Rev*. 2008;72:457–470.
7. Masenga SK, Mweene BC, Luwaya E, Muchaili L, Chona M, Kirabo A. HIV–host cell interactions. *Cells*. 2023;12:1351.
8. Blood GAC. Human immunodeficiency virus (HIV). *Transfus Med Hemotherapy*. 2016;43:203.
9. Kanki PJ, Koofhethile CK. *HIV/AIDS Global Epidemic, Infectious Diseases*. Springer; 2023:221–250.
10. Saag MS, Gandhi RT, Hoy JF, et al. Rio, Antiretroviral drugs for treatment and prevention of HIV infection in adults: 2020 recommendations of the International Antiviral Society–USA Panel. *Jama*. 2020;324:1651–1669.
11. Hussein M, Molina MA, Berkhout B, Herrera-Carrillo E. A CRISPR-Cas cure for HIV/AIDS. *Int J Mol Sci*. 2023;24:1563.

12. Maslennikova A, Mazurov D. Application of CRISPR/Cas genomic editing tools for HIV therapy: toward precise modifications and multilevel protection. *Front Cell Infect Microbiol*. 2022;12:880030.
13. Eggleton J.S., Nagalli S. Highly Active Antiretroviral Therapy (HAART). StatPearls. Treasure Island (FL) 2024.
14. Alkhatib G. The biology of CCR5 and CXCR4. *Curr OpHIV AIDS*. 2009;4:96–103.
15. Khamaikawin W, Saisawang C, Tassaneetrithep B, Bhukhai K, Phanthong P, Borwornpinyo S, et al. CRISPR/Cas9 genome editing of CCR5 combined with C46 HIV-1 fusion inhibitor for cellular resistant to R5 and X4 tropic HIV-1. *Sci. Rep*. 2024;14(1):10852.
16. Zotova A, Pichugin A, Atemasova A, et al. Isolation of gene-edited cells via knock-in of short glycophosphatidylinositol-anchored epitope tags. *Sci Rep*. 2019;9:3132.
17. Maslennikova A, Kruglova N, Kalinichenko S, et al. Engineering T-cell resistance to HIV-1 infection via knock-in of peptides from the heptad repeat 2 domain of gp41. *MBio*. 2022;13:e03589-21.
18. Brown R, Goulder P, Matthews PC. Sexual dimorphism in chronic hepatitis B virus (HBV) infection: evidence to inform elimination efforts. *Wellcome Open Res*. 2022;7.
19. Tu T, Budzinska MA, Vondran FW, Shackel NA, Urban S. Hepatitis B virus DNA integration occurs early in the viral life cycle in an in vitro infection model via sodium taurocholate cotransporting polypeptide-dependent uptake of enveloped virus particles. *J Virol*. 2018;92. https://doi.org/10.1128/jvi.02007-17
20. Kanda T, Goto T, Hirotsu Y, Moriyama M, Omata M. Molecular mechanisms driving progression of liver cirrhosis towards hepatocellular carcinoma in chronic hepatitis B and C infections: a review. *Int J Mol Sci*. 2019;20:1358.
21. Liang TJ. Hepatitis B: the virus and disease. *Hepatology*. 2009;49:S13–S21.
22. Najafi S, Tan SC, Aghamiri S, et al. Therapeutic potentials of CRISPR-Cas genome editing technology in human viral infections. *Biomed Pharmacother*. 2022;148:112743.
23. Li H, Yang Y, Hong W, Huang M, Wu M, Zhao X. Applications of genome editing technology in the targeted therapy of human diseases: mechanisms, advances and prospects. *Signal Transduct Target Ther*. 2020;5:1.
24. Guo H, Jiang D, Zhou T, Cuconati A, Block TM, Guo J-T. Characterization of the intracellular deproteinized relaxed circular DNA of hepatitis B virus: an intermediate of covalently closed circular DNA formation. *J Virol*. 2007;81:12472–12484.
25. Bhat SA, Kazim SN. HBV cccDNA— A culprit and stumbling block for the hepatitis B virus infection: its presence in hepatocytes perplexed the possible mission for a functional cure. *ACS Omega*. 2022;7:24066–24081.
26. Kostyushev D, Kostyusheva A, Brezgin S, et al. Depleting hepatitis B virus relaxed circular DNA is necessary for resolution of infection by CRISPR-Cas9. *Mol Therapy-Nucleic Acids*. 2023;31:482–493.
27. Mahmood F, Xu R, Awan MUN, et al. HBV vaccines: advances and development. *Vaccines*. 2023;11:1862.
28. Stone D, Long KR, Loprieno MA, et al. CRISPR-Cas9 gene editing of hepatitis B virus in chronically infected humanized mice. *Mol Ther Methods Clin Dev*. 2021;20:258–275.
29. Yang Y-C, Yang H-C. Recent progress and future prospective in HBV cure by CRISPR/Cas. *Viruses*. 2021;14:4.
30. Lebbink RJ, de Jong DC, Wolters F, et al. A combinational CRISPR/Cas9 gene-editing approach can halt HIV replication and prevent viral escape. *Sci Rep*. 2017;7:41968.
31. Dong C, Qu L, Wang H, Wei L, Dong Y, Xiong S. Targeting hepatitis B virus cccDNA by CRISPR/Cas9 nuclease efficiently inhibits viral replication. *Antivir Res*. 2015;118:110–117.

32. Martinez MG, Combe E, Inchauspe A, et al. CRISPR-Cas9 targeting of hepatitis B virus covalently closed circular DNA generates transcriptionally active episomal variants. *MBio.* 2022;13:e02888-21.
33. Bartosh UI, Dome AS, Zhukova NV, Karitskaya PE, Stepanov GA. CRISPR/Cas9 as a new antiviral strategy for treating hepatitis viral infections. *Int J Mol Sci.* 2023;25:334.
34. McCoullough LC, Fareh M, Hu W, et al. CRISPR-Cas13b-mediated suppression of hepatitis B virus replication and protein expression. *J Hepatol.* 2024.
35. Luria L., Cardoza-Favarato G. Human Papillomavirus. StatPearls. Treasure Island (FL) 2024.
36. Narisawa-Saito M, Kiyono T. Basic mechanisms of high-risk human papillomavirus-induced carcinogenesis: roles of E6 and E7 proteins. *Cancer Sci.* 2007;98:1505–1511.
37. Pal A, Kundu R. Human papillomavirus E6 and E7: the cervical cancer hallmarks and targets for therapy. *Front Microbiol.* 2020;10:3116.
38. Tian R, Liu J, Fan W, et al. Gene knock-out chain reaction enables high disruption efficiency of HPV18 E6/E7 genes in cervical cancer cells. *Mol Therapy-Oncolytics.* 2022;24:171–179.
39. Suzuki K, Tsunekawa Y, Hernandez-Benitez R, et al. In vivo genome editing via CRISPR/Cas9 mediated homology-independent targeted integration. *Nature.* 2016;540:144–149.
40. Li C, Guo L, Liu G, et al. Reprogrammed CRISPR-Cas13a targeting the HPV16/18 E6 gene inhibits proliferation and induces apoptosis in E6-transformed keratinocytes. *Exp Ther Med.* 2020;19:3856–3860.
41. Cascella M, Rajnik M, Aleem A, Dulebohn SC, Di Napoli R. *Features, evaluation, Treat coronavirus (COVID-19).* 2020.
42. Mackenzie JS, Smith DW. COVID-19: a novel zoonotic disease caused by a coronavirus from China: what we know and what we don't. *Microbiol. Aust.* 2020;41:45–50.
43. Brant AC, Tian W, Majerciak V, Yang W, Zheng Z-M. SARS-CoV-2: from its discovery to genome structure, transcription, and replication. *Cell Biosci.* 2021;11:1–17.
44. Rando HM, Brueffer C, Lordan R, et al. Molecular and serologic diagnostic technologies for SARS-CoV-2. *ArXiv.* 2022.
45. Broughton JP, Deng X, Yu G, et al. CRISPR–Cas12-based detection of SARS-CoV-2. *Nat Biotechnol.* 2020;38:870–874.
46. Chen JS, Ma E, Harrington LB, et al. CRISPR-Cas12a target binding unleashes indiscriminate single-stranded DNase activity. *Science.* 2018;360:436–439.
47. Ebrahimi S, Khanbabaei H, Abbasi S, et al. CRISPR-Cas system: a promising diagnostic tool for Covid-19. *Avicenna J Med Biotechnol.* 2022;14:3.
48. Wang W, Wang S, Meng X, et al. A virus-like particle candidate vaccine based on CRISPR/Cas9 gene editing technology elicits broad-spectrum protection against SARS-CoV-2. *Antivir Res.* 2024;225:105854.

CHAPTER FOUR

Advances in CRISPR-Cas systems for gut microbiome

Namra Ali[a,1], Chaitali Vora[b,1], Anshu Mathuria[c], Naina Kataria[c], and Indra Mani[a,*]

[a]Department of Microbiology, Gargi College, University of Delhi, New Delhi, India
[b]Department of Biosciences and Biomedical Engineering, Indian Institute of Technology, Indore, India
[c]Department of Biochemistry, Sri Venkateswara College, University of Delhi, New Delhi, India
*Corresponding author. e-mail address: indra.mani@gargi.du.ac.in; indramanibhu@gmail.com

Contents

Abstract

CRISPR-Cas technology has revolutionized microbiome research by enabling precise genetic manipulation of microbial communities. This review explores its diverse applications in gut microbiome studies, probiotic development, microbiome diagnostics, pathogen targeting, and microbial community engineering. Engineered bacteriophages and conjugative probiotics exemplify CRISPR-Cas's capability for targeted bacterial manipulation, offering promising strategies against antibiotic-resistant infections and other gut-related disorders. CRISPR-Cas systems also enhance probiotic efficacy by improving stress tolerance and colonization in the gastrointestinal tract. CRISPR-based techniques in

[1] *Namra Ali and Chaitali Vora contributed equally the first authors.*

Progress in Molecular Biology and Translational Science, Volume 208
ISSN 1877-1173, https://doi.org/10.1016/bs.pmbts.2024.07.008

diagnostics enable early intervention by enabling fast and sensitive pathogen identification. Furthermore, CRISPR-mediated gene editing allows tailored modification of microbial populations, mitigating risks associated with horizontal gene transfer and enhancing environmental and health outcomes. Despite its transformative potential, ethical and regulatory challenges loom large, demanding robust frameworks to guide its responsible application. This chapter highlights CRISPR-Cas's pivotal role in advancing microbiome research toward personalized medicine and microbial therapeutics while emphasizing the imperative of balanced ethical deliberations and comprehensive regulatory oversight.

1. Introduction

Trillions of bacteria living in the gastrointestinal (GI) tract make up the human gut microbiome, which is crucial to both host health and disease. Recent discoveries have revealed the complex interactions that these commensal bacteria have with human physiology, affecting everything from immunological responses to metabolism and even brain processes. The gut microbiota interacts with its host through a wide range of processes, and among them, CRISPR-Cas systems have become potent tools with significant consequences. The development of CRISPR-Cas technologies has fundamentally changed our capacity to modify genetic material with previously unheard-of efficiency and precision.[1]

CRISPR-Cas systems show extraordinary versatility and functionality inside the gut microbiota. They mediate connections among members of the microbiota and with the host environment, in addition to giving bacteria defensive mechanisms against viral predation. Clarifying microbial ecology, evolution, and their consequences for human health and disease requires an understanding of the dynamics of CRISPR-Cas systems in the gut microbiome. As a result, increasing attention is being paid to using CRISPR-Cas technologies to clarify and maybe modify these microbial communities.[2]

Both archaea and bacteria have defense mechanisms called CRISPR and Cas proteins. By enclosing fragments of the genetic material of these invaders within their own DNA, they provide protection against plasmid and viral DNA. It functions as a genetic memory of previous infections thanks to this stored data. In nature, CRISPR systems are present in most archaea and approximately 47% of bacteria, including both benign and pathogenic varieties. Different bacterial strains can be distinguished from one another using the distinctive stored sequences known as spacers. This makes them useful for classifying bacteria found in probiotics, as well as in identifying disease-causing strains.[3] *Bacteroides fragilis* is important in the "hot spot" of horizontal gene transfer (HGT) between microorganisms.[4] They represent one of the human gut's densest populations of resistance gene repositories.[5] The effectiveness of many antimicrobial treatments has been compromised by the spread of resistance and virulence genes due to the transmission of genetic material across distinct bacterial species.[6,7] Resistance genes can move between various pathogens, between commensal organisms and potential infections, and even among pathogens and probiotics in the gut ecosystem.[8,9]

Genetic studies of clinical isolates of *B. fragilis* and isolates from the GI tract reveal extensive HGT activity, including the procurement of genes for resistance from gram-positive bacteria with extreme divergence.[10,11] Because *B. fragilis* CRISPR-Cas systems can regulate or record HGT, which involves the absorption of genes linked to virulence and antibiotic resistance, these systems are important.

Recent pioneering investigations on harmful bacteria in the gut justify the usage of modified bacteriophages with an exogenous CRISPR-Cas system. This method is a flexible tool for manipulating bacteria since it can be steered toward any point of interest. In particular, transcriptional repression and lysogenic phages have been used to great effect recently.[12] Nevertheless, more research is required to allow stable modification of the gene content and structure of microbial communities without requiring the integration of a viral genome.

Bacteria have conserved DNA repair mechanisms that can be used to inducing targeted genomic deletions. This makes CRISPR-Cas9 an improved tool for the job. Colonizing the epidermis (eliminating a single genetic area) is an approach that has the advantage of not having as much of an influence on the remainder of the microbiota in the gut as eradicating the strain.[13] The range of removal sizes (379–68,321 bp) is quite large, indicating the capacity of bacteria to withstand large deletions. This makes it possible to eradicate whole biosynthetic gene clusters or pathogenicity islands in vivo. Consequently, it could be able to introduce more intricate genetic circuits to *E. coli* in order to modify immunological function or enhance metabolic pathways that are advantageous to its mammalian host. Furthermore, co-administration of a DNA-repair template with CRISPR-Cas9 may constrain the size of the resultant chromosomal deletions.[14]

Unexpectedly, Neil et al. found an extensive variety of removal sizes (379–68,321 bp), suggesting that bacteria are capable of withstanding significant deletions. This makes it possible to eliminate whole biosynthetic gene clusters or pathogenicity islands in vivo. This work has demonstrated that *E. coli* can be given more intricate genetic circuits to enhance metabolic pathways advantageous to its mammal host or modify immune function. Moreover, supplying CRISPR-Cas9 with a DNA repair template may lessen the severity of the chromosomal deletions that are produced.[15]

This novel technique offers the way for new treatment strategies that modify the gut microbiota, potentially targeting antibiotic resistance and pathogenic infections with precision and efficacy. Using CRISPR-Cas9 in a conjugative probiotic system allows us to carefully edit or eliminate detrimental bacterial genes while maintaining helpful microbiota members. This technique also opens up new possibilities for improving probiotic effects, such as increasing metabolic pathways that benefit the host or modifying the immunological response. As a result, modified conjugative probiotics may become a significant tool in personalized medicine, providing individualized treatments for several gut-related diseases and ailments.[16]

2. CRISPR-Cas for probiotic development

Fermented food manufacturing has long benefited from the use of probiotics, live bacteria that are beneficial to health when taken in the right doses. Lately, focus has switched to probiotic metabolites, like fatty acids with short chains, which maintain gut integrity by lowering pH to limit the growth of harmful bacteria and giving colonic cells energy.[17,18] These bioactive compounds also show promise in anti-inflammatory and neurological roles[19] expanding probiotics' applications in therapeutic and animal health sectors. *Bifidobacterium*, a common gut bacterium, produces exopolysaccharides that aid immune modulation and protect against pathogens.[20] Additionally, probiotics exhibit antitumor effects attributed to their immunomodulatory and antiproliferative activities.[21] Live biotherapeutic products, considered as living microbial drugs, are increasingly explored as alternatives in clinical settings.[22]

However, probiotics vary significantly in their tolerance and ability to colonize the gut, which imposes limitations on their applications.[23] Genome engineering techniques, involving modifications to introduce, remove, or alter genetic traits, offer a pathway to enhance probiotics. These changes have the potential to increase their probiotic efficacy, strengthen their ability to survive in the GI tract, and increase their level of stress resistance during food production. Technological developments in genome manipulation and artificial biology have made it easier to develop novel probiotic strains with targeted functionalities. Treatments for inflammatory conditions, metabolic problems, pathogenic infections, and maybe cancer have all benefited from this progress.[24]

Numerous CRISPR-based technologies have developed as effective tools for genome editing as the mechanics behind CRISPR-Cas systems have become more understood. Notably, because of their remarkable specificity, effectiveness, and adaptability, Cas9 and Cas12-based genome editing approaches have been extensively studied and extensively used.[25] Fig. 1 shows probiotic genome editing methods based on CRISPR.[26]

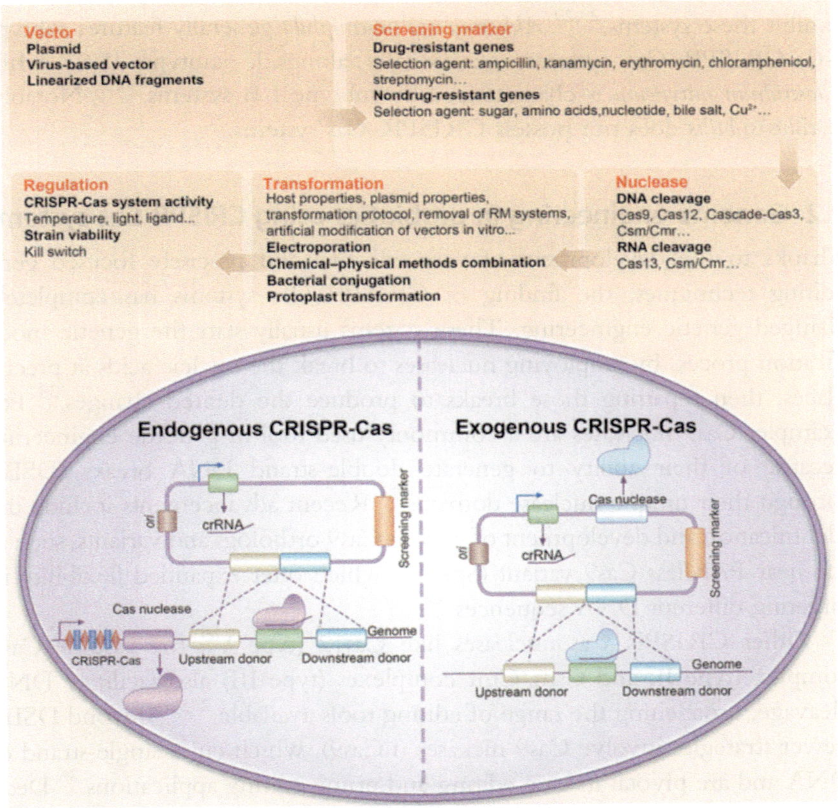

Fig. 1 Probiotic genome editing techniques based on CRISPR. Top section: Crucial elements that CRISPR-based genome editing should take into account in its entirety. Bottom section: Endogenous CRISPR-Cas system-based genome editing schematic (left); exogenous CRISPR-Cas system-based genome editing schematic (right). Restriction-modification, or RM. *Adapted from Ling L, Shimaa Elsayed H, Nan P. CRISPR-Cas-based engineering of probiotics. BioDesign Res. 2023;5:0017. https://doi.org/10.34133/bdr.0017.*

2.1 Variety of CRISPR-Cas systems found in probiotics

CRISPR-Cas systems are widespread among various microbial species, albeit with varying prevalence. Whole CRISPR-Cas systems have been found in the *Lactobacillus genus* surpasses 90% in species like *Lactobacillus crispatus* and *Lactobacillus delbrueckii*, whereas *Lactobacillus acidophilus* only harbors CRISPR arrays devoid of associated cas genes.[27] In *Bifidobacterium*, 57% of strains have CRISPR-Cas systems, with subtype I-E being the predominant type, similar to *Lactobacillus*. Usually, *Streptococcus thermophilus* can support one or more types I, II, and III CRISPR-Cas systems. On the other hand, some species have less CRISPR-Cas systems than others; for instance, only 17% of *Limosilactobacillus reuteri* strains and around 30% of *Pediococcus acidilactici* strains exhibit these systems.[28,29] *Akkermansia muciniphila* generally features subtype I-C CRISPR-Cas systems, occasionally alongside subtype II-C, while *Clostridium butyricum* is characterized by subtype I-B systems.[30,31] Notably, *Bacillus subtilis* does not possess CRISPR-Cas systems.[32]

2.2 Genome engineering in probiotics using CRISPR-Cas systems

Thanks to the development of very effective and precisely focused gene editing techniques, the finding of CRISPR-Cas systems has completely changed genetic engineering. These systems usually start the genetic modification process by employing nucleases to break the nucleic acids at precise places, then repairing those breaks to produce the desired changes.[33] For example, Cas9 nucleases are a commonly used tool in genome engineering because of their ability to generate double-strand DNA breaks (DSBs) through their unique nuclease domains.[34] Recent advancements include the identification and development of various Cas9 orthologs and variants, such as the near-PAMless Cas9 variant (SpRY), which offer expanded flexibility in targeting different DNA sequences.[35]

Other CRISPR-Cas nucleases like Cas12 (type V), the cascade-Cas3 complex (type I), and Csm/Cmr complexes (type III) also facilitate DNA cleavage, broadening the range of editing tools available.[28,29] Beyond DSBs, newer strategies involve Cas9 nickases (nCas9), which cut a single strand of DNA and are pivotal in base editing and prime editing applications.[36] Dead Cas9 (dCas9), lacking nuclease activity, is utilized for precise genetic modifications such as CRISPR activation (CRISPRa), interference (CRISPRi), and epigenetic alterations.

Genome editing using CRISPR tools has revolutionized genetic manipulation, offering versatile capabilities like gene deletion, insertion, and replacement.[37]

Microhomology-mediated end joining, which depends on microhomologous sequences for genome modification, offers another error-prone alternative to non-homologous end joining (NHEJ), which is more common in eukaryotes than prokaryotes.[38] Homology-directed repair (HDR), which introduces desired alterations using repair templates, is recommended for precise adjustments.[39] HDR-based editing optimization options include reducing NHEJ activity, boosting donor template stability, employing high activity Cas nucleases, and increasing recombination efficiency. Additionally, for manipulating large genomic segments in cells with low HDR efficiency, transposon-associated CRISPR systems offer promising alternatives.[40]

2.3 Application of CRISPR-engineered probiotics in therapeutics

Human gut microbiota microorganisms have been demonstrated to impact human health via a number of different pathways, including the gut-organ axis.[41] Probiotic use is becoming more widely acknowledged as a successful method for curing or preventing human ailments. Genetically modified strains of probiotics have unique or improved qualities compared to other strains, making them more valuable for study and use.[42] These days, a rising number of genetically modified probiotics—mostly created via homologous recombination (HR) or conventional plasmid expression—are being used to cure or prevent a variety of illnesses.[43]

For example, gene deletion and integration were carried out in *E. coli* Nissle 1917 (EcN) via lambda red recombination, producing an engineered strain that could produce L-arginine from ammonia, a metabolic byproduct of tumor cells. This change has an anti-tumor effect by increasing the number of T cells that infiltrate tumors.[44] The creation of antimicrobial C-type lectin regenerating islet-derived 3 gamma in the gut is further encouraged by modified *L. reuteri* expressing the anti-inflammatory cytokine interleukin-22 via a plasmid, which helps to mitigate alcohol-induced liver damage.[45]

As a result of the quick developments in synthetic biology and CRISPR-based editing technologies, probiotics with CRISPR-Cas systems are being researched as microbial treatments. Antibiotic-resistant bacteria are becoming a major worldwide issue, which is why CRISPR-Cas technologies are being used to study antibiotic resistance pathways and fight infectious diseases. For example, a highly effective approach for antimicrobial therapy has been devised using CRISPR-Cas9, wherein selected resistant to antibiotics *E. coli* strains were nearly completely removed from the gut microbiome by use of modified probiotic EcN.

Furthermore, a CRISPR-Cas9-engineered vaccine strain of EcN was created to protect against F4+/F18+ enterotoxigenic *E. coli* (ETEC) infections. This required chromosomal integration of genes from the F4 and F18 fimbriae clusters. Research revealed that vaccinated mice and piglets' antibodies dramatically decreased the F4+ and/or F18+ ETEC strains' adhesion to porcine intestinal cells in vitro.[15]

A complicated, long-term inflammatory gastrointestinal (GI) ailment is called inflammatory bowel disease (IBD). CRISPR-Cas9 was used to develop BS016, a self-regulating modified strain of *S. cerevisiae*, in order to solve issue. This strain helps prevent intestinal inflammation in mice models of IBD by expressing a human P2Y2 purinergic receptor and secreting an enzyme that breaks down ATP. Recently, BsS-RS06551, a strain of *B. subtilis* that produces butyrate, was modified using CRISPR-Cas9 to improve metabolic control and obesity in mice given a high-fat diet.[46] These advancements demonstrate the potential of probiotics designed with CRISPR to address a range of health issues.

2.4 Strategies and challenges for CRISPR based genome editing of probiotics

Effective genome manipulation of probiotics relies on delivering an editing toolbox into cells, often achieved through methods like electroporation for lactic acid bacteria (LAB) transformation.[47] Optimization of transformation efficiency involves considering and refining factors such as host properties (cell wall composition, growth stage, etc.), plasmid characteristics (source, concentration, etc.), and transformation conditions (electrical parameters, buffer composition, etc.).[48]

Additional strategies like in vitro vector modifications, removing host defense mechanisms, and iterative transformation cycles can enhance electroporation efficiency. In addition to electroporation, other potential vector delivery techniques include physicochemical methods, conjugation of bacteria, protoplast modification, and natural competence. To reduce the unintended consequences of DNA breakage during genome editing, new nucleases like as dCas9 and dCpf1 have been developed. These enzymes do not induce DSBs, thereby enhancing the precision of genome editing processes.[49]

Despite advances in CRISPR-based genome editing, concerns persist regarding engineered probiotics. Key challenges include ensuring microbial biocontainment, addressing safety issues such as immunogenicity and potential inflammation, and regulating CRISPR-Cas system activity through external stimuli like temperature, light, and ligands. Ensuring stability, elucidating mechanisms of action, and facilitating market application are essential to reaching one's full potential of engineered probiotics in promoting animal and human health industries.

3. CRISPR-Cas for microbiome diagnostics

CRISPR-Cas-based functional microbiome analysis entails modifying microbial genomes to investigate gene functions and interactions. CRISPRi and CRISPRa can be used to suppress or activate gene expression, respectively. This technique aids in the identification of genes required for microbial survival, pathogenicity, and metabolic activity. Furthermore, CRISPR-Cas allows for the production of synthetic microbial communities with predetermined genetic features, which aids the study of microbial dynamics and interactions.[50,51]

Numerous investigations have shown how useful CRISPR-Cas is for studying microbiomes. For example, CRISPR-Cas9 was employed to reduce the spread of antibiotic resistance in clinical settings by selectively removing antibiotic-resistance genes from the gut microbiota.[52] Another use was to develop probiotics with therapeutic benefits by sensing and reacting to environmental stimuli in the gut through the use of CRISPR-based systems. These illustrations show how adaptable and useful CRISPR-Cas technology is for advancing our understanding of microbiomes.[50] The specificity of CRISPR-based diagnostics is due to the precise targeting mechanism of guide RNAs (sgRNAs), which ensures that only the required DNA or RNA sequences are identified. This high specificity reduces false positives while increasing the dependability of the data. The amplification of target signals by collateral cleavage activity improves the sensitivity of these diagnostics, allowing them to detect low-abundance infections. Furthermore, the speed of CRISPR-based diagnostics allows for timely clinical decision-making, which is critical in controlling infectious diseases.[53] The capacity to monitor microbiome alterations at high resolution allows for a better understanding of disease mechanisms. CRISPR-based diagnostics can reveal how microbial diversity and function changes contribute to disease pathogenesis, giving insights into possible treatment targets.[54] Identifying specific microbial biomarkers associated with disease allows for targeted therapeutic interventions. CRISPR-Cas tools not only enable the detection of disease-associated microbes but also offer potential strategies for precision medicine by modulating microbial communities to restore health.[55]

Several real-world applications show how well CRISPR-Cas technology works for diagnosing microbiomes. One significant example was the utilization of CRISPR-Cas9 to create a bacterium capable of detecting and digesting environmental contaminants, highlighting the potential for environmental monitoring and bioremediation. A different case study showed how a CRISPR-based assay was developed to identify tuberculosis in clinical specimens, providing a quick and precise diagnostic choice.[56] Fig. 2 illustrates the utilization of the microbiome of humans as an indicator of disease indicators for diagnostics.[57]

3.1 Applications in microbiome diagnostics

3.1.1 Detection of pathogenic bacteria in clinical samples

Clinical samples have been successfully utilized to identify harmful germs using CRISPR-Cas technology. For example, the CRISPR-based

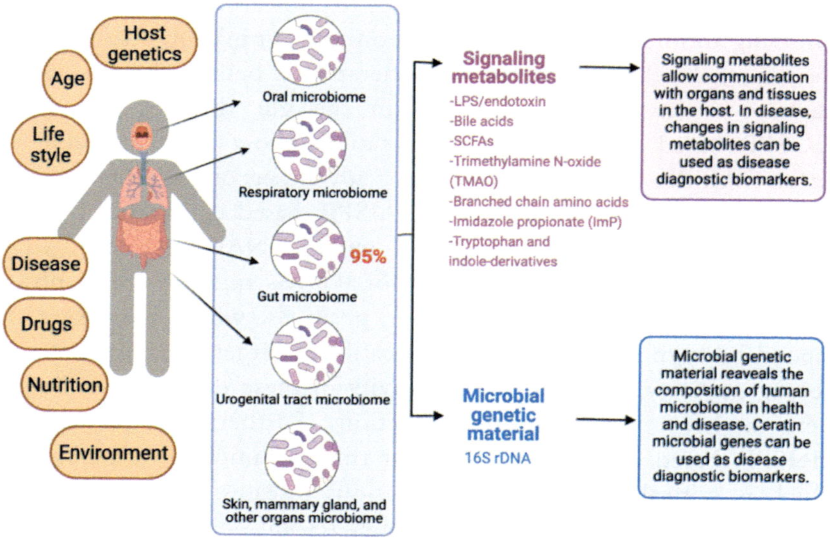

Fig. 2 Utilizing the human microbiome as a source of illness indicators for diagnosis *Adapted from Hajjo R, Sabbah DA, Al Bawab AQ. Unlocking the potential of the human microbiome for identifying disease diagnostic biomarkers. Diagnostics (Basel). 2022;12(7):1742. https://doi.org/10.3390/diagnostics12071742.*

DETECTR platform was successfully employed to identify SARS-CoV-2 in respiratory samples, demonstrating its potential for diagnosing viral infections. Similar approaches can be applied to detect bacterial pathogens in blood, urine, and other clinical specimens, providing rapid and accurate diagnostics.[50]

3.1.2 Identification of antibiotic resistance genes

The public's health is seriously threatened by the rise in resistance to antibiotics. CRISPR-Cas systems can help find and describe antibiotic resistance genes in microbial communities. This application involves targeting and sequencing resistance genes, enabling the detection of multidrug-resistant strains and informing appropriate treatment strategies.[56]

3.1.3 CRISPR-based diagnostics for microbiome monitoring

Because CRISPR-Cas systems can recognize specific DNA or RNA sequences, they can be employed for diagnostic purposes. This approach involves designing guide RNAs that target microbial biomarkers associated with health or disease states. When the Cas enzyme recognizes the target sequence, it generates a detectable signal that shows the target microbe's number and existence.[55]

3.1.4 Tracking microbiome shifts

Longitudinal monitoring of microbiome changes using CRISPR-based diagnostics allows researchers to track shifts in microbial populations over time. By analyzing these shifts, researchers can identify biomarkers that correlate with disease progression or therapeutic responses. For example, fluctuations in gut microbiota composition can provide early indicators of conditions like IBD, guiding early intervention strategies.[54]

3.2 Technological advancements and innovations

Recent developments in the CRISPR-Cas system have led to in the creation of more effective and diverse diagnostic instruments. Innovations such as CRISPR-Cas12 and CRISPR-Cas13 have expanded the range of detectable targets, including both DNA and RNA sequences. These advancements enhance the diagnostic capabilities of CRISPR-based platforms, making them more adaptable to various clinical and environmental applications.[50,58] Next-generation sequencing (NGS) and biosensors are two examples of other technologies that can be combined with CRISPR-Cas systems, has further improved diagnostic accuracy and functionality. For example, combining CRISPR diagnostics with NGS allows for high-throughput screening and comprehensive analysis of microbial communities. Biosensors equipped with CRISPR components enable real-time monitoring of pathogens and environmental contaminants, providing immediate feedback for timely interventions.[50] A major advancement in healthcare is the creation of point-of-care (POC) diagnostics with CRISPR-Cas technology. POC diagnostics offer rapid, on-site testing without the need for specialized laboratory equipment. This is particularly beneficial in remote or resource-limited settings where access to healthcare infrastructure is limited. CRISPR-based POC tests can provide quick and reliable results, facilitating early diagnosis and treatment.[50]

3.3 Challenges and limitations

Despite the promising applications, CRISPR-Cas diagnostics face several technical challenges. These include issues related to the delivery and stability of CRISPR components, the efficiency of target recognition and cleavage, and the potential for off-target effects. Addressing these challenges requires ongoing research and optimization of CRISPR systems.[50] Even while CRISPR-based diagnostics are very specific, there is still a chance that unintentional sequences will be targeted and cleaved, leading to off-target consequences. This can lead to false positives or negatives, affecting the accuracy of diagnostics. Strategies to enhance the fidelity of guide RNAs and improve target recognition are essential to minimize these off-target effects.[50]

Developments in CRISPR science, coupled with bioinformatics tools, hold promise for advancing microbiome research and clinical applications. Integrating multi-omics approaches and machine learning algorithms can further enhance our ability to interpret microbiome data and translate findings into personalized medicine.[55,59]

4. CRISPR-Cas for pathogen targeting

CRISPR-Cas technology simplifies and accelerates pathogen detection with on-site lateral flow assays, yielding results in hours. It combines CRISPR systems with fluorophores, quenchants, and nanoparticles to improve sensitivity and specificity over qPCR and allows for simultaneous detection of numerous targets. Cas9 cleaves double-stranded DNA (dsDNA) using sgRNA interactions. FLASH detects antimicrobial resistance (AMR) sequences, which are critical for diagnosing methicillin-resistant *Staphylococcus aureus* (MRSA) and vancomycin-resistant infections. Combining optical DNA mapping with CRISPR-Cas9 reveals resistance genes (e.g., *bla*CTX-M, *bla*NDM) quickly, which aids in infection detection. CRISPR-Cas9 DNA labeling increases marker detection by focusing on specific 20-bp regions, increasing structural variation localization accuracy. Using *H. influenzae* as a paradigm, it detects single alleles and single-base variations in conserved sequences, allowing strain distinction.[60]

ssRNA is recognized by CRISPR-Cas13, which then chops it to reveal the target RNA's existence. CRISPR-mediated surface-enhanced Raman spectroscopy (SERS) detects antimicrobial resistance genes via CRISPR-Cas9 and Au-coated magnetic nanoparticles (Au MNPs). This method combines SERS's sensitivity to accurately distinguish bacteria, including MDR strains like *S. aureus, A. baumannii,* and *K. pneumoniae,* and allows for on-site detection with a 3D nanopillar array swab.[60]

Most research focuses on Cas13a-based nucleic acid detection; however, it can also detect non-nucleic acid targets. In order to stimulate Cas13a and detect human IL-6 and VEGF with high sensitivity, researchers created the CRISPR/Cas13a signal enhancement linked immunosorbent assay (CLISA), which substituted dsDNA with a T7 promoter for traditional horseradish peroxidase.[61] The scientists created a Cas13a-based detection assay called TITAC-Cas, which combines the trans-cleavage activity of Cas13a with target-induced transcription amplification. Alkaline phosphatase (ALP) dephosphorylation was used to shield dsDNA from λ exonuclease digestion, resulting in accurate ALP activity detection. A fluoride riboswitch was utilized with a LOD of 1.7 μM to regulate transcription and activate Cas13a's collateral cleavage, resulting in a fluorescence signal. Additionally, we utilize the SHERLOCK-based analyzing of in vitro transcription approach[62] that combines Cas13a with ligand-dependent transcription activation for high-throughput analysis, enabling the simultaneous quantification of various chemicals based on ligand-dependent transcription. RNA methylation is also detected by Cas13a. Through the creation of mismatches in crRNA, studies have detected and measured RNA methylation, including N1-methyladenosine (m1A) and N6-methyladenosine (m6A).[63]

dCas9, which lacks enzymatic cleavage but can precisely bind to target dsDNA, has been used to identify pathogens. Reconnection of split luciferase's N- and C-terminal halves (NFluc and CFluc) results in the restoration of its enzymatic activity. When dCas9 attaches to two successive target sites in the substrate DNA, the fusing of NFluc and CFluc to the

N-terminus of dCas9 causes luciferase to generate light in the presence of D-luciferin.[64] Two sgRNAs that target sequences roughly 20 bp apart are used in this method to detect *Mycobacterium tuberculosis* (Mtb) with great specificity. However, its sensitivity needs improvement due to the lack of additional signal amplification. dCas9 has also been used to detect MRSA[65] and SARS-CoV-2.[66]

On the other hand, target binding-induced "collateral cleavage" activity is employed in CRISPR-Cas12/Cas13/Cas14-based assays.[67] Two RNA-guided DNases, Cas12 and Cas14, have different specificities for dsDNA and single-stranded DNA (ssDNA).[68] Both sequences that target and non-target ssDNA reporters are cleaved by these proteins. Technologies based on Cas12, such HOLMES and DETECTR, are able to reliably identify viruses like Japanese encephalitis virus and human papillomavirus (HPV). In order to facilitate rapid SARS-CoV-2 detection, SARS-CoV-2 DETECTR was created. It combines reverse transcription, isothermal amplification via loop-mediated amplification (RT-LAMP), and "collateral cleavage" by Cas12. Another method, called POC-CRISPR, uses droplet magnetofluidics in conjunction with Cas12 and reverse transcription and recombinase polymerase amplification (RT-RPA) to identify SARS-CoV-2 at the point of care. Within 30 min, this automated, portable device finds raw clinical nasopharyngeal swab eluates.[66]

5. CRISPR-Cas for microbiome engineering

5.1 Targeting specific microbial genes

Targeting specific microbial genes is a cornerstone of microbiome engineering, enabling precise manipulation of microbial functions and behaviors. By editing genes within microbial genomes, scientists can control metabolic pathways, influence microbial interactions, and tailor microbiomes for desired applications. With its unmatched accuracy and adaptability, the CRISPR-Cas system has completely changed this sector by making it possible to suppress, activate, or wipe out particular genes. These capabilities open up a wide array of possibilities, from enhancing microbial production of valuable compounds to improving health outcomes by modulating the human microbiome.[56,69] In this section, we will explore the primary strategies for targeting microbial genes using CRISPR-Cas, including gene knockouts and gene activation/repression, highlighting their mechanisms and applications in microbiome engineering.

5.1.1 Gene knockouts

Gene knockout involves the complete disruption of a gene's function, effectively nullifying its expression. In order to accomplish this, CRISPR-Cas9 introduces DSBs at particular genomic locations. NHEJ, a prone to mistakes process that frequently results in insertions or deletions (indels) that change the gene's sequence, is then used to repair these breaks. Numerous microbial species have been subjected to this technique in order to study metabolic pathways and gene function, leading to the creation of microbial strains with desired characteristics like increased biofuel or pharmaceutical output.[56]

5.1.2 Gene activation and repression

Furthermore, CRISPR-Cas systems can modify the expression of genes without altering the DNA sequence, in contrast to knockouts. CRISPRa and CRISPRi are the methods used to do this. Through the use of the dCas9 protein, which attaches to DNA without cutting it, CRISPRi represses the expression of genes by inhibiting transcription factors. Conversely, CRISPR uses dCas9 fused to transcriptional activators to upregulate target gene expression. These methods allow precise control over microbial metabolic pathways, facilitating the optimization of microbial communities for specific functions, such as waste degradation or nutrient cycling.[70,71]

5.2 Modulating microbial communities

5.2.1 Editing microbial population dynamics

It is possible to modify the dynamics and makeup of microbial communities by using CRISPR-Cas technologies. By targeting specific species within a community, researchers can selectively promote or inhibit the growth of particular microbes. This approach is useful in environments such as the human gut, where maintaining a balanced microbial community is crucial for health. For instance, CRISPR-based strategies have been proposed to selectively remove pathogenic bacteria or enhance the growth of beneficial ones, thereby restoring or maintaining a healthy microbiome balance.[72]

5.2.2 Synthetic biology approaches

In synthetic biology, new biological components, apparatus, and systems are designed and built by combining CRISPR-Cas technology with additional genetic engineering tools. In microbiome engineering, this can involve the creation of synthetic microbial consortia with engineered traits tailored for specific applications. For example, researchers can design synthetic communities capable of efficient bioremediation by engineering each member to perform a distinct but complementary role in degrading pollutants. This modular approach allows for the customization of microbial communities to perform complex, multi-step processes more efficiently than single strains.[73,74]

5.3 Horizontal gene transfer (HGT) and CRISPR-Cas

5.3.1 Preventing unwanted gene transfer

Since HGT can disperse unwanted features, such antibiotic resistance, throughout microbial communities, it presents a serious barrier to microbiome engineering. By focusing on and cleaving incoming plasmid DNA or phage genomes that facilitate gene transfer, Using CRISPR-Cas systems, HGT can be prevented. Through the utilization of CRISPR sequences that identify and eliminate these mobile genetic components, scientists can curtail the proliferation of deleterious genes and preserve the stability of modified microbial communities.[75,76]

5.3.2 Harnessing HGT for beneficial traits

Conversely, HGT can be harnessed for beneficial purposes. It is possible to engineer CRISPR-Cas systems to promote the spread of advantageous genes amongst microorganisms. To improve the general resilience and effectiveness of the microbiome, genes that impart metabolic capacities or resistance to environmental challenges, for instance, may be transferred to additional community members. This approach leverages the natural mechanisms of gene transfer while providing the precision and control of CRISPR technology.[77,78]

6. Ethical and regulatory issues related to use of CRISPR-Cas

Over the past 20 years, biotechnology has advanced rapidly, making it possible to create scientific instruments that were previously only seen in science fiction. Genome editing, especially with the aid of the CRISPR-Cas system, constitutes one of the most significant advances in science of our time. It permits precise modification of any living organism's genetic composition, including human DNA.[79] The CRISPR-Cas system offers a number of moral, ethical, political, and scientific issues that may result in a ban on this kind of research despite its efficacy in modifying genomic DNA.[80] Therefore, it is crucial to acknowledge these issues, make sense of them, and promote communication among all relevant parties—including society, academic institutions, and scientific and technological organizations—in order to develop the greatest ethical and legal framework possible.[79–81]

Developers, researchers, and ethicists debated the ethical, legal, and biological ramifications of CRISPR-Cas9 at a 2015 symposium in Napa Valley.[82] There were lengthy lectures across the entire conference. To examine the application of this method on humans, the Chinese Academy of Sciences and the Royal Society of the UK were invited by the US National Academies of Sciences, Engineering, and Medicine (NASEM) to an international meeting on human genome editing.[83] In reaction to these approaches, which enable precise gene alterations in living cells and have consequences for biomedicine, reproduction, animals and plants, industrial biotechnology, and ecology, the Nuffield Council on Bioethics promptly issued a request for proof.[84,85]

Early genome editing studies on human embryos brought up significant concerns regarding security, ethics, and morality. It is recommended that researchers cease their use of CRISPR-Cas9 for genome editing on human germ cells to address safety concerns and allow a full public discussion of the ethical and sociological implications of these modifications across a range of society groups.[82,86] The legal framework and guiding principles for the medical use of genome editing were thoroughly described in the NASEM report.[87] However, there are still some unclear aspects of the regulations. Furthermore, not enough is being said about the worldwide regulatory framework for CRISPR-Cas genome-edited plants, particularly in regards to the controversy surrounding the classification of the transformed plants or crops as genetically modified organisms (GMOs). There are continuing international conversations centered on figuring out how genome editing might differ from other biotechnologies in theory and execution.[88]

The manipulation of the gut microbiome by CRISPR-Cas technology poses distinct ethical and regulatory difficulties. The potential to alter microbial communities within the human body raises concerns about unintended health consequences, ecological impacts, and the potential for creating antibiotic-resistant bacteria. The consequences of altering organisms that are essential to the environment and human health are among the ethical issues to be taken into account. Regulatory issues focus on ensuring the safety and efficacy of such interventions, requiring rigorous oversight and comprehensive risk assessments. Strong ethical norms and open public participation are necessary to negotiate these challenges and guarantee the proper application of CRISPR-Cas in gut microbiome research.

Some contend that to tackle ethical difficulties, an entity should supervise the ethical use of genome editing. To provide such guidance, European scientists founded the Association for Responsible Research and Innovation in Genome Editing (ARRIGE).[89,90] In addition, researchers have suggested creating a "global observatory for gene editing" to incorporate various viewpoints in moral discussions, akin to the networks set up for human rights and climate change.[91] As CRISPR technology advances, so too will the discourse surrounding its moral and legal parameters. National and international organizations must develop research and ethical guidelines with input from diverse societal disciplines. To reduce risks and enhance benefits, funding from government organizations and boards of institutional review will need to Institutional Review Boards (IRBs) adhere to these recommendations. Law enforcement agencies, main investigators, and IRBs will probably be responsible for upholding ethical research rules in the end.

7. Conclusion and future perspectives

The development of CRISPR-Cas technology has fundamentally changed how we do research on the gut microbiota, opening up a plethora of possibilities for both scientific inquiry and therapeutic applications. The accuracy and adaptability of this technique allow for the modification of microbial communities, the insertion of targeted microbial genes, and the creation of improved probiotics. Furthermore, CRISPR-Cas systems provide novel approaches to pathogen identification and microbiome diagnostics, enabling hitherto unattainable precision in the targeting and eradication of dangerous microbes. However, with these scientific breakthroughs come significant ethical and regulatory challenges. The ability to alter microbial communities within the human body necessitates a thorough examination of potential health risks, ecological impacts, and ethical considerations. Thorough regulatory frameworks and rigorous risk evaluations are required to ensure the safety and efficacy of CRISPR-Cas therapies in the gut microbiome. Public involvement and ethical standards must underpin the creation and application of this technology. Future CRISPR-Cas research on gut microbiomes has the potential to fundamentally alter our understanding of human health, agriculture, and environmental sustainability. Advancements in precision and control of genetic modifications will enhance the safety and efficacy of microbiome engineering. Personalized medicine will benefit from tailored CRISPR-based interventions that target specific pathogens and modulate beneficial microbes. Therapeutic applications will expand, offering new treatments for infections, metabolic disorders, and inflammatory diseases. In agriculture, CRISPR will improve crop yields and pest resistance, while environmental applications will focus on bioremediation. Ethical and regulatory frameworks will be essential to ensure responsible use, and integration with technologies like metagenomics and synthetic biology will deepen our understanding of microbial interactions. Overall, CRISPR-Cas will unlock innovative solutions, improving health and sustainability.

References

1. Barrangou R, Marraffini LA. CRISPR-Cas systems: prokaryotes upgrade to adaptive immunity. *Mol Cell*. 2014;54(2):234–244. https://doi.org/10.1016/j.molcel.2014.03.011
2. Waters JL, Ley RE. The human gut bacteria Christensenellaceae are widespread, heritable, and associated with health. *BMC Biol*. 2019;17(1):83. https://doi.org/10.1186/s12915-019-0701-8
3. Barrangou R, Fremaux C, Deveau H, et al. CRISPR provides acquired resistance against viruses in prokaryotes. *Science*. 2007;315:1709–1712. https://doi.org/10.1126/science.1138140
4. Kurokawa K, Itoh T, Kuwahara T, et al. Comparative metagenomics revealed commonly enriched gene sets in human gut microbiomes. *DNA Res*. 2007;14:169–181. https://doi.org/10.1093/dnares/dsm018
5. Salyers AA, Gupta A, Wang Y. *Human intestinal bacteria as reservoirs for antibiotic resistance genes*. *Trends Microbiol*. 12. 2004; 2004:412–416. https://doi.org/10.1016/j.tim.2004.07.004
6. Barlow M. What antimicrobial resistance has taught us about horizontal gene transfer. *Methods Mol Biol*. 2009;532:397–411. https://doi.org/10.1007/978-1-60327-853-9_23
7. Mani I, Singh V, Khalid JA, Dinh-Toi C. *Microbial Genomic Islands in Adaptation and Pathogenicity*. Springer Nature Singapore; 2023:356 ISBN: 9789811993411.
8. Capozzi V, Spano G. Horizontal gene transfer in the gut: is it a risk? *Food Res Int*. 2009;42:1501–1502. https://doi.org/10.1016/j.foodres.2009.08.001
9. Gupta A, Singh V, Mani I. *Dysbiosis of human microbiome and infectious diseases*. *Progress in Molecular Biology and Translational Science*. vol. 192. Academic Press, Elsevier; 2022:33–51 *ISSN 1877-1173*. *ISBN: 9780323912105*.
10. Husain F, Veeranagouda Y, Boente R, Tang K, Mulato G, Wexler HM. The Ellis Island effect: a novel mobile element in a multi-drug resistant Bacteroides fragilis clinical isolate includes a mosaic of resistance genes from Gram-positive bacteria. *Mob Genet Elem*. 2014;4:E29801–E29812. https://doi.org/10.4161/mge.29801
11. Saini A, Mani I, Rawal MK, Verma C, Singh V, Mishra S. An introduction of microbial genomic islands for evolutionary adaptation and pathogenicity. In: Mani I, Singh V, Alzahrani KJ, Chu DT, eds. *Microbial Genomic Islands in Adaptation and Pathogenicity*. Singapore: Springer Nature; 2023:1–15 ISBN: 9789811993411.
12. Hsu BB, Plant IN, Lyon L, Anastassacos FM, Way JC, Silver PA. In situ reprogramming of gut bacteria by oral delivery. *Nat Commun*. 2020;11:5030.

13. Lam KN, Spanogiannopoulos P, Soto-Perez P, et al. Phage-delivered CRISPR-Cas9 for strain-specific depletion and genomic deletions in the gut microbiome. *Cell Rep.* 2021;37(5):109930. https://doi.org/10.1016/j.celrep.2021.109930
14. Jiang W, Bikard D, Cox D, Zhang F, Marraffini LA. RNA-guided editing of bacterial genomes using CRISPR-Cas systems. *Nat Biotechnol.* 2013;31:233–239.
15. Neil K, Allard N, Roy P, et al. High-efficiency delivery of CRISPR-Cas9 by engineered probiotics enables precise microbiome editing. *Mol Syst Biol.* 2021;17(10):e10335. https://doi.org/10.15252/msb.202110335
16. Javed MU, Hayat MT, Mukhtar H, Imre K. CRISPR-Cas9 system: a prospective pathway toward combatting antibiotic resistance. *Antibiotics (Basel).* 2023;12(6):1075. https://doi.org/10.3390/antibiotics12061075 PMID: 37370394; PMCID: PMC10295005.
17. Gupta A, Mani I. Beneficial effects of psychobiotic bacteria, cyanobacteria, algae and modified yeast in various food industries. In: Kumar A, Patruni K, Singh V, eds. *Recent Advances in Food Biotechnology.* Singapore: Springer Nature; 2022:161–173 *ISBN: 9789811681240.*
18. Kelly CJ, Zheng L, Campbell EL, Saeedi B, Scholz CC, Bayless AJ. Crosstalk between microbiota-derived short-chain fatty acids and intestinal epithelial HIF augments tissue barrier function. *Cell Host Microbe.* 2015;17(5):662–671.
19. Lavelle A, Sokol H. Gut microbiota-derived metabolites as key actors in inflammatory bowel disease. *Nat Rev Gastro Hepat.* 2020;17(4):223–237.
20. Fanning S, Hall LJ, Cronin M, Zomer A, MacSharry J, Goulding D. Bifidobacterial surface-exopolysaccharide facilitates commensal-host interaction through immune modulation and pathogen protection. *Proc Natl Acad Sci U S A.* 2012;109(6):2108–2113.
21. Garbacz K. Anticancer activity of lactic acid bacteria. *Semin Cancer Biol.* 2022;86(Pt 3):356–366.
22. O'Toole PW, Marchesi JR, Hill C. Next-generation probiotics: the spectrum from probiotics to live biotherapeutics. *Nat Microbiol.* 2017;2:17057.
23. Suez J, Zmora N, Segal E, Elinav E. The pros, cons, and many unknowns of probiotics. *Nat Med.* 2019;25(5):716–729.
24. Chung Y, Ryu Y, An BC, Yoon YS, Choi O, Kim TY. A synthetic probiotic engineered for colorectal cancer therapy modulates gut microbiota. *Microbiome.* 2021;9(1):122.
25. Doudna JA, Charpentier E. Genome editing. The new frontier of genome engineering with CRISPR-Cas9. *Science.* 2014;346(6213):1258096.
26. Ling L, Shimaa Elsayed H, Nan P. CRISPR-Cas-based engineering of probiotics. *BioDesign Res.* 2023;5:0017. https://doi.org/10.34133/bdr.0017
27. Crawley AB, Henriksen JR, Barrangou R. CRISPRdisco: an automated pipeline for the discovery and analysis of CRISPR-Cas systems. *CRISPR J.* 2018;1(2):171–181.
28. Liu G, Lin Q, Jin S, Gao C. The CRISPR-Cas toolbox and gene editing technologies. *Mol Cell.* 2021;82(2):333–347.
29. Liu L, Yang D, Zhang Z, Liu T, Hu G, He M. High-efficiency genome editing based on endogenous CRISPR-Cas system enhances cell growth and lactic acid production in Pediococcus acidilactici. *Appl Environ Microbiol.* 2021;87(20):e0094821.
30. Karcher N, Nigro E, Puncochar M. Genomic diversity and ecology of human-associated Akkermansia species in the gut microbiome revealed by extensive metagenomic assembly. *Genome Biol.* 2021;22(1):209.
31. Zhou X, Wang X, Luo H, Wang Y, Wang Y, Tu T. Exploiting heterologous and endogenous CRISPR-Cas systems for genome editing in the probiotic Clostridium butyricum. *Biotechnol Bioeng.* 2021;118(7):2448–2459.
32. Su F, Xu P. Genomic analysis of thermophilic Bacillus coagulans strains: efficient producers for platform bio-chemicals. *Sci Rep-UK.* 2014;4:3926.

33. Gao C. Genome engineering for crop improvement and future agriculture. *Cell.* 2021;1840(6):1621–1635.
34. Gasiunas G, Barrangou R, Horvath P, Siksnys V. Cas9-crRNA ribonucleoprotein complex mediates specific DNA cleavage for adaptive immunity in bacteria. *Proc Natl Acad Sci U S A.* 2012;109(39):E2579–E2586.
35. Walton RT, Christie KA, Whittaker MN, Kleinstiver BP. Unconstrained genome targeting with near-PAMless engineered CRISPR-Cas9 variants. *Science.* 2020;368(6488):290–296.
36. Komor AC, Kim YB, Packer MS, Zuris JA, Liu DR. Programmable editing of a target base in genomic DNA without double-stranded DNA cleavage. *Nature.* 2016;533(7603):420–424.
37. Zhao B, Rothenberg E, Ramsden DA, Lieber MR. The molecular basis and disease relevance of non-homologous DNA end joining. *Nat Rev Mol Cell Bio.* 2020;21(12):765–781.
38. Martinez-Galvez G, Joshi P, Friedberg I, Manduca A, Ekker SC. Deploying MMEJ using MENdel in precision gene editing applications for gene therapy and functional genomics. *Nucleic Acids Res.* 2021;49(1):67–78.
39. Ran FA, Hsu PD, Wright J, Agarwala V, Scott DA, Zhang F. Genome engineering using the CRISPR-Cas9 system. *Nat Protoc.* 2013;8:2281–2308.
40. Klompe SE, Vo PLH, Halpin-Healy TS, Sternberg SH. Transposon-encoded CRISPR-Cas systems direct RNA-guided DNA integration. *Nature.* 2019;571.
41. Wang B, Shen J. NF-kappaB inducing kinase regulates intestinal immunity and homeostasis. *Front Immunol.* 2022;13:895636.
42. Sola-Oladokun B, Culligan EP, Sleator RD. Engineered probiotics: applications and biological containment. *Annu Rev Food Sci Technol.* 2017;8:353–370.
43. Aggarwal N, Breedon AME, Davis CM, Hwang IY, Chang MW. Engineering probiotics for therapeutic applications: recent examples and translational outlook. *Curr Opin Biotech.* 2020;65:171–179.
44. Canale FP, Basso C, Antonini G, Perotti M, Li N, Sokolovska A. Metabolic modulation of tumours with engineered bacteria for immunotherapy. *Nature.* 2021;598(7882):662–666.
45. Hendrikx T, Duan Y, Wang Y. Bacteria engineered to produce IL-22 in intestine induce expression of REG3G to reduce ethanol-induced liver disease in mice. *Gut.* 2019;68(8):1504–1515.
46. Wang L, Cheng X, Bai L, et al. Positive interventional effect of engineered butyrate-producing bacteria on metabolic disorders and intestinal flora disruption in obese mice. *Microbiol Spectr.* 2022;10(2):e0114721.
47. Wang C, Cui Y, Qu X. Optimization of electrotransformation (ETF) conditions in lactic acid bacteria (LAB). *J Microbiol Methods.* 2020;174:105944.
48. Zuo F, Marcotte H. Advancing mechanistic understanding and bioengineering of probiotic lactobacilli and bifidobacteria by genome editing. *Curr Opin Biotech.* 2021;70:75–82.
49. Gao Y, Han M, Shang S, Wang H, Qi LS. Interrogation of the dynamic properties of higher-order heterochromatin using CRISPR-dCas9. *Mol Cell.* 2021;81:4287–4299.e5.
50. Jolany vangah S, Katalani C, Boone HA, et al. CRISPR-based diagnosis of infectious and noninfectious diseases. *Biol Proced Online.* 2020;22:22. https://doi.org/10.1186/s12575-020-00135-3
51. Zhang R, Xu W, Shao S, Wang Q. Gene silencing through CRISPR interference in bacteria: current advances and future prospects. *Front Microbiol.* 2021;12:635227. https://doi.org/10.3389/fmicb.2021.635227
52. Duan C, Cao H, Zhang LH, Xu Z. Harnessing the CRISPR-Cas systems to combat antimicrobial resistance. *Front Microbiol.* 2021;12:716064. https://doi.org/10.3389/fmicb.2021.716064 PMID: 34489905; PMCID: PMC8418092.
53. Puig-Serra P, Casado-Rosas MC, Martinez-Lage M, et al. CRISPR approaches for the diagnosis of human diseases. *Int J Mol Sci.* 2022;23:1757. https://doi.org/10.3390/ijms23031757

54. Qin J, Li R, Raes J, et al. A human gut microbial gene catalogue established by metagenomic sequencing. *Nature*. 2010;464(7285):59–65.
55. Gootenberg JS, Abudayyeh OO, Kellner MJ, et al. Multiplexed and portable nucleic acid detection platform with Cas13, Cas12a, and Csm6. *Science*. 2018;360(6387):439–444.
56. Hsu PD, Lander ES, Zhang F. Development and applications of CRISPR-Cas9 for genome engineering. *Cell*. 2014;157(6):1262–1278. https://doi.org/10.1016/j.cell.2014.05.010 PMID: 24906146; PMCID: PMC4343198.
57. Hajjo R, Sabbah DA, Al Bawab AQ. Unlocking the potential of the human microbiome for identifying disease diagnostic biomarkers. *Diagnostics (Basel)*. 2022;12(7):1742. https://doi.org/10.3390/diagnostics12071742
58. Bhattacharjee, et al. A paper-based assay for detecting hypervirulent Klebsiella pnuemoniae using CRISPR-Cas13a system. *Microchem J*. 2024;203:110931.
59. Mani I, Singh V. *Multi-Omics Analysis of the Human Microbiome—From Technology to Clinical Applications*. Springer Nature Singapore; 2024:354 ISBN: 9789819718436.
60. Wu Y, Battalapalli D, Hakeem MJ, et al. Engineered CRISPR-Cas systems for the detection and control of antibiotic-resistant infections. *J Nanobiotechnol*. 2021;19(1):401. https://doi.org/10.1186/s12951-021-01132-8
61. Chen Q, Tian T, Xiong E, Wang P, Zhou X. CRISPR/Cas13a signal amplification linked immunosorbent assay for Femtomolar protein detection. *Anal Chem*. 2020;92:573–577. https://doi.org/10.1021/acs.analchem.9b04403
62. Iwasaki RS, Batey RT. SPRINT: a Cas13a-based platform for detection of small molecules. *Nucleic Acids Res*. 2020;48:e101. https://doi.org/10.1093/nar/gkaa673
63. Yang S, Wu F, Peng S, et al. A m(6) a sensing method by its impact on the stability of RNA double helix. *Chem Biodivers*. 2020;17:e2000050. https://doi.org/10.1002/cbdv.202000050
64. Zhang Y, Qian L, Wei W, et al. Paired design of dCas9 as a systematic platform for the detection of featured nucleic acid sequences in pathogenic strains. *ACS Synth Biol*. 2017;6(2):211–216.
65. Guk K, Keem JO, Hwang SG, et al. A facile, rapid and sensitive detection of MRSA using a CRISPR-mediated DNA FISH method, antibody-like dCas9/sgRNA complex. *Biosens Bioelectron*. 2017;95:67–71.
66. Huang T, Zhang R, Li J. CRISPR-Cas-based techniques for pathogen detection: retrospect, recent advances, and future perspectives. *J Adv Res*. 2023;50:69–82. https://doi.org/10.1016/j.jare.2022.10.011 Epub 2022 Oct 30.
67. Aman R, Mahas A, Mahfouz M. Nucleic acid detection using CRISPR/Cas biosensing technologies. *ACS Synth Biol*. 2020;9(6):1226–1233.
68. van Dongen JE, Berendsen JTW, Steenbergen RDM, Wolthuis RMF, Eijkel JCT, Segerink LI. Point-of-care CRISPR/Cas nucleic acid detection: recent advances, challenges and opportunities. *Biosens Bioelectron*. 2020;166:112445.
69. Jiang W, Marraffini LA. CRISPR-Cas: new tools for genetic manipulations from bacterial immunity systems. *Annu Rev Microbiol*. 2015;69:209–228. https://doi.org/10.1146/annurev-micro-091014-104441 Epub 2015 Jul 22.
70. Gilbert LA, Larson MH, Morsut L, et al. CRISPR-mediated modular RNA-guided regulation of transcription in eukaryotes. *Cell*. 2013;154(2):442–451. https://doi.org/10.1016/j.cell.2013.06.044
71. Qi LS, Larson MH, Gilbert LA, et al. Repurposing CRISPR as an RNA-guided platform for sequence-specific control of gene expression. *Cell*. 2013;152(5):1173–1183. https://doi.org/10.1016/j.cell.2013.02.022 Erratum in: Cell. 2021 Feb 4;184(3):844. https://doi.org/10.1016/j.cell.2021.01.019.
72. Mimee M, Tucker AC, Voigt CA, Lu TK. Programming a human commensal bacterium, Bacteroides thetaiotaomicron, to sense and respond to stimuli in the murine gut microbiota. *Cell Syst*. 2015;1(1):62–71. https://doi.org/10.1016/j.cels.2015.06.001

73. Johns NI, Blazejewski T, Gomes AL, Wang HH. Principles for designing synthetic microbial communities. *Curr Opin Microbiol*. 2016;31:146–153. https://doi.org/10.1016/j.mib.2016.03.010
74. Smanski MJ, Bhatia S, Zhao D, et al. Functional optimization of gene clusters by combinatorial design and assembly. *Nat Biotechnol*. 2014;32(12):1241–1249. https://doi.org/10.1038/nbt.3063
75. Koonin EV, Wolf YI. Evolution of microbes and viruses: a paradigm shift in evolutionary biology? *Front Cell Infect Microbiol*. 2012;2:119. https://doi.org/10.3389/fcimb.2012.00119
76. Yosef I, Manor M, Kiro R, Qimron U. Temperate and lytic bacteriophages programmed to sensitize and kill antibiotic-resistant bacteria. *Proc Natl Acad Sci U S A*. 2015;112(23):7267–7272. https://doi.org/10.1073/pnas.1500107112
77. Bennett GM, Moran NA. Heritable symbiosis: the advantages and perils of an evolutionary rabbit hole. *Proc Natl Acad Sci U S A*. 2015;112(33):10169–10176. https://doi.org/10.1073/pnas.1421388112 Epub 2015 Feb 23. PMID: 25713367; PMCID: PMC4547261.
78. Shintani M, Sanchez ZK, Kimbara K. Genomics of microbial plasmids: classification and identification based on replication and transfer systems and host taxonomy. *Front Microbiol*. 2015;6:242. https://doi.org/10.3389/fmicb.2015.00242
79. de Lecuona I, Casado M, Marfany G, Baroni ML, Escarrabill M. Focus: genome editing: gene editing in humans: towards a global and inclusive debate for responsible research. *Yale J Biol Med*. 2017;90:673.
80. Lyon J. Bioethics panels open door slightly to germline gene editing. *JAMA*. 2017;318:1639–1640.
81. Khambhati K, Bhattacharjee G, Gohil N, et al. Phage engineering and phage-assisted CRISPR-Cas delivery to combat multidrug-resistant pathogens. *Bioeng Transl Med*. 2022;8(2):e10381. https://doi.org/10.1002/btm2.10381
82. Baltimore D, Berg P, Botchan M, et al. A prudent path forward for genomic engineering and germline gene modification. *Science*. 2015;348:36–38.
83. Brokowski C, Adli M. CRISPR ethics: moral considerations for applications of a powerful tool. *J Mol Biol*. 2019;431:88–101.
84. Mulvihill JJ, Capps B, Joly Y, Lysaght T, Zwart HA, Chadwick R. Ethical issues of CRISPR technology and gene editing through the lens of solidarity. *Br Med Bull*. 2017;122:17–29.
85. Nuffield Council on Bioethics. Genome editing: an ethical review. http://nuffieldbioethics.org/wp-content/uploads/Genome-editing-an-ethical-review.pdf; 2016. Last assessed April 7, 2019.
86. Olson S. *Committee on science, and national academies of sciences, engineering, and medicine. International Summit on Human Gene Editing: A Global Discussion*. Washington, DC: National Academies Press; 2016.
87. National Academies of Sciences, Engineering and Medicine (NASEM). Human Genome Editing: Science, Ethics, and Governance. Washington, DC: The National Academies Press; 2017.
88. Fears R, Ter Meulen V. Point of view: how should the applications of genome editing be assessed and regulated? *Elife*. 2017;6:e26295.
89. Enserink M. Interested in responsible genome editing? Join the new club. *Science*. 2018. https://doi.org/10.1126/science.aat7183
90. Lluis M, Jennifer M, François H, et al. ARRIGE arrives: toward the responsible use of genome editing. *CRISPR J*. 2018;1(2):128–129.
91. Jasanoff S, Hurlbut JB. A global observatory for gene editing. *Nature*. 2018;555:435–437.

CHAPTER FIVE

Advances in CRISPR-Cas systems for fungal infections

Avinash Singh[a], Monisa Anwer[b], Juveriya Israr[c], and Ajay Kumar[b,*]
[a]Department of Biotechnology, Axis Institute of Higher Education, Kanpur, Uttar Pradesh, India
[b]Department of Biotechnology, Faculty of Engineering and Technology Rama University, Mandhana, Kanpur, Uttar Pradesh, India
[c]Institute of Biosciences and Technology, Shri Ramswaroop Memorial University, Lucknow, Barabanki, Uttar Pradesh, India
*Corresponding author. e-mail address: ajaymtech@gmail.com

Contents

Abstract

Fungi contain a wide range of bioactive secondary metabolites (SMs) that have numerous applications in various fields, including agriculture, medicine, human health, and more. It is common for genes responsible for the production of secondary metabolites (SMs) to form biosynthetic gene clusters (BGCs). The identification and analysis of numerous unexplored gene clusters (BGCs) and their corresponding substances (SMs) has been significantly facilitated by the recent advancements in genomic and genetic technologies. Nevertheless, the exploration of secondary metabolites with commercial value is impeded by a variety of challenges. The emergence of modern CRISPR/*Cas* technologies has brought about a paradigm shift in fungal genetic engineering, significantly streamlining the process of discovering new bioactive compounds. This study begins with an examination of fungal

Progress in Molecular Biology and Translational Science, Volume 208
ISSN 1877-1173, https://doi.org/10.1016/bs.pmbts.2024.07.006

biosynthetic gene clusters (BGCs) and their interconnections with the secondary metabolites (SMs) they generate. Following that, a brief summary of the conventional methods employed in fungal genetic engineering is provided. This study explores various sophisticated CRISPR/Cas-based methodologies and their utilization in examining the synthesis of secondary metabolites (SMs) in fungi. The chapter provides an in-depth analysis of the limitations and obstacles encountered in CRISPR/*Cas*-based systems when applied to fungal genetic engineering. It also proposes promising avenues for future research to optimize the efficiency of these systems.

Abbreviations

BGCs	Biosynthetic Gene Clusters
CRISPR/Cas	Clustered Regularly Interspaced Short Palindromic Repeats/CRISPR-associated
dCas	Deactivated Cas
DNA	Deoxyribonucleic Acid
GA	Gibberellic Acid
HDR	Homology-Directed Repair
INDELs	Insertions and Deletions
miRNA	MicroRNA
NHEJ	Non-Homologous End Joining
PKS	Polyketide Synthase
RNPs	Protospacer-Adjacent Motif-PAMRibonucleoproteins
SMs	Secondary Metabolites
sgRNA	Single Guide RNA

1. Introduction

Fungal secondary metabolism involves a range of physiologically active compounds that can potentially be used to tackle issues in the environment, agriculture, and medication development.[1] Fungi, found in many different settings, generate a diverse range of secondary metabolites (SMs) as part of their complex ecological network.[2] Fungi possess the capacity to generate bioactive compounds, yet comprehending and harnessing their secondary metabolism presents difficulties.[3] Researchers have difficulties in cultivating specific fungi and understanding cryptic biosynthetic gene clusters (BGCs), which impedes the complete use of fungal secondary metabolites (SMs) in the industry.[1] Despite encountering numerous challenges, recent breakthroughs possess the capacity to revolutionize the domains of natural product discovery and fungal genetic engineering.[4] Various obstacles impede the extensive exploration of secondary metabolites for industrial applications. Fungal secondary metabolism is highly diverse and enables the synthesis of bioactive chemicals. Nevertheless, the extensive investigation and application of these compounds are impeded by numerous obstacles.[2] Secondary metabolites produced by fungi are notoriously difficult to identify and characterize because of cryptic biosynthetic gene clusters (BGCs) that are activated and the difficulties of growing fungi that cannot be cultured[1,5] Clustered Regularly Interspaced Short Palindromic Repeats (CRISPR-associated protein) could solve these problems. Gene editing techniques have greatly advanced fungal genetic engineering, leading to the discovery of novel bioactive compounds that hold promise for commercial applications.[5] states that CRISPR/*Cas* technology can be used to modify the genomes of fungi. Alterations to biosynthetic pathways, enhanced bioactivity and productivity, and higher secondary metabolite synthesis are all possible outcomes. One of the many benefits of using CRISPR/*Cas* systems for fungal genetic engineering is the ability to identify bioactive compounds. Our understanding of infections caused by fungal metabolites and fungal pathogenesis has significantly improved. Fungal secondary metabolites encompass mycotoxins and phytotoxins. The pathogenicity of fungi, as well as the interactions between hosts and pathogens, and the occurrence of fungal infections in other organisms, rely on these compounds.[3] By illuminating the genomic basis of fungal pathogenicity, CRISPR/*Cas* technology may improve disease prevention and management.[4] There is optimism that CRISPR/*Cas*-based approaches will contribute to the reduction of the prevalence of metabolic disorders and fungal infections, which are both on the rise. Two approaches are developing fungal strains with less toxic effects and blocking mechanisms that lead to toxins.[2] Fungi, which inhabit terrestrial and aquatic environments, synthesize several secondary metabolites (SMs). Given their wide range of structures and important roles in the ecology, physiology, and relationships of fungi, these chemicals are essential. Fungal secondary metabolites (SMs) contribute to defense, communication, competition, and symbiosis, whereas primary metabolites are crucial for cellular processes. SM synthesis is frequently triggered by stress and environmental variables. Fungal biosynthetic gene clusters (BGCs) regulate intricate metabolic pathways responsible for the manufacture of secondary metabolites (SMs).[6] Genes involved in SM biosynthesis are physically organized into gene clusters known as GSCs. Transporters, regulatory elements, essential biosynthetic enzymes, modifying enzymes, and other components are frequently found in general signaling complexes (GSCs). The expression of fungal secondary metabolism is greatly influenced by pH, temperature, and species interactions, as these factors are always changing.[7,8] The vast number of fungal species indicates a significant capacity for discovering new secondary metabolites. Although there have been advancements in fungal genetics and bioinformatics, our understanding of most fungal species and their secondary metabolites (SMs) remains limited. The primary technical challenges that impede our thorough examination of fungal secondary metabolism are the incapacity to culture some fungi, the induction of hidden biosynthetic gene clusters (BGCs), and the isolation of bioactive compounds from intricate mixtures. Fungal secondary metabolites (SMs) encompass a wide range of chemical compounds, including terpenoids, alkaloids, polyketides, non-ribosomal peptides, and hybrids. This leads to a significant level of variation in their molecular structures. These compounds possess useful features such as immunosuppressive, antibacterial, anticancer, antiviral, insecticidal, and cytotoxic effects, which make them highly valuable in the sectors of medicine, biotechnology, agriculture, and environmental remediation. Fungal secondary metabolites (SMs) have a significant impact on the wide range of diversity, mainly due to their therapeutic importance.[9] Polyketide synthases (PKSs) and non-ribosomal peptide synthetases (NRPSs) are enzymatic catalysts responsible for the biosynthesis of polyketides and non-ribosomal peptides, respectively. These chemicals play a crucial role in the advancement of antifungals, antibiotics, and cancer medicines.[10] Notable examples are the antibiotic penicillin and the immunosuppressant cyclosporine. Penicillium spp. and *Tolypocladium inflatum* produce penicillin and cyclosporine, as stated by Makarova et al.[11] Terpenoids, which are a group of chemicals produced by fungi, have important use in medicine and economics. They originate from isoprenoid precursors. *Streptomyces nodosus* synthesizes amphotericin B, a medicine with antifungal properties, whereas *Artemisia annua* generates artemisinin, a compound with antiviral activity.[12] Fungal alkaloids possess cytotoxic, antiviral, and antimalarial effects as a result of their nitrogen-containing heterocyclic structures. The utilization of fungi's secondary metabolites in biotechnology and drug development necessitates the application of transdisciplinary methodologies such as genomics, metabolomics, synthetic biology, and bioinformatics. Contemporary techniques such as synthetic biology, pathway engineering, heterologous expression, and genome mining can be employed to discover, produce, and enhance fungal secondary metabolites (SMs) with enhanced bioactivity, pharmacokinetics, and therapeutic potential.

2. Common methods for changing the genes of fungi

Random DNA fusion, gene-targeting technology, and RNA technology were the more traditional approaches to altering fungal genomes and manipulating gene expression before CRISPR/*Cas*. As per Wang et al.,[13] three methods for random DNA integration are called REMI, ATMT, and TAGKO.

Despite the time investment, random mixing yields excellent results. Genomic editing and insertion rely on homologous recombination (HR), a typical technique for targeted gene editing and insertion when a DNA template is available. Some have hypothesized that gene-targeted technologies struggle to effectively combat filamentous fungus and other species.[7,8] Foreign DNA integration into filamentous fungi needs a long, homologous sequence, which results in poor HR efficiency.[14] On the other hand, filamentous mushrooms were determined to have a lower HR efficiency than yeast, according to Huang et al.[7,8] many fungal species prefer the NHEJ approach for DNA damage repair because it is highly conserved and reduces the HR frequency of gene delivery.[7,8] To improve HR efficiency and encourage more precise genetic modifications, it may be possible to block DNA ligase IV, KU70, and KU80 in the NHEJ pathway of filamentous fungus.[15] Many individuals have employed conventional methods to produce various types of bioactive small molecules by modifying biosynthetic gene clusters (BGCs) in fungus, particularly using model organisms and strains that hold significance in the industry. However, these techniques are plagued by numerous issues, including limited utility, prolonged duration, and a scarcity of precise genetic markers.[16–18] They have also been hard to use on older fungal strains because they are hard to change and screen, and they don't have a vector system. Recent developments in gene editing technology, including as the CRISPR/Cas system, have enabled the successful application of high-efficiency genetic engineering to fungus, thereby overcoming these obstacles. As a result, a fresh strategy for discovering and fostering powerful social media connections has emerged. Clustering Regularly Interspaced Short Palindromic Repeats is a type of genetic coding that exists. A number of bacteria and archaea first developed the CRISPR system to defend themselves against viruses.[9,10] Classification of CRISPR/*Cas* systems into "class I" and "class II" has been further refined according to recent genetic guidelines. In each group, there are six distinct sorts, as stated by Makarova et al.[11] There have been a plethora of investigations and uses of CRISPR/*Cas9* technology in gene editing. The endonuclease *Cas9*, crRNA, and tracer RNA are the components of this system, as stated by Swartzjes et al.[19] These kinds of CRISPR RNA are trans-activating. *Cas9*can precisely target the necessary DNA sequence with the help of crRNA and tracer RNA. The arrival of *Cas9* and its cutting of the DNA's double strands causes a double-strand break (DSB).[12] Multiple pathways are utilized by the DNA repair mechanism of cells to fix the double-strand break.[20] Unlike the old system, which was dependent on non-template-dependent and error-prone methods, the HR system uses error-free templates. Both mechanisms referred to as non-homologous end joining (NHEJ) are primarily responsible for cellular DNA repair.[21–23] Another option is single-strand annealing (SSA) or microhomology-mediated end joining (MMEJ), although all of these methods have significant drawbacks.[24] Although non-homologous end joining (NHEJ) is a common method for repairing double-strand breaks (DSBs), it does introduce mutations at specific areas. The use of a DNA donor template allows for precise genome editing by homologous recombination (HR), on the other hand. The single RNA chimera called Dual-tracrRNA is one of the crRNA versions that has been developed. Motivating this change is a desire to make the CRISPR/*Cas9* technology more user-friendly. In addition, this chimera can be utilized to direct *Cas9* to break DNA at particular sites.[9] Several CRISPR-based approaches have been successfully employed to modify the SM pathway, including altering fungal genomes, regulating transcription, and modifying epigenetics. Fig. 1 displays various CRISPR/*Cas* technologies, together with their structures and functionalities. Table 1 provides an explanation of the feature, its functionality, and the advantages it offers. Editing fungal genomes and generating innovative therapeutic molecules are areas that could greatly benefit from the efficient, adaptive, and customizable CRISPR/*Cas* technology. Another enzyme that can alter fungal genes is TALEN or transcription activator-like effector nuclease. A domain inherited from *Fok*I endonucleases and another from TALE endonucleases make up the structure.[21] A filamentous fungus known as *Trichoderma reesei* was the initial host for the gene editing and transcription control methods involving the TALEN and TALE transcription factor fusion proteins. *Trichoderma reesei* is particularly noteworthy for its extraordinary protein production capabilities. The study was conducted by Liu et al.[25] and Tsuboi et al.[26] created TALENs and exonuclease overexpression to efficiently modify genes in Rhizopus oryzae. There are many benefits to using TALEN instead of CRISPR, even though it is challenging to make TALE repeats. It is more effective in altering the genome within heterochromatin regions, has a smaller range of non-specific effects, and can reach more target locations, according to Yee et al.[27] and Jain et al.[28]

3. Utilization of CRISPR/*Cas* systems in fungal genetic engineering

3.1 Classification of CRISPR/*Cas* systems

To stay up with the constantly growing number of CRISPR/Cas systems that have been discovered, a lot of work has gone into establishing classification algorithms. Nature Journals Microbiology released an article in

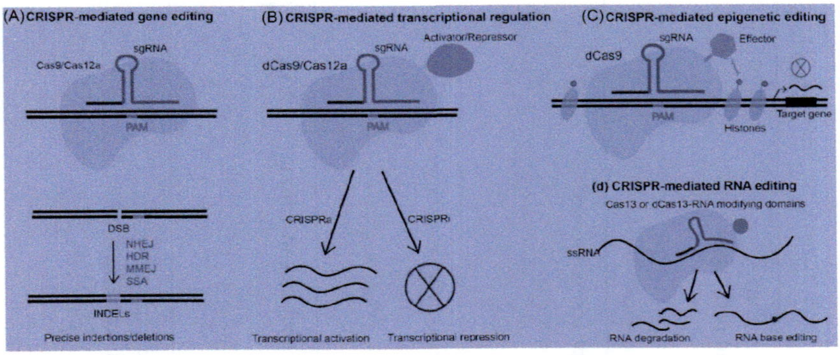

Fig. 1 The procedures of the CRISPR/Cas tools based on fungus are illustrated in Fig. 1. (A) In CRISPR-mediated gene editing, small guiding RNAs (sgRNAs) direct the proteinases Cas9 and Cas12a to specific sites on the genome, where they cleave the DNA to produce a double-stranded break. The DSB can be fixed via NHEJ, HDR, MMEJ, and SSA. In addition to large deletions or insertions, DSB repair can also introduce little indels. (B) CRISPR-mediated transcriptional regulation, in which dCas9/dCas12a, or deactivated Cas9 coupled with activation or repression domains, is utilized to modify gene expression by targeting the promoter region of a specific gene. CRISPRi blocks CRISPRa from acting. (C) Epigenetic editing using CRISPR alters target gene expression by modifying histones through dCas9-effector fusion. (D) Either a specifically engineered dCas13-RNA-modifying domains-fusion protein edits specific nucleotide residues, or RNA-targeting Cas13 degrades messenger RNA (mRNA) to suppress gene expression. (ssRNA) stands for single-strand RNA; PAM stands for protospacer adjacent motif.

2015 and provided further explanations of the CRISPR/*Cas* classification. Recently published research on CRISPR/Cas systems has cast doubt on the existing taxonomy and offered practical replacements. The CRISPR/*Cas* system now includes two new classifications, Class II and Class I. The CRISPR/*Cas* system has 33 distinct subtypes, with six variants in each subtype, according to Makarova et al.[36] There are sixteen distinct ways in which the latest class I CRISPR/*Cas* system differs from its 2015 predecessor. Class II CRISPR/*Cas* systems are very new, however, there are already seventeen subtypes, including classes II, V, and VI.[36] The most widely used CRISPR/*Cas9* gene editing tools today are based on the type II-A CRISPR/*Cas* system discovered in *Streptococcus pyogenes*. The CRISPR array, *cas1*, *cas2*, *cas9*, and *csn2* are encoded by a single operon in the type II CRISPR/*Cas9* system. Transcription of CRISPR arrays results in two molecules: pre-crRNA, a long precursor to mature crRNAs and tracrRNA, a short trans-activating RNA that enhances the CRISPR repeat. By using short crRNA, the *Cas9* protein inserts double-strand

Table 1 CRISPR/Cas aspect its mechanism of action and advantages.

CRISPR/Cas aspect	Mechanism of action	Advantages	References
DNA-Based CRISPR/Cas9 System	Utilizes DNA-based strategies for delivering Cas9 and sgRNA expression cassettes into the fungal nucleus. Successful in editing fungal genomes for SMs and metabolic pathways.	Precise editing of fungal genomes, Modification of metabolic pathways	29,30
CRISPR/Cas9 RNPs	Involves introducing pre-assembled Cas/sgRNA complexes (RNPs) into fungal cells. Offers ease of use and compatibility across different species/strains.	Ease of use, Compatibility across species/strains	29,31
Combination of *In Vitro* and *In Vivo* Expression	Combines both in vitro and in vivo methods for efficient gene targeting. Enhances CRISPR/Cas component delivery into fungal cells.	Enhanced gene editing efficiency, Flexibility in delivery methods	29,32
CRISPR/Cas12a-Based Editing	Cas12a (Cpf1) provides an alternative to Cas9, recognizing different PAM sequences and cleaving DNA differently. Expands CRISPR-based editing scope and flexibility.	Expanded target scope, Flexibility in editing methods	29,33
CRISPR/Cas Transcriptional Regulation	Enables gene editing and transcriptional regulation in fungi using CRISPRa and CRISPRi techniques. Allows precise control over gene expression.	Precise control over gene expression, Facilitates study of fungal secondary metabolism	34,35

breaks (DSBs) at predefined places, mediating interference. A protospacer-adjacent motif (PAM) is a short sequence that is required for *Cas9* to operate. This component of the CRISPR array prevents self-DNA cleavage, according to an article published by Helfer et al.[37] Numerous CRISPR-based methods have proliferated since the creation of CRISPR/*Cas* systems. In contrast to older methods such as homologous activation and heterologous expression, these new technologies allow for the efficient and extensive alteration of fungal genes. By manipulating fungal genomes with CRISPR/*Cas*-based methods, scientists have enhanced secondary metabolite synthesis.

3.2 The CRISPR/*Cas9* system that relies on DNA

Utilizing DNA-based methods, scientists have introduced *Cas9* and sgRNA expression cassettes into the nucleus of fungus cells in an effort to modify fungal genomes through the use of CRISPR/*Cas9*. Before the *Cas9*/sgRNA expression cassettes are transcribed, the fungus starts the synthesis of the *Cas9* RNPs complex. Genetic editing techniques based on CRISPR/*Cas9* are not applicable in all cases due to their reliance on fungal strains that are dependent on transformation and expression vectors that are particular to certain species. According to Shi et al.,[38] this was the standard method for fungal secondary metabolite production in the past. The CRISPR/Cas system was tested in an *in vivo* environment with the filamentous fungus *Fusarium fujikuroi*. An NLS from histone H2B was used to construct a *Cas9* delivery mechanism. The sgRNA expression promoters came from either 5S rRNA or U6 small nuclear RNA. The method changed the GA product profiles and GA metabolic pathways in *F. fujikuroi*, as reported by Shi et al.[39] The esterase genes IAH1 and TIP1 in *Saccharomyces cerevisiae* were rendered inactive by CRISPR/*Cas9* technology, according to Dank et al.[40] Ester synthesis and the manufacture of aromatic compounds were both boosted as a result. Using CRISPR/*Cas9* technology to render enzymes inactive could reduce the rate of sterol and fatty acid synthesis by filamentous fungus. This in turn can lead to enhanced production of the most potent, popular drugs lovastatin and taxol. As mentioned by El-Sayed et al.,[5] the effectiveness of these drugs is increased when compared to the conventional methods. The metabolic route for the synthesis of alternariol and its derivatives, which are potent secondary metabolites that cause disease, was elucidated in *Alternaria alternata*, using the CRISPR/*Cas9*knock-out technique.[41] Thus, using CRISPR/*Cas* systems in the study of fungal SM biosynthesis and metabolism, respective approaches can be developed to promote the synthesis of beneficial bioactive SMs while reducing the production of undesirable ones. Using CRISPR/*Cas9*-expression plasmid to replace gloF with ap-htyE in a recent study of *Glarea lozoyensis*, researchers achieved a decrease in pneumocandin C0 yield and an increase in the production of caspofungin-pneumocandin B0.[42] The CRISPR/Cas9 system is highly effective in creating genetically modified fungal strains, which can be applied to produce new secondary metabolites. For instance, by using CRISPR/Cas9, we managed to eliminate four biosynthetic gene clusters (BGCs) in a *Penicillium rubens* strain. Therefore, this strain fails to produce a particular secondary metabolite (SM). Later, the SM-lacking strain was exploited to integrate the foreign gene cluster Calbistrin, thus leading to the development of a novel strain with higher levels of decumbenone and a more improved SM profile due to decreased interference from native SMs.[43] An overabundance of double-strand breaks (DSBs) would hinder the viability capability of fungus, hence leading to a decrease in the effectiveness of CRISPR/*Cas* multi-gene editing.[44] To correct this issue, exogenous DNA and *Cas9*/sgRNAs can be administered to the fungus.[45] Recently, a more advanced CRISPR/*Cas9* system, coupled with a DNA repair template, was used to eliminate numerous sorbicillinoid biosynthesis genes in *Acremonium chrysogenum*. In turn, the outcome was a decrease in sorbicillinoids and an increase in the synthesis of cephalosporin C.[44] Recently, it was demonstrated that in the industrial fungus *A. chrysogenum* FC3–5-23, using CRISPR/*Cas9* technology to eliminate the Acaxl2 gene, which controls the development of arthrospores, increased cephalosporin C synthesis. This finding proves that there is a high correlation between fungal mycelium structure and production ability.[46,47] The majority of fungal species strongly favor non-homologous end joining (NHEJ) as the primary mechanism for the repair of double-strand breaks (DSBs), instead of homology-directed repair (HDR).[48] The CRISPR/*Cas* systems based on DNA were then used for the development of mutant fungus strains with disrupted NHEJ processes in a bid to enhance HDR for more accurate genetic modifications. The yeast *Scheffersomyces stipitis* was used to create a mutant strain called ku70Ŕku80Ŕ by deleting the KU70 and KU80 genes using CRISPR. The findings proved that the change improved the efficiency of HDR-based genome editing in comparison to the initial strain. Genome editing using CRISPR/*Cas9* was performed on the NHEJ-deficient *Shiraia bambusicola* strain to target the hypocrellin pathway. Using this mutant strain,[49] found that hypocrellin yield increased twelvefold.

3.3 CRISPR/Cas9 ribonucleoproteins (RNPs)

Several fungi can't express *Cas* endonuclease or sgRNA DNA expression cassettes. Alternatively, the Cas/sgRNA combination could be introduced into the fungal nucleus after pre-assembled RNPs have been transformed *in vitro*. A major advantage of RNP-based CRISPR technology over its DNA-based counterpart is that it is compatible with nearly all species and strains and does not require any strain-specific modifications.[50] detailed a novel experimental setup that outperformed traditional gene replacement methods in identifying target genes in *Aspergillus fumigatus*. The *in vitro* assembly of the system made use of *Cas9* ribonucleoprotein (RNP) and micro homology repair templates. Numerous clinical isolates of *Acinetobacter fumigatus* were swiftly and entirely cleared of drug resistance and pathogenicity using this RNP-based gene editing approach. To increase the variety of secondary metabolites generated by wild-type *Aspergillus wentii*, scientists have developed a novel strategy based on CRISPR/*Cas9* technology. Based on the results demonstrated by Oakley et al.[51] and Yuan et al.,[52] this technique is capable of removing the mcrA negative transcriptional regulator. Several secondary metabolites' activity and production are regulated by this regulator. The use of RNP complexes, which consist of a modified *Cas9* nuclease and two single guide RNAs, allows the Epichloë species to eliminate a large group of ergot alkaloid synthesizers.[53] found that our method significantly decreased the production of these secondary metabolites, which can harm cow health. The RNP method allows for the independent modification of endophytes, eliminating the need for transgenic approaches. Better yet, it makes it easier to create mutant strains that do not have toxin genes, which could lead to feed crop improvements. Some economically significant plant diseases have been engineered with highly focused genetic alterations using CRISPR RNP-based approaches. Researchers focused their CRISPR RNP-based tool development efforts on the rice blast fungus *Magnaporthe oryzae*. This application allows users to quickly and accurately change base pairs, substitute genes, and tweak many genes simultaneously.[54]

3.4 Utilizing both *in vitro* and *in vivo* methods to express the *Cas*/sgRNA complex

Because fungi lack the necessary promoters to generate small guide RNAs (sgRNAs), it is now possible to use laboratory-grown gRNAs to target specific genes within fungal cells. This not only gets around the problem of insufficient sgRNA promoters, but it also gets rid of the necessity to make sgRNA expression cassettes. Using the filamentous fungus *T. reesei*, Liu and colleagues enhanced a CRISPR-based method in 2015. They succeeded in producing sgRNA by *in vitro* transcription by employing an in vivo codon-optimized*Cas9* version. Thanks to this method, targeted mutations and homologous recombination are now within reach. Using a U6 promoter to regulate sgRNA expression in Nodulisporium resulted in rather ineffective CRISPR/*Cas9* gene editing.[55] found that mutagenesis was greatly amplified in a strain that had a lab-generated sgRNA in addition to a linear marker gene cassette. The increasing evidence suggests that the strategies used to propagate CRISPR/*Cas* components may affect the effectiveness of gene editing. The lab-synthesized sgRNA was integrated with an organism-expressed *Cas9* plasmid using a CRISPR system to extract galactaric acid from *Aspergillus niger*. we do this by editing genes involved in galactaric acid digestion. This strategy for gene deletion outperformed its predecessors by a wide margin, as reported by Kuivanen et al.[56]

3.5 Gene editing using CRISPR/*Cas12a*

A certain PAM sequence is required for accurate DNA identification under sgRNA guidance and for CRISPR/*Cas* systems to generate the necessary cuts. The 5′-NGG-3′ PAM sequence is necessary for the commonly used *Staphylococcus pyogenes* Sp*Cas9*.[57] Due to its reliance on a specific PAM (protospacer adjacent motif) at the target site, CRISPR/*Cas9* based gene editing has not found extensive use. For CRISPR-based genomic editing to be more versatile, cas nucleases are required with a broader range of PAM sequence detecting capabilities. Class II systems were the go-to for molecular biologists because of how easy they were to operate. The Sp*Cas9* enzyme from *Staphylococcus pyogenes* and the *Cas12a* type V enzyme from *Francisella novicida*, Acidaminococcus sp., or Lachnospiraceae, among other bacteria, were discovered as components of this system by Oue-draogo et al.[58] *Cpf1* (or *Cas12a*) is highly specific for DNA cleavage and is dependent on the PAM sequence, in contrast to *Cas9*. In particular, *Cas12a* aims for a consensus protospacer adjacent motif (PAM) close to the 5′ end of the displaced strand. In situations where N can be any nucleotide other than T, the 5′-TTN-3′ PAM is preferable to the 5′-NTN PAM. Staggered double-strand breaks are created when specific locations on the DNA molecule are cleaved by the *Cas12a* protein's only RuvC domain. In connected type II CRISPR systems, crRNA directs the *Cas9* nuclease cleavage activity. The *Cas9* nuclease and the crRNA it is attached to can be guided to the exact location of the cleavage by use of tracrRNA. In contrast to the type II CRISPR system, the mature crRNA produced by the *Cas12a*-associated CRISPR array is quite short, clocking in at a mere 42 or 44 nucleotides in length. There seems to be no function for tracrRNA in this pathway. Following a series of 19 extremely similar nucleotides, the crRNA advances to a group of 23–25 distinct nucleotides. Scientists have used the CRISPR/*Cas12a* method to modify many strains of industrial fungus. This group of Aspergillus species includes variants such as nidulans, aculeatus, oryzae, and sojae, as shown in several studies[59–62] showed that four genes in *Ashbya gossypii* can be eliminated using a multiplexed CRISPR/*Cas12a* system. The plasmid containing several CRISPR/*Cas12a* genes allowed this to happen. Numerous studies have examined the effectiveness of *Cas* nucleases in destroying both individual and clusters of fungal genes. To find the most effective, scalable, and generally applicable CRISPR methods for usage with *Thermothelomyces thermophilus*. Gene editing efficiency was investigated by the authors using Sp*Cas9*, Fn*Cas12a*, and As*Cas12a*. It is possible that one or more genes are the targets of these nucleases. The methods of delivery described by Kwon et al.[63] encompassed both plasmids and RNPs. Editing efficiency is determined by the targeted region and the *Cas* nuclease, according to the results. A number of DNA changes, including large deletions and insertions (LDIs) and small insertions and deletions (INDELs), can be induced in the *Musella oryzae* genome by ribonucleoproteins (RNPs) based on *Cas12a*, according to recent findings employing sanger and nanopore sequencing. Both Huang and Cook[24] and Huang et al.[64] showed that several DNA-repair pathways are involved in fixing the *Cas12a*-induced double-strand staggered break. Some DNA repair processes may have preferences, according to the study, which is based on their observation of biased DNA mutations. If the epigenome has the ability to influence this hierarchy, it could have significant implications for genome engineering as well as evolution.[24]

3.6 Transcriptional regulation with CRISPR/*Cas*

Some fungal biosynthetic gene clusters (BGCs) have their expression drastically decreased or eliminated due to strict regulation. Numerous mechanisms exist for the activation of biosynthetic gene clusters (BGCs). Some of these methods include using foreign expression systems, altering global regulators, boosting TF expression, and changing promoters. The most recent method for activating genes is by using CRISPR. Using CRISPR/*Cas9* to knock-in the promoter of the *Thermomyces dupontii* gene is one novel way to produce the PKS-NRPS gene, as described by Huang et al.[8] By utilizing the CRISPR/*Cas9* promoter exchange approach, it is possible to activate many BGCs.[65] used CRISPR/*Cas9* to activate a native promoter in a group of physiologically weak biosynthetic genes. The TAR resource and the single-marker multiplex CRISPR/*Cas9* technology were used to construct the mCRISTAR gene editing tool.[65] states that this approach could simultaneously replace many native promoters in BGC. The mCRISTAR method improved gene expression by utilizing many promoters situated immediately upstream of BGC. It becomes more difficult to synthesize DNA when employing a single CRISPR array with numerous target sites, and it becomes even more difficult to target certain gene combinations. This limitation is not encountered by the mpCRISTAR approach because it is based on several plasmids and employs CRISPR/*Cas9* and TAR. It is possible to replace numerous sets of biosynthetic genes simultaneously using this strategy.[66] found that by combining BGC constructs of diverse designs into one assembly, the technique improved promoter engineering while being more cost-effective than mCRISTAR.[67] state that the CRISPR activation (CRISPRa) mechanism activates silent biosynthetic gene clusters (BGCs) using trans-acting effectors and a deactivated *Cas* (*dCas*) protein. By controlling the expression of fungal BGCs using CRISPRa technology, the hunt for bioactive small molecules has been accelerated. Using CRISPRa systems like CRISPR/dLb*Cas12a*-VPR and CRISPR/dSp*Cas9*-VPR has been considered as a potential method to stimulate transcription in *A. nidulans*.[68] Scientists discovered that d*Cas12a* activated genes far more effectively than d*Cas9*. Microperfuranone production showed a remarkable rise following the activation of the NRPS-like gene utilizing CRISPR/dLb*Cas12a*-VPR. By inserting d*Cas9*-VPR and a sgRNA module into *P. rubens* through the non-integrative AMA1 vector, we were able to create a CRISPRa system that does not integrate into the genome. This is the process that activated the BGC cryptic macrophorin, according to Mózsik et al.[69,70] Using d*Cas9* with multiple activator domains is one potential strategy to improve transcription activation, as stated by Román et al.[71] Pickar-Oliver and Gersbach describe the CRISPR interference (CRISPRi) approach to transcriptional suppression in their 2019 publication. They state that d*Cas9* and repressors work together to achieve this goal. The promoter regions of *Candida albicans* were silenced by the researchers using d*Cas9* with a repressor domain and a CRISPRi-based mechanism. Wensing et al.[72] and Román et al.[71] found that the extent of transcriptional suppression was regulated by the specific location inside the promoter region that was targeted by the CRISPR complex. Heterologous expression is an additional mechanism for reawakening sets of inactive genes involved in biosynthesis. Using this approach, we can manipulate the fungus genome in a lab setting. After that, we shall clone the BGC. Copying large amounts of DNA in a controlled setting with this method has been a bit of a challenge. According to Xu et al.,[73] the full complement of filamentous fungus biosynthetic gene clusters (BGCs) was generated during the first round of using a novel CRISPR/*Cas9* method. In this approach, RNA-guided *Cas9* endonuclease was used to controllably cleave the genomic DNA of the bacteria. Enriching the DNA fragments into live yeast cells was the initial step in an innovative approach of employing vectors to construct entire biosynthetic gene clusters (BGCs).

3.7 Epigenetic editing using CRISPR/*Cas*

It is possible to control gene expression by means of epigenetic regulation. It changes how transcription factors make genes accessible and readable. The epigenetic regulatory controls, which include DNA methylation, histone modifications, chromatin remodeling, and microRNA, are affected by environmental stresses and particular signals. Environmental signals or genetic manipulation of master regulators can cause epigenetic fingerprints to undergo global alterations. But for site-specific epigenetic marker alteration, it's required to rearrange nearby epigenetic markers. There is strong evidence from multiple studies linking epigenetic alterations to SM's secondary metabolism. The epigenetic profile of Aspergillus spp. has been altered by chemically inducing DNA methyl transferase and inhibiting histone deacetylase. Because of this, the SM profiles of these species are now well-known. Conventional methods can allow for epigenome rewriting by epigenetic remodelers modification. The CRISPR/*Cas* system has allowed for remarkable advancements in epigenetic editing in a variety of organisms, including fungus, plants, mammals, and bacteria. The CRISPR/d*Cas9*-mediated epigenetic modification approach was used to apply histone marks to many secondary metabolic genes in *A. niger*. The *breF, fuml,* and *fwnA* genes were among these. In addition to d*Cas9*, this technique made use of histone acetyltransferase and histone deacetylase, two epigenetic regulators. In each case, the intended gene induction or repression was accomplished. Through the use of CRISPR/*Cas9* technology, the histone deacetylase-coding *rpd3* gene was successfully removed from a marine fungus. The elimination method led to the isolation of a novel class of compounds. These findings strongly suggest that epigenomic editing using CRISPR could open up new avenues for understanding fungal secondary metabolite use. Editing fungal epigenomes with *Cas12a* and other types of *Cas*, in addition to *Cas9*, might theoretically enhance the number of potential target sites.

3.8 Gene editing system utilizing CRISPR/*Cas9* technology without the need for genetic markers

The inadequateness of selection markers poses a major obstacle to genetic alteration in fungi. A plasmid containing the AMA1 sequence and other critical components can be used to make genetic modifications, thus circumventing this restriction. Plasmids like this eliminate the need for markers. The AMA1 sequence was first found to reside autonomously upon transformation in *A. nidulans*, rather than being integrated into the homologous chromosome.[74] After inserting the AMA1 gene into the plasmid, it becomes completely autonomous from the fungal genome, making it highly transformable and retaining its ability to replicate on its own. According to Wang and Coleman,[75] the AMA plasmid can be utilized for several sub-culturing cycles of the fungus without a selection agent, and it also allows for the repeated use of a marker or markers during transformation. Several strains of fungi that are important to the industry have been modified using CRISPR/*Cas9* technology. *Aspergillus terreus*, *Cordygillus oryzae*, *Cordyceps militaris*, and *Aspergillus niger* are some of these types. These methods are based on an AMA1-plasmid that replicates on its own. Publications by Liu et al.,[76] Meng et al.,[77] Yao et al.,[78] and Li et al.[79] have proved the methods' effectiveness. In one study, strains of *Paecilomyces variotii* and *Penicillium roquefortii* were generated using a genome-editing technique based on AMA1. These strains lost their ability to produce melanin due to the loss of the PKS genes.[80] ran the targeted deletion on fungi that cause food spoilage to determine if it altered their resilience to heat and UV-C radiation, as conidia are prevalent in these organisms. One alternative is to edit the *AowA* and *sC* genes in *A. oryzae* without using markers by using genome editing vectors based on AMA1 and codon-optimized *Cas12a* expression cassettes. Furthermore,[81] noted that *A. sojae* did not contain any deletion markers in the *AswA* gene. The AMA1 plasmid has allowed for the development of novel CRISPR-associated editing tools that do not require markers. It makes use of a telomere vector, or one that is improved upon it. According to Leisen et al.[82] and Leisen et al.,[83] these have been designed to suppress the production of harmful chemicals by *Botrytis cinerea* fungus by turning particular genes dormant. Minichromosomes, which are plasmids produced from telomeres and may replicate autonomously, are not susceptible to natural selection since they lack centromeres. These plasmids provide a novel, promising method of marker-less gene editing in fungi. Constant *Cas9* production during plasmid replication can have deleterious effects on cell metabolism and development. Scientists have started using a synthetic *Cas9* protein instead of the naturally occurring one to avoid this problem. To create a marker-less gene substitution technique, the citric acid-hyper-producing *Aspergillus tubingensis* WU-2223L strain was used. In this method, a DNA fragment was utilized to initiate the CRISPR/*Cas9* system-mediated *in vitro* expression of *Cas9* and sgRNAs. Improving gene replacement in gene-deficient situations is made possible by creating sgRNAs to target both the marker gene and the gene of interest. Inadequate selection markers can be addressed in part by making use of endogenous genes as a screening marker. One species of Monascus recently served as a screening marker for a naturally occurring gene that confers resistance to this organism; nevertheless, there are fewer of these genes available for genetic engineering. Genetic engineering can only employ a limited set of genes to create resistance to this specific bacterium. By utilizing mutant strains deficient in many endogenous genes (mrpyrG),[46,47] created a marker-less approach to multigene modification.

4. Current constraints and future potential of CRISPR/*Cas*-mediated fungi genome engineering

The CRISPR/*Cas* system allows certain fungal species to edit and express their genes. Having said that, they are rendered useless due to a handful of notable constraints. The efficiency of gene editing is affected by a number of factors, including cas enzyme kinetics, sgRNA structure, repair template, editing method, quantity of gene copies, and similar parameters. When developing techniques for CRISPR/*Cas* gene targeting, design-by-design methods fail to take into account the significance of sgRNA secondary structure. Improved models for secondary structure prediction are necessary to enable sgRNA synthesis. The target location, accessibility of chromatin, nucleosome occupancy, and transcription factor occupancy are some of the important elements that must be considered for CRISPR gene editing to be effective. Changing the colored parts changes the exact spots where the *Cas* protein or *Cas*-effector complex is used. To create sgRNA, we need novel methods to evaluate the density and structure of the key chromatin sites. Assessing the responsiveness of various genomic regions to sgRNA using high-throughput methods enhances the design. A recent study successfully targeted many loci using a high-throughput CRISPR-based approach, gRNA libraries, and *in vivo* expressed *Cas9*. This approach eliminates the need to sub-clone sgRNA expression cassettes. CRISPR/*Cas*-induced non-homologous end joining repair might not be the best option for fixing double-strand fractures. Unexpected mutagenesis at the target region has been observed in recent research that utilized long-read sequencing and long-range PCR genotyping. A human differentiated cell line, hematopoietic progenitors, and embryonic stem cells from mice were subjected to complicated rearrangements and large deletions with the application of single-guide RNA/*Cas9* at specified loci.[84] Scientists[85] utilized *Sclerotinia sclerotiorum* to introduce plasmids into DNA repair pathways including non-homologous end joining (NHEJ) and DSSBs produced by CRISPR/*Cas9*. Using CRISPR/*Cas12a*,[24] discovered that *M. oryzae* exhibited substantial insertions, deletions, and combinations of the two. Multiple complex on-target effects show that numerous natural DNA repair pathways can patch double-strand breaks (DSBs) caused by CRISPR/*Cas*. Mutual influences are exerted by these systems. It is critically important to study the factors that influence the selection of cure mechanisms for double-strand breaks (DSBs). The possibility that gene editing could have unintended consequences is another cause for concern. Several papers have proposed methods to lessen the occurrence of unintended biological consequences. Using shorter sgRNAs, especially Tru-RNA with 17 or 18 nucleotides,[86] demonstrated reduced specificity of gene editing in human cells without affecting target activity. Due of its lower size compared to the 20-nucleotide sgRNA, the nuclease RNAdirected complex could be more susceptible to mismatching. Compared to truRNAs with 17 or 18 nucleotides, those with shorter or longer complementarity sequences had little to no effect. This finding amply supports the need for additional inquiry, as stated by Fu et al.[86] Cho et al.[87] found that by combining two sgRNAs with *Cas9* nickases, it is feasible to produce two SSBs or nicks on separate strands of DNA. This helps maintain the ability to alter the genome while reducing off-target consequences. The decrease in off-targeting and the improvement of the regulatory mechanism may be explained by the fact that cas nickases reduced the formation of off-target indels. A number of studies have claimed to have enhanced the specificity and accuracy of the SpRY Sp*Cas9* mutant; they include,[88] eSp*Cas9,* Hypa*Cas9, Cas9*-NG,[89] and.[90] According to Walton et al.,[91] these variations reduce off-target effects by lowering the demand for PAM in human cells. There are less unintended effects on the non-targeted areas when customized *Cas* is used. Due to the *Cas* modifications, PAM compatibility was much improved, and the DNA-targeting CRISPR enzymes were utilized less frequently. Several PAM-*Cas9* proteins modified fungal genes. The possibility of destabilizing the genome of *S. cerevisiae* and performing nucleotide alterations with Sp*Cas9*-NG was demonstrated in an experiment conducted in 2021. The effective modification of nucleotides was demonstrated by Tan et al.[92] using *Cas9*-VQR, VRER, x*Cas9,* and Sp*Cas9*-NG in yeast. For CRISPR to be useful in more contexts, more studies with different types of fungi and *Cas* proteins are required. Genome engineering allows us to pinpoint the specific genes and clusters involved in SM biosynthesis. Fungal mechanisms and regulatory networks that contribute to the production of secondary metabolites whose precise identities remain unknown. Maybe we might look into the latest research in molecular biology, bioinformatics, and omics. Gene circuits can be precisely and efficiently manufactured using synthetic biology modules.[69,70] It is now possible to reprogram fungal genomes quickly and effectively, all because of developments in synthetic biology. Based on the findings of Liu et al.[93] and Ning et al.,[94] genetic modification is currently reserved for model strains and industrial backgrounds. For CRISPR-based genetic editing of fungus to work, optimization of transformation pathways for different strains is essential. There is evidence that CRISPR/Cas can modify RNA. All three types of *Cas9*—Type II, Type III Cmr/Csm, and Type VI *Cas13*—target unique RNA sequences. The RNA-targeting mechanisms evolved by *Neisseria meningitidis* and *Leptotrichia shahii* from Type II, III, and VI of the CRISPR gene family allow these bacteria to alter, track, and inactivate RNA. The messenger RNA (mRNA) that disrupts yeast genes can be degraded by specialized tools such as *Cas13a, Lwa, Rfx,* and Nme1*Cas9.* The bases of fission yeast can be precisely targeted utilizing a crRNA/pRNA in conjunction with the catalytic domain of hADAR2d, as reported by Jing et al.[95] We can study faulty genes using RNA-targeting, as opposed to CRISPR, which allows us to alter inheritance. Fungi that actually employ CRISPR to alter their DNA are quite rare. Our knowledge of fungal secondary metabolism will be substantially improved by future CRISPR-RNA-targeting approaches.

5. Conclusion

Through this approach, scientists have been able to manipulate the expression of specific genes in various fungal species, greatly expanding our understanding of the molecular biology of these organisms. Scientists are utilizing the CRISPR-*Cas9* technology to make alterations to the genetic makeup of fungi, particularly Aspergillus and Candida species that pose a threat to human health. This cutting-edge technique enables efficient gene manipulation, providing a deeper understanding of fungus pathogenesis, potential antifungal targets, and microbial pathogenicity factors. Advancements in technology have been instrumental in gaining insights into the biology, pathophysiology, and mechanisms of fungus resistance to antifungal medications. As a result, there are exciting prospects for further research and practical applications related to the manipulation of fungal genes. In the realm of fungal research, the application of CRISPR-*Cas* systems may shed light on new perspectives regarding fungal diseases, genetic engineering, and disease preventive strategies. The advancements in genome editing have greatly enhanced the field of fungal molecular biology, opening up exciting possibilities for investigating fungal virulence factors, medicinal targets, and host-pathogen interactions.

References

1. Singh P, Jayashree T, Reddy MS. Plant secondary metabolites as defenses, regulators, and primary metabolites: the blurred functional trichotomy. *Plant Physiol.* 2020;184:39–52.
2. Hawksworth DL, Lücking R. Fungal diversity revisited: 2.2 to 3.8 million species. *Microbiology.* 2017;5(4).
3. Woloshuk CP, Shim WB. Aflatoxins, fumonisins, and trichothecenes: a convergence of knowledge. *FEMS Microbiol Rev.* 2013;37:94–109 Fungal BGCs and their relationships with associated SMs.
4. Mózsik L, Iacovelli R, Bovenberg RAL, Driessen AJM. Transcriptional Activation of biosynthetic gene clusters in filamentous fungi. *Front Bioeng Biotechnol.* 2022;10:901037.
5. El-Sayed ASA, Abdel-Ghany SE, Ali GS. Genome editing approaches: manipulating of lovastatin and taxol synthesis of filamentous fungi by CRISPR/Cas9 system. *Appl Microbiol Biotechnol.* 2017;101:3953–3976.
6. Wang H, Xu X. Microhomology-mediated end joining: new players join the team. *Cell Biosci.* 2017;7:6.
7. Huang PW, Yang Q, Zhu YL, et al. The construction of CRISPR-Cas9 system for endophytic Phomopsis liquidambaris and its PmkkA-deficient mutant revealing the effect on rice. *Fungal Genet Biol.* 2020;136:103301.
8. Huang WP, Du YJ, Yang Y, et al. Two CRISPR/Cas9 systems developed in Thermomyces dupontii and characterization of key gene functions in thermolide biosynthesis and fungal adaptation. *Appl Env Microbiol.* 2020;86:e01486-20. ([Google Scholar] [CrossRef] [PubMed]).
9. Jinek M, Chylinski K, Fonfara I, Hauer M, Doudna JA, Charpentier EA. programmable dual-RNA-guided DNA endonuclease in adaptive bacterial immunity. *Science.* 2012;337:816–821. ([Google Scholar] [CrossRef]).

10. Barrangou R, Fremaux C, Deveau H, et al. CRISPR provides acquired resistance against viruses in prokaryotes. *Science*. 2007;315(5819):1709–1712. https://doi.org/10.1126/science.1138140
11. Makarova KS, Haft DH, Barrangou R, et al. Evolution and classification of the CRISPR-Cas systems. *Nat Rev Microbiol*. 2011;9:467–477. ([Google Scholar] [CrossRef] [PubMed] [Green Version]).
12. Gasiunas G, Barrangou R, Horvath P. Cas9-crRNA ribonucleoprotein complex mediates specific DNA cleavage for adaptive immunity in bacteria. *Proc Natl Acad Sci U S A*. 2012;109:E2579–E2586. ([Google Scholar] [CrossRef] [Green Version]).
13. Wang H, Xu X. Homology-mediated end joining-based targeted integration using CRISPR/Cas9. *Cell Res*. 2017;27:801–814.
14. Hua SB, Qiu M, Chan E, Zhu L, Luo Y. Minimum length of sequence homology required for in vivo cloning by homologous recombination in yeast. *Plasmid*. 1997;38:91–96. ([Google Scholar] [CrossRef] [PubMed]).
15. Ninomiya Y, Suzuki K, Ishii C, Inoue H. Highly efficient gene replacements in Neurospora strains deficient for nonhomologous end-joining. *Proc Natl Acad Sci USA*. 2014;101:12248–12253. ([Google Scholar] [CrossRef] [Green Version]).
16. Krappmann S. Gene targeting in filamentous fungi: the benefits of impaired repair. *Fungal Biol Rev*. 2007;21:25–29. ([Google Scholar] [CrossRef]).
17. Kück U, Hoff B. New tools for the genetic manipulation of filamentous fungi. *Appl Microbiol Biotechnol*. 2010;86:51–62. ([Google Scholar] [CrossRef]).
18. Shapiro RS, Chavez A, Collins JJ. CRISPR-based genomic tools for the manipulation of genetically intractable microorganisms. *Nat Rev Microbiol*. 2018;16:333–339. ([Google Scholar] [CrossRef]).
19. Swartjes T, Staals RHJ, van der Oost J. Editor's cut: DNA cleavage by CRISPR RNA-guided nucleases Cas9 and Cas12a. *Biochem Soc Trans*. 2020;48:207–219. ([Google Scholar] [CrossRef] [Green Version]).
20. Xue C, Greene EC. DNA repair pathway choices in CRISPR-Cas9-mediated genome editing. *Trends Genet*. 2021;37:639–656. ([Google Scholar] [CrossRef]).
21. Joung JK, Sander JD. TALENs: a widely applicable technology for targeted genome editing. *Nat Rev Mol Cell Biol*. 2013;14:49–55. ([Google Scholar] [CrossRef] [Green Version]).
22. Chang HHY. Non-homologous DNA end joining and alternative pathways to double-strand break repair. *Nat Rev Mol Cell Biol*. 2017;18:495–506. ([Google Scholar] [CrossRef]).
23. Sung P, Klein H. Mechanism of homologous recombination: mediators and helicases take on regulatory functions. *Nat Rev Mol Cell Biol*. 2006;7:739–750. ([Google Scholar] [CrossRef]).
24. Huang J, Cook DE. CRISPR-Cas12a ribonucleoprotein-mediated gene editing in the plant pathogenic fungus Magnaporthe oryzae. *STAR Protoc*. 2021;3:101072. ([Google Scholar] [CrossRef] [PubMed]).
25. Liu P, Wang W, Wei D. Use of transcription activator-like effector for efficient gene modification and transcription in the filamentous fungus Trichoderma reesei. *J Ind Microbiol Biotechnol*. 2017;44:1367–1373. ([Google Scholar] [CrossRef]).
26. Tsuboi Y, Sakuma T, Yamamoto T, et al. Gene manipulation in the Mucorales fungus Rhizopus oryzae using TALENs with exonuclease overexpression. *FEMS Microbiol Lett*. 2022;369:fnac010. ([Google Scholar] [CrossRef]).
27. Yee JK. Off-target effects of engineered nucleases. *FEBS J*. 2016;283:3239–3248. ([Google Scholar] [CrossRef] [Green Version]).
28. Jain S, Shukla S, Yang C, et al. TALEN outperforms Cas9 in editing heterochromatin target sites. *Nat Commun*. 2021;12:606. ([Google Scholar] [CrossRef]).
29. Norton EL, Sherwood RK, Bennett RJ. Development of a CRISPR-Cas9 system for efficient genome editing of Candida lusitaniae. *mSphere*. 2017;2:e00217. https://doi.org/10.1128/mSphere.00217-17 ([PMC free article] [PubMed] [CrossRef] [Google Scholar]).

30. Song N, Chu Y, Tang J, Yang D. Lipid-, inorganic-, polymer-, and DNA-based nano-carriers for delivery of the CRISPR/Cas9 system. *Chembiochem.* 2023;24(16):e202300180. https://doi.org/10.1002/cbic.202300180 Epub 2023 Jul 25. PMID: 37183575.
31. Wichmann M, Maire CL, Nuppenau N, et al. Deep characterization and comparison of different retrovirus-like particles preloaded with CRISPR/Cas9 RNPs. *Int J Mol Sci.* 2023;24(14):11399. https://doi.org/10.3390/ijms241411399 PMID: 37511168; PMCID: PMC10380221.
32. Griffin MF, Borrelli MR, Garcia JT, et al. JUN promotes hypertrophic skin scarring via CD36 in preclinical in vitro and in vivo models. *Sci Transl Med.* 2021;13(609):eabb3312. https://doi.org/10.1126/scitranslmed.abb3312 Epub 2021 Sep 1. PMID: 34516825; PMCID: PMC8988368.
33. Zhao J, Zuo S, Huang L, Lian J, Xu Z. CRISPR-Cas12a-based genome editing and transcriptional repression for biotin synthesis in Pseudomonas mutabilis. *J Appl Microbiol.* 2023;134(3):lxad049. https://doi.org/10.1093/jambio/lxad049 PMID: 36914213.
34. Wright AV, Nuñez JK, Doudna JA. Biology and applications of CRISPR systems: harnessing nature's toolbox for genome engineering. *Cell.* 2016;164(1-2):29–44. https://doi.org/10.1016/j.cell.2015.12.035 PMID: 26771484.
35. Mahas A, Neal Stewart C, Jr, Mahfouz MM. Harnessing CRISPR/Cas systems for programmable transcriptional and post-transcriptional regulation. *Biotechnol Adv.* 2018;36(1):295–310. https://doi.org/10.1016/j.biotechadv.2017.11.008 Epub 2017 Nov 29. PMID: 29197619.
36. Makarova KS, Wolf YI, Iranzo J, et al. Evolutionary classification of CRISPR-Cas systems: a burst of class 2 and derived variants. *Nat Rev Microbiol.* 2020;18:67–83. ([Google Scholar] [CrossRef] [PubMed]).
37. Heler R, Samai P, Modell JW, et al. Cas9 specifies functional viral targets during CRISPR-Cas adaptation. *Nature.* 2015;519:199–202. ([Google Scholar] [CrossRef] [PubMed] [Green Version]).
38. Shi TQ, Liu GN, Ji RY, et al. CRISPR/Cas9-based genome editing of the filamentous fungi: the state of the art. *Appl Microbiol Biotechnol.* 2017;101:7435–7443. ([Google Scholar] [CrossRef]).
39. Shi TQ, Gao J, Wang WJ, et al. CRISPR/Cas9-based genome editing in the filamentous fungus Fusarium fujikuroi and its application in strain engineering for gibberellic acid production. *ACS Synth Biol.* 2019;8:445–454. ([Google Scholar] [CrossRef]).
40. Dank A, Smid EJ, Notebaart RA. CRISPR-Cas genome engineering of esterase activity in Saccharomyces cerevisiae steers aroma formation. *BMC Res Notes.* 2018;11:682. ([Google Scholar] [CrossRef]).
41. Wenderoth M, Garganese F, Schmidt-Heydt M, et al. Alternariol as virulence and colonization factor of Alternaria alternata during plant infection. *Mol Microbiol.* 2019;112:131–146. ([Google Scholar] [CrossRef]).
42. Wei TY, Wu YJ, Xie QP, et al. CRISPR/Cas9-based genome editing in the filamentous fungus glarea lozoyensis and its application in manipulating gloF. *ACS Synth Biol.* 2020;9:1968–1977. ([Google Scholar] [CrossRef] [PubMed]).
43. Pohl C, Polli F, Schütze T, et al. A Penicillium rubens platform strain for secondary metabolite production. *Sci Rep.* 2020;10:7630. ([Google Scholar] [CrossRef] [PubMed]).
44. Gnügge R, Symington LS. Efficient DNA double-strand break formation at single or multiple defined sites in the Saccharomyces cerevisiae genome. *Nucleic Acids Res.* 2020;48:e115. ([Google Scholar] [CrossRef]).
45. Chen C, Liu J, Duan C, Pan Y, Liu G. Improvement of the CRISPR-Cas9 mediated gene disruption and large DNA fragment deletion based on a chimeric promoter in *Acremonium chrysogenum. Fungal Genet Biol.* 2020;134:103279. ([Google Scholar] [CrossRef] [PubMed]).

46. Xu N, Li L, Chen F. Construction of gene modification system with highly efficient and markerless for Monascus ruber M7. *Front Microbiol.* 2022;13:952323. ([Google Scholar] [CrossRef]).
47. Xu Y, Liu L, Chen Z, Tian X, Chu J. The arthrospore-related gene Acaxl2 is involved in cephalosporin C production in industrial Acremonium chrysogenum by the regulatory factors AcFKH1 and CPCR1. *J Biotechnol.* 2022;347:26–39. ([Google Scholar] [CrossRef]).
48. Mladenov E, Iliakis G. Induction and repair of DNA double strand breaks: the increasing spectrum of non-homologous end joining pathways. *Mutat Res.* 2011;711:61–72. ([Google Scholar] [CrossRef]).
49. Deng H, Liang W, Fan TP, Zheng X, Cai Y. Modular engineering of Shiraia bambusicola for hypocrellin production through an efficient CRISPR system. *Int J Biol Macromol.* 2020;165:796–803. ([Google Scholar] [CrossRef]).
50. Al Abdallah Q, Ge W, Fortwendel JR. A simple and universal system for gene manipulation in Aspergillus fumigatus: in vitro-assembled Cas9-guide RNA ribonucleoproteins coupled with microhomology repair templates. *mSphere.* 2017;2:e00446-17. ([Google Scholar] [CrossRef] [Green Version]).
51. Oakley CE, Ahuja M, Sun WW, et al. Discovery of McrA, a master regulator of Aspergillus secondary metabolism. *Mol Microbiol.* 2017;103:347–365. ([Google Scholar] [CrossRef] [PubMed] [Green Version]).
52. Yuan B, Keller NP, Oakley BR, Stajich JE, Wang CCC. Manipulation of the global regulator mcrA upregulates secondary metabolite production in Aspergillus wentii using CRISPR-Cas9 with in vitro assembled ribonucleoproteins. *ACS Chem Biol.* 2022;17:2828–2835. ([Google Scholar] [CrossRef]).
53. Florea S, Jaromczyk J, Schardl CL. Non-transgenic CRISPR-mediated knockout of entire ergot alkaloid gene clusters in slow-growing asexual polyploid fungi. *Toxins.* 2021;13:153. ([Google Scholar] [CrossRef]).
54. Foster AJ, Martin-Urdiroz M, Yan X, Wright HS, Soanes DM, Talbot NJ. CRISPR-Cas9 ribonucleoprotein-mediated co-editing and counterselection in the rice blast fungus. *Sci Rep.* 2018;8:14355. ([Google Scholar] [CrossRef] [Green Version]).
55. Zheng YM, Lin FL, Gao H, et al. Development of a versatile and conventional technique for gene disruption in filamentous fungi based on CRISPR-Cas9 technology. *Sci Rep.* 2017;7:9250. ([Google Scholar] [CrossRef]).
56. Kuivanen J, Wang YJ, Richard P. Engineering Aspergillus niger for galactaric acid production: elimination of galactaric acid catabolism by using RNA sequencing and CRISPR/Cas9. *Microb Cell Fact.* 2016;15:210. ([Google Scholar] [CrossRef] [Green Version]).
57. Anders C, Niewoehner O, Duerst A, Jinek M. Structural basis of PAM-dependent target DNA recognition by the Cas9 endonuclease. *Nature.* 2014;513:569–573. ([Google Scholar] [CrossRef] [PubMed] [Green Version]).
58. Ouedraogo JP, Tsang A. CRISPR_Cas systems for fungal research. *Fungal Biol Rev.* 2020;34:189–201. ([Google Scholar] [CrossRef]).
59. Abdulrachman D, Eurwilaichitr L, Champreda V, Chantasingh D, Pootanakit K. Development of a CRISPR/Cpf1 system for targeted gene disruption in Aspergillus aculeatus TBRC 277. *BMC Biotechnol.* 2021;21:15. ([Google Scholar] [CrossRef]).
60. Katayama T, Maruyama JI. CRISPR/Cpf1-mediated mutagenesis and gene deletion in industrial filamentous fungi *Aspergillus oryzae* and *Aspergillus sojae*. *J Biosci Bioeng.* 2022;133:353–361. ([Google Scholar] [CrossRef]).
61. Vanegas KG, Jarczynska ZD, Strucko T, Mortensen UH. Cpf1 enables fast and efficient genome editing in Aspergilli. *Fungal Biol Biotechnol.* 2019;6:6. ([Google Scholar] [CrossRef] [Green Version]).
62. Jiménez A, Hoff B, Revuelta JL. Multiplex genome editing in Ashbya gossypii using CRISPR-Cpf1. *N Biotechnol.* 2020;57:29–33. ([Google Scholar] [CrossRef]).

63. Kwon MJ, Schütze T, Spohner S, Haefner S, Meyer V. Practical guidance for the implementation of the CRISPR genome editing tool in filamentous fungi. *Fungal Biol Biotechnol.* 2019;6:15. ([Google Scholar] [CrossRef] [Green Version]).
64. Huang J, Rowe D, Subedi P, et al. CRISPR-Cas12a induced DNA double-strand breaks are repaired by multiple pathways with different mutation profiles in Magnaporthe oryzae. *Nat Commun.* 2022;13:7168. ([Google Scholar] [CrossRef]).
65. Kang HS, Charlop-Powers Z, Brady SF. Multiplexed CRISPR/Cas9- and TAR-mediated promoter engineering of natural product biosynthetic gene clusters in yeast. *ACS Synth Biol.* 2016;5:1002–1010. ([Google Scholar] [CrossRef] [PubMed] [Green Version]).
66. Kim H, Ji CH, Je HW, Kim JP, Kang HS. mpCRISTAR: multiple plasmid approach for CRISPR/Cas9 and TAR-mediated multiplexed refactoring of natural product biosynthetic gene clusters. *ACS Synth Biol.* 2020;9:175–180. ([Google Scholar] [CrossRef] [PubMed]).
67. Pickar-Oliver A, Gersbach CA. The next generation of CRISPR-Cas technologies and applications. *Nat Rev Mol Cell Biol.* 2019;20:490–507. ([Google Scholar] [CrossRef]).
68. Roux I, Woodcraft C, Hu J, Wolters R, Gilchrist CLM, Chooi YH. CRISPR-mediated activation of biosynthetic gene clusters for bioactive molecule discovery in filamentous fungi. *ACS Synth Biol.* 2020;9:1843–1854. ([Google Scholar] [CrossRef]).
69. Mózsik L, Hoekzema M, de Kok NA, Bovenberg RA, Nygård Y, Driessen AJ. CRISPR-based transcriptional activation tool for silent genes in filamentous fungi. *Sci Rep.* 2021;11:1118. ([Google Scholar] [CrossRef] [PubMed]).
70. Mózsik L, Pohl C, Meyer V, Bovenberg RAL, Nygård Y, Driessen AJM. Modular synthetic biology toolkit for filamentous fungi. *ACS Synth Biol.* 2021;10:2850–2861. ([Google Scholar] [CrossRef]).
71. Román E, Coman I, Prieto D, Alonso-Monge R, Pla J. Implementation of a CRISPR-based system for gene regulation in Candida albicans. *mSphere.* 2019;4:e00001-19. ([Google Scholar] [CrossRef] [Green Version]).
72. Wensing L, Sharma J, Uthayakumar D, Proteau Y, Chavez A, Shapiro RS. A CRISPR interference platform for efficient genetic repression in Candida albicans. *mSphere.* 2019;4:e00002-19. ([Google Scholar] [CrossRef] [PubMed] [Green Version]).
73. Xu X, Feng J, Zhang P, Fan J, Yin WB. A CRISPR/Cas9 cleavage system for capturing fungal secondary metabolite gene clusters. *J Microbiol Biotechnol.* 2021;31:8–15. ([Google Scholar] [CrossRef]).
74. Gems D, Johnstone IL, Clutterbuck AJ. An autonomously replicating plasmid transforms Aspergillus nidulans at high frequency. *Gene.* 1991;98:61–67. ([Google Scholar] [CrossRef] [PubMed]).
75. Wang Q, Coleman JJ. Progress and challenges: development and implementation of CRISPR/Cas9 technology in filamentous fungi. *Comput Struct Biotechnol J.* 2019;17:761–769. ([Google Scholar] [CrossRef] [PubMed]).
76. Liu D, Liu Q, Guo W, et al. Development of genetic tools in glucoamylase-hyper-producing industrial Aspergillus niger strains. *Biology.* 2022;11:1396. ([Google Scholar] [CrossRef]).
77. Meng G, Wang X, Liu M, Wang F, Liu Q, Dong C. Efficient CRISPR/Cas9 system based on autonomously replicating plasmid with an AMA1 sequence and precisely targeted gene deletion in the edible fungus, Cordyceps militaris. *Microb Biotechnol.* 2022;15:2594–2606. ([Google Scholar] [CrossRef] [PubMed]).
78. Yao G, Chen X, Han Y, Zheng H, Wang Z, Chen J. Development of versatile and efficient genetic tools for the marine-derived fungus Aspergillus terreus RA2905. *Curr Genet.* 2022;68:153–164. ([Google Scholar] [CrossRef]).
79. Li Y, Zhang H, Chen Z, et al. Construction of single, double, or triple mutants within kojic acid synthesis genes kojA, kojR, and kojT by the CRISPR/Cas9 tool in Aspergillus oryzae. *Folia Microbiol.* 2022;67:459–468. ([Google Scholar] [CrossRef]).

80. Seekles SJ, Teunisse PPP, Punt M, et al. Preservation stress resistance of melanin deficient conidia from Paecilomyces variotii and Penicillium roqueforti mutants generated via CRISPR/Cas9 genome editing. *Fungal Biol Biotechnol.* 2021;8:4. ([Google Scholar] [CrossRef]).
81. Katayama T, Nakamura H, Zhang Y, Pascal A, Fujii W, Maruyama JI. Forced recycling of an AMA1-based genome-editing plasmid allows for efficient multiple gene deletion/integration in the industrial filamentous fungus Aspergillus oryzae. *Appl Environ Microbiol.* 2019;85:e01896-18. ([Google Scholar] [CrossRef] [Green Version]).
82. Leisen T, Bietz F, Werner J, et al. CRISPR/Cas with ribonucleoprotein complexes and transiently selected telomere vectors allows highly efficient marker-free and multiple genome editing in Botrytis cinerea. *PLoS Pathog.* 2020;16:e1008326. ([Google Scholar] [CrossRef] [PubMed]).
83. Leisen T, Werner J, Pattar P, et al. Multiple knockout mutants reveal a high redundancy of phytotoxic compounds contributing to necrotrophic pathogenesis of Botrytis cinerea. *PLoS Pathog.* 2022;18:e1010367. ([Google Scholar] [CrossRef] [PubMed]).
84. Kosicki M, Tomberg K, Bradley A. Repair of double-strand breaks induced by CRISPR-Cas9 leads to large deletions and complex rearrangements. *Nat Biotechnol.* 2018;36:765–771. ([Google Scholar] [CrossRef]).
85. Li J, Zhang Y, Zhang Y, Yu PL, Pan H, Rollins JA. Introduction of large sequence inserts by CRISPR-Cas9 to create pathogenicity mutants in the multinucleate filamentous pathogen Sclerotinia sclerotiorum. *mBio.* 2018;9:e00567-18. ([Google Scholar] [CrossRef] [Green Version]).
86. Fu Y, Sander J, Reyon D. Improving CRISPR-Cas nuclease specificity using truncated guide RNAs. *Nat Biotechnol.* 2014;32:279–284. ([Google Scholar] [CrossRef] [Green Version]).
87. Cho SW, Kim S, Kim Y, et al. Analysis of off-target effects of CRISPR/Cas-derived RNA-guided endonucleases and nickases. *Genome Res.* 2014;24:132–141. ([Google Scholar] [CrossRef] [Green Version]).
88. Slaymaker IM, Gao L, Zetsche B, Scott DA, Yan WX, Zhang F. Rationally engineered Cas9 nucleases with improved specificity. *Science.* 2016;351:84–88. ([Google Scholar] [CrossRef] [Green Version]).
89. Zhong Z, Sretenovic S, Ren Q, et al. Improving plant genome editing with high-fidelity xCas9 and non-canonical PAM-targeting Cas9-NG. *Mol Plant.* 2019;12:1027–1036. ([Google Scholar] [CrossRef]).
90. Hu JH, Miller SM, Geurts MH, et al. Evolved Cas9 variants with broad PAM compatibility and high DNA specificity. *Nature.* 2018;556:57–63. ([Google Scholar] [CrossRef]).
91. Walton RT, Christie KA, Whittaker MN, Kleinstiver BP. Unconstrained genome targeting with near-PAMless engineered CRISPR-Cas9 variants. *Science.* 2020;368:290–296. ([Google Scholar] [CrossRef]).
92. Tan J, Zhang F, Karcher D, Bock R. Expanding the genome-targeting scope and the site selectivity of high-precision base editors. *Nat Commun.* 2020;11:629. ([Google Scholar] [CrossRef] [Green Version]).
93. Liu Z, Friesen TL. Polyethylene glycol (PEG)-mediated transformation in filamentous fungal pathogens. *Methods Mol Biol.* 2012;835:365–375. ([Google Scholar]).
94. Ning YD, Hu B, Yu HB, Liu XY, Jiao BH, Lu XL. Optimization of protoplast preparation and establishment of genetic transformation system of an arctic-derived fungus Eutypella sp. *Front Microbiol.* 2022;13:769008. ([Google Scholar] [CrossRef]).
95. Jing X, Xie B, Chen L, et al. Implementation of the CRISPR-Cas13a system in fission yeast and its repurposing for precise RNA editing. *Nucleic Acids Res.* 2018;46:e90. ([Google Scholar] [CrossRef] [Green Version]).

CHAPTER SIX

Recent development in CRISPR-Cas systems for human protozoan diseases

Utkarsh Gangwar[a], Himashree Choudhury[a], Risha Shameem[a], Yashi Singh[b], and Abhisheka Bansal[a,*]
[a]School of Life Sciences, Jawaharlal Nehru University, New Delhi, India
[b]Department of Biosciences & Biomedical Engineering, Indian Institute of Technology, Indore, India
*Corresponding author. e-mail address: abhisheka@jnu.ac.in; bansal.abhisheka@gmail.com

Contents

Progress in Molecular Biology and Translational Science, Volume 208
ISSN 1877-1173, https://doi.org/10.1016/bs.pmbts.2024.07.010

Abstract

Protozoan parasitic diseases pose a substantial global health burden. Understanding the pathogenesis of these diseases is crucial for developing intervention strategies in the form of vaccine and drugs. Manipulating the parasite's genome is essential for gaining insights into its fundamental biology. Traditional genomic manipulation methods rely on stochastic homologous recombination events, which necessitates months of maintaining the cultured parasites under drug pressure to generate desired transgenics. The introduction of mega-nucleases (MNs), zinc-finger nucleases (ZFNs), and transcription activator-like effector nucleases (TALENs) greatly reduced the time required for obtaining a desired modification. However, there is a complexity associated with the design of these nucleases. CRISPR (Clustered regularly interspaced short palindromic repeats)/Cas (CRISPR associated proteins) is the latest gene editing tool that provides an efficient and convenient method for precise genomic manipulations in protozoan parasites. In this chapter, we have elaborated various strategies that have been adopted for the use of CRISPR-Cas9 system in *Plasmodium*, *Leishmania* and *Trypanosoma*. We have also discussed various applications of CRISPR-Cas9 pertaining to understanding of the parasite biology, development of drug resistance mechanism, gene drive and diagnosis of the infection.

Abbreviations

CRISPR	Clustered Regularly Interspaced Short Palindromic Repeats
Cas	CRISPR Associated Proteins
DSB	Double-Stranded Break
NHEJ	Non-Homologous End-Joining
HDR	Homology-Directed Repair
MMEJ	MicroHomology End Joining
SSA	Single Strand Annealing
SgRNA	Single guide RNA
PAM	Protospacer Adjacent Motif
cNHEJ	Canonical non-homologous end joining
hDHFR	Human dihydrofolate reductase
T7 RNAP	T7 RNA Polymerase
BSD	Blasticidin S deaminase
GAPDH	Glyceraldehyde 3-phosphate dehydrogenase
Hsp70	Heat shock protein 70
SRB	A suicide-rescue-based
ChIP	Chromatin ImmunoPrecipitation
IP	ImmunoPrecipitation
IFA	Immunofluorescence Assay
RNP	RiboNucleoProtein
ssODN	Single stranded Oligodeoxynucleotide
glmS ribozyme	Glucosamine-6-phosphate riboswitch ribozyme
DiCre	Dimerisable Cre
loxP	Locus of X-over P1
dCas9	Dead Cas9

SHERLOCK RT-RPA	Specific High-sensitivity enzymatic reporter unlocking Reverse transcriptase recombinase polymerase amplification
CrRNA	CRISPR RNA
VL	Visceral leishmaniasis
SpCas9	Streptococcus pyogenes Cas9
SaCas9	Staphylococcus aureus Cas9
DHFR-TS	DiHydroFolate Reductase Thymidylate Synthase
HDV ribozyme	Hepatitis delta virus
LdMT	Leishmania donovani miltefosine transporter gene
MLF	Miltefosine
eGFP	Enhanced Green Fluorescent protein
RCA	Rolling Circle Amplification
rel gene	Relish gene
RNAi	RNA Interference

1. Introduction

Protozoan parasites cause infectious diseases that affect human health across the globe. *Plasmodium*, *Leishmania* and *Trypanosoma* are some of the major protozoan parasites that cause important human diseases and are responsible for millions of deaths annually. The lifecycle of these parasites is highly complex and there are thousands of genes that are yet to be functionally annotated. Understanding the functional roles of the un-characterized genes in the complex parasite biology requires target gene manipulation through available gene editing tools. From the discovery of CRISPR-Cas9 as a self-defense mechanism in bacteria against invading viruses and DNA[1] to its adaptation as a genetic tool for genome modification across all the spectrum of living organisms[2] is one of the greatest and most impactful discoveries in the history of Science. CRISPR-Cas9 gene editing tool has provided huge opportunity for genome editing of protozoan parasites that are genetically more difficult to manipulate compared to other eukaryotic organisms. The Cas9 endonuclease is a site-specific endonuclease that makes a double-stranded break (DSB) at the target site that is complementary to the guide RNA sequence.[2] The DSB is repaired through either non-homologous end-joining (NHEJ) or homology-directed repair (HDR).[3–5] In some protozoan parasites microhomology end joining (MMEJ) and single strand annealing (SSA) are also used for repair of the DSB.[6,7] NHEJ is not functional in many protozoan parasites and hence the DSB repair is repaired either through HDR or MMEJ pathway.[8] The ability of Cas9 endonuclease to target specific nucleic acid sequences is exploited for targeted genomic manipulations including gene knock-out, endogenous tagging with epitope tags, allelic replacements, gene drive and detection of the protozoan parasites in infected patient samples.[9] CRISPR-Cas systems have greatly revolutionized protozoan parasite research, and provided a much-needed tool for genome manipulation in these genetically intractable organisms. Here, we have discussed three diseases caused by protozoan parasites viz malaria, leishmaniasis and trypanosomiasis. We have elaborated various systems that have been utilized for incorporating desired modification in the parasite genomes for each disease. Applications of CRISPR-Cas9 for each of the three diseases are also discussed.

2. Plasmodium

Plasmodium falciparum causes the most serious type of human malaria. Despite continued efforts to eradicate malaria, approximately 249 million malaria cases reported across 85 malaria endemic nations and regions in 2022, as stated in World Malaria Report 2023. Due to increasing resistance, currently used antimalarials including the front-line drug, artemisinin are gradually losing their effectiveness. Therefore, there is an urgent need for the development of effective intervention strategies in the form of highly efficient drugs or vaccines against malaria. *P. falciparum* exhibits digenetic lifecycle with asexual phase in humans and sexual phase in female *Anopheles* mosquito. The infective sporozoites are released in human skin by the mosquito bite that enter the nearby capillaries and reach liver sinusoids through blood flow. The sporozoites eventually come out of the liver sinusoids and invade hepatocytes. Over a period of 4–5 days they get transformed into a multinucleated schizont. Schizonts rupture, releasing merozoites that invade erythrocytes through a complex process involving interactions between parasite ligands and receptors on the erythrocyte surface. The merozoites undergo several morphological changes in the erythrocytes as they pass through ring, trophozoite and schizont stages. A mature schizont rupture by a process called egress, releasing merozoites that invade fresh erythrocytes. After repeated schizogony, some of the merozoites undergo gametocytogenesis to produce gametocytes which are ingested by another mosquito. Gametogony in the alimentary canal of mosquito produces male and female gametes. The gametes upon fertilization leads to the production of a zygote which elongates and pierces through the epithelial lining of mosquito stomach, stops and begins to encyst. The encysted zygote, also known as the oocyst undergoes sporogony to produce sporozoites which when liberated into the haemocoel begins penetrating the mosquito's salivary glands, where they can be transmitted to humans.

The processes of immune evasion, malaria progression, transmission and antimalarial resistance need to be understood in greater details at the molecular level. The advent of Cas-mediated genome editing techniques has offered a much-needed toolkit for the genetic engineering of difficult to manipulate malaria parasite. CRISPR-Cas gene editing technique has contributed to deeper insights into parasite biology, identify and functionally characterize essential genes, and develop gene drive and diagnostic systems.

2.1 First developed CRISPR-Cas9 systems for gene editing in *P. falciparum*

CRISPR-Cas9 approach of gene manipulation in *P. falciparum* depends on three essential components, a Cas9 endonuclease, a single-guide RNA (sgRNA) and a donor DNA having homologous regions for repair of the double stranded break (DSB) site. *P. falciparum* does not possess canonical non-homologous end joining (cNHEJ) method for repair of endogenous DNA. Initially, a two-plasmid system was used for the generation of desired transgenics wherein the components of CRISPR-Cas9 were distributed in two different plasmids that are co-transfected into the parasite.[10,11] (1) *Streptococcus pyogenes* Cas9 was expressed using plasmodial regulatory elements from a plasmid (pUF1-Cas9) with y*dhodh* drug-selectable marker, and sgRNA was expressed using U6 snRNA promoter (*P. falciparum* RNA Polymerase III promoter) from another plasmid (pL7) which also contains a homologous region for DSB repair (Fig. 1A). The pL7 plasmid contains a positive selection marker gene (h*dhfr*) along with a negative selection marker gene (y*fcu*), enabling the utilization of 5-fluorocytosine (5-FC) to eliminate parasites harboring copies of pL7 construct after desired transgenics are obtained. To enhance system efficiency, a linearized pL7 was used for gene disruption instead of a circular plasmid. Since linear DNA introduced into *P. falciparum* is lost within four days, there is no requirement for negative selection against parasites that retain the plasmid. The use of linear DNA also facilitates approaches that do not require molecular cloning for supplying donor DNA. To introduce a single point mutation without integrating a selection marker, a donor DNA containing the intended mutation along with an extra alteration in the form of shield or silent mutation at the target site was designed to safeguard the modified locus from recurrent Cas9 cleavage.[10] (2) The first plasmid expressing the Cas9 and sgRNA using T7 promoter, and the second plasmid expressing the T7 RNAP (T7 RNA Polymerase) and carry a homologous region to repair the DSB (Fig. 1B).[11]

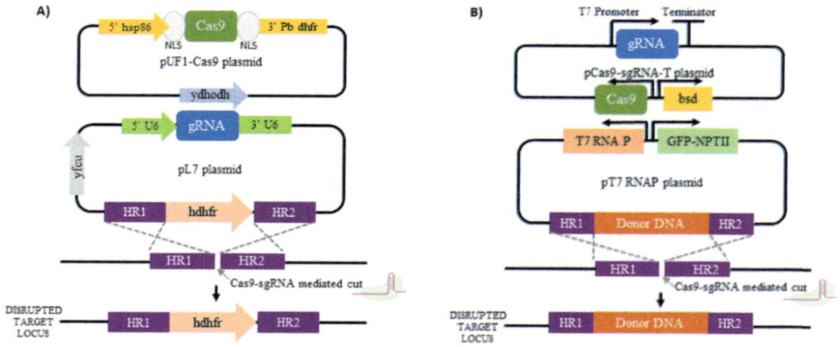

Fig. 1 General schematics of CRISPR-Cas9 based systems for genome editing in *P. falciparum*. (A) The Cas9 endonuclease, which includes nuclear localization signals (NLS), is expressed from pUF1-Cas9 plasmid using regulatory elements from *Plasmodium*. The pL7 plasmid contains the cassette for expressing single-guide RNA (sgRNA) and donor DNA. The sgRNA is transcribed from the promoter of *P. falciparum* U6 snRNA by RNA polymerase III (5′ U6). To achieve gene knockout, homology regions 1 (HR1) and 2 (HR2) of the gene of interest (GoI) must flank a drug selectable marker (*hdhfr*, human dihydrofolate reductase gene). 5′ hsp86, heat shock protein 86 promoter region; 3′ Pbdhfr, 3′ region of *P. berghei* dhfr; *ydhodh*, yeast dihydroorotate dehydrogenase gene; *yfcu*, yeast cytosine deaminase and uridyl phosphoribosyl transferase. (B) The Cas9 endonuclease and T7 promoter driven sgRNA are expressed from the pCas9-sgRNA-T plasmid. To facilitate homology-directed repair following the induced double-stranded break, pT7 RNAP plasmid was engineered to incorporate homology regions flanking the donor DNA. T7 RNAP, T7 RNA polymerase; *bsd*, blasticidin S deaminase.

2.2 Enhanced CRISPR-Cas9 systems for gene editing in *P. falciparum*

A new approach was designed for rapid generation of transgenic lines using CRISPR-Cas9, wherein first plasmid contains Cas9 gene with blasticidin S deaminase (BSD) selection marker, and the second plasmid contains sgRNA and donor DNA with a dual selection marker cassette, i.e. a positive selection marker gene fused to a negative selection marker gene (Fig. 2A). The positive-negative selection marker cassette is used to select marker-free transgenics without integrating into the parasite genome. Three GFP expressing reporter lines using promoters of three stably transcribing genes *(calmodulin, gapdh and hsp70)* of *P. falciparum* were generated using this approach. These transgenic lines showed comparable blood-stage growth rates and drug susceptibility as the WT (wild-type).[12]

A suicide-rescue based (SRB) system using CRISPR-Cas9 was established to overcome the challenges of restricted pool of available selection

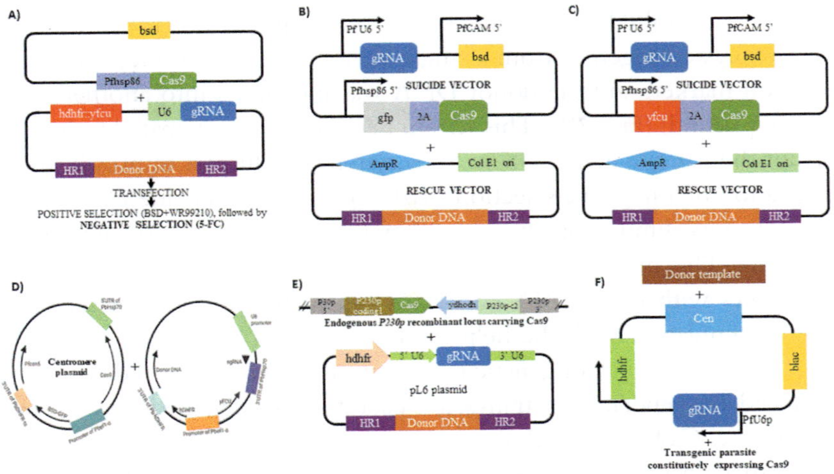

Fig. 2 General schematics of enhanced CRISPR-Cas9 based systems for genome editing of *P. falciparum*. (A) Parasites are transfected with a Cas9 construct carrying the *bsd* selectable marker and an sgRNA/ donor construct carrying a fusion of the positive selectable marker *hdhfr* and the negative selectable marker *yfcu*, along with homology regions targeting the gene of interest to introduce donor DNA via homologous recombination. Double positive selection employing both BSD and WR99210 results in selection of parasites harboring both the plasmids. Following positive selection, cultures are maintained without drugs for 2–4 days before negative selection using 5-FC to select plasmid-DNA free parasites. (B) Suicide-rescue-based (SRB) approach for desired gene editing. Parasites are transfected with two plasmids; the first plasmid (suicide vector) contains cassettes for expression of Cas9 and sgRNA with a *bsd* selectable marker and the second plasmid (rescue vector) contains homology regions flanking the modifications without any drug selectable marker. This strategy can be used for large DNA knock-ins with the use of only single selectable marker gene. PfCAM5′, Pf calmodulin promoter region; GFP, green fluorescent protein; AmpR, ampicillin resistance gene; ori, origin of replication. (C) This strategy is an improvement of the SRB CRISPR-Cas9 system. Negative selectable marker *yfcu* is added to the suicide plasmid for the generation of marker free parasites after negative selection using 5-FC. (D) A parasite line stably expressing Cas9 endonuclease from centromere plasmid is generated first, followed by transfection with another plasmid carrying sgRNA expressing cassette and donor DNA. (E) The Cas9 and *ydhodh* containing plasmid is integrated into the endogenous locus of *P. falciparum* P230p gene via single crossover recombination first, followed by transfection with plasmid containing sgRNA expression cassette and donor DNA. (F) The constitutively expressing Cas9 parasite is generated first via double crossover recombination, followed by transfection with linear donor DNA and sgRNA expressing plasmid.

markers and limited size of knock-in DNA sequences. Cas9 and sgRNA were expressed together from a single vector (suicide plasmid) with single selection marker, and the donor DNA was integrated into another vector (rescue plasmid) (Fig. 2B). This system performed as anticipated to insert or disrupt the parasite genes.[13] The SRB system was further improved by the integration of a negative selection marker gene in the suicide plasmid so as to mediate its elimination after the generation of desired transgenic parasites (Fig. 2C). This modified SRB system is capable of producing marker-free transgenics with large DNA knock-ins (up to 6.3 kb) and for sequential gene manipulation with the help of same selection marker.[14]

The low efficiency of genetic modification with two-plasmid system was enhanced by generating a transgenic parasite, *Pf*CAS9, designed to stably express the Cas9 endonuclease via the centromere plasmid (Fig. 2D).[15] Centromere plasmid has the ability to segregate efficiently, leading to its stable maintenance across numerous nuclear divisions. The centromere plasmid, which carries the Cas9 endonuclease, and another plasmid containing sgRNA and donor DNA (Fig. 2D), were predicted to coexist with greater frequency in *Pf*CAS9 parasites, owing to the greater modification efficiency provided by the *Pf*CAS9 parasite. To evaluate the efficacy of genetic alteration with *Pf*CAS9 parasite, site-directed mutagenesis of kelch13 gene was done and the desired mutation was successfully achieved with nearly 100% efficiency when the transfected parasites underwent dual drug treatment to retain both the sgRNA-containing plasmid and the centromere plasmid.[15] Furthermore, an integrative CRISPR-Cas9 (Cas9i) system was generated wherein a Cas9 endonuclease expressing parasite was developed by single-crossover recombination and intermittent drug treatment (Fig. 2E). This system not only improved the efficiency of gene editing by reducing the time required to generate transgenic parasite, but multigenic editing to study the interactions between different genes also became feasible.[16] The Cas9 parasite generated previously by single crossover recombination may lose its Cas9 through intra-chromosomal recombination, regenerating the wild-type.[16,17] The shortcomings of unwanted recombination and off-target effects using CRISPR-Cas9 approach could be resolved by transfecting a constitutively expressing Cas9 parasite with a linear donor DNA (Fig. 2F). Here, the Cas9 endonuclease gene was inserted into the parasite genome via double-crossover recombination using a centromere-based plasmid.[15,17] This strategy is also helpful in the simultaneous genetic manipulation of two genes on different chromosomes by transfecting two sgRNA expressing plasmid and two linear donor DNAs into the schizont stage parasites. Single nucleotide insertion and green fluorescent protein tagging were achieved using this method with 85–100% efficiency, no unwanted recombination or off-target mutations.[17]

2.3 CRISPR-Cas9 based system for tagging endogenous genes of *P. falciparum*

CRISPR-Cas9 approach is also used to append commercial epitope tag DNA sequences to endogenous genes of *P. falciparum* with greater efficiency than was possible using the conventional gene editing techniques. Epitope tags such as HA or HA-TY1 were appended at the 3′ end of *ck2α*, *ck2β1* and *stk* genes using two-plasmid system (Fig. 3A). Parasite transfection followed by positive selection using drug markers incorporated within the CRISPR plasmids resulted in resistant parasites within three weeks of drug pressure. Successful integration of the epitope tag was verified using various molecular techniques including PCR, DNA sequencing, Immunoblotting and Immunofluorescence Assay (IFA). Performing ChIP, IP and IFAs became easier using the tagged proteins.[18]

2.4 CRISPR-Cas9 based system to explore drug resistance in *P. falciparum*

A method was devised to study the molecular basis of drug resistance in malaria parasite using vector-free CRISPR-Cas9 approach. This method involves using recombinant Cas9 protein complexed with a synthetic sgRNA, and 200-nucleotide single-stranded oligodeoxynucleotide (ssODN) donor DNA template for transfection into the parasite (Fig. 3B). To illustrate this method, two single-edits were incorporated into the sodium efflux channel, *Pf*ATP4 (ATPase4) and the resulting transgenic parasite developed resistance to the antimalarial compound SJ733. Scar-free nature and no requirement of molecular cloning makes this method an appealing gene editing tool.[19]

2.5 CRISPR-Cas9 based conditional knockdown and knockout systems to better characterize essential genes of *P. falciparum*

A glmS-ribozyme based glucosamine-inducible knockdown system was generated using CRISPR-Cas9 approach. This system involves the integration of a glmS ribozyme to the 3′ end of targeted gene via double-crossover recombination. The glmS-ribozyme located at the 3′ end of mRNAs is activated by glucosamine treatment which results in the degradation of associated mRNAs and lowering of protein levels. Three-plasmid based approach was used wherein the first plasmid was used for

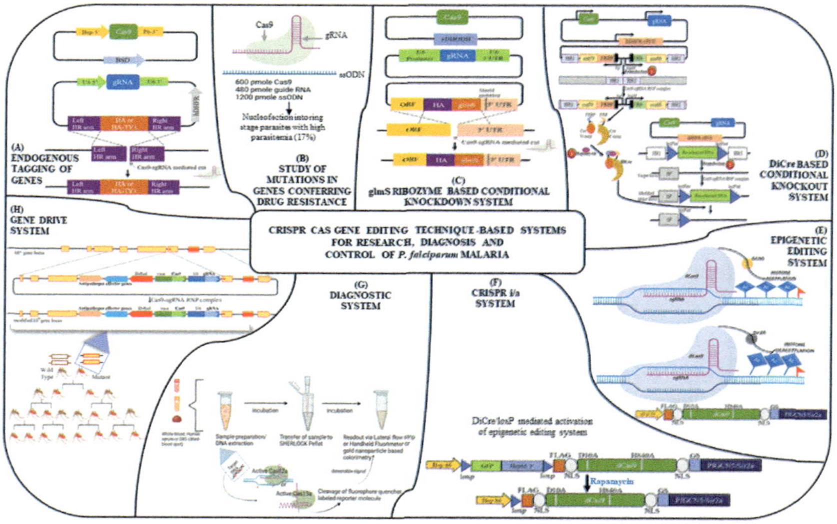

Fig. 3 General schematics of different CRISPR-Cas9 gene editing technique-based systems. (A) Two-plasmid CRISPR-Cas9 based system for tagging endogenous genes of *P. falciparum* is depicted. The epitope tag sequences are inserted immediately preceding the stop codon of the targeted gene through homologous recombination. (B) Plasmid-free strategy to introduce drug resistance mutations in the *P. falciparum* genes using CRISPR-Cas9 is depicted. To introduce edits into PfATP4, synchronized ring stage parasites with 17% parasitemia were transfected using nucleofection with Cas9 protein, guide RNA and a template ssODN. Post-transfection, cultures were subjected to drug pressure using 500 nM SJ733 starting from day 2. (C) Design of conditional knockdown system is depicted. To introduce glmS ribozyme coding sequence at the 3′ end of the targeted gene via double crossover homologous repair, three-plasmid CRISPR-Cas9 system is depicted. The repair plasmid contains homologous regions to the open reading frame (ORF) and the 3′ UTR of the targeted gene sequence while the other two plasmids contain expression cassettes for Cas9 and sgRNA. (D) Design of conditional gene excision system is depicted. Two-plasmid CRISPR-Cas9 strategy is employed first for the generation of parasite line with modified endogenous locus of *P. falciparum* carrying the expression cassettes for two halves of Cre recombinase. The N-terminal and C-terminal segments of Cre recombinase are linked to either FK506-binding protein (FKBP12) or the FKBP12-rapamycin binding (FRB) domain of FKBP12-rapamycin associated protein. CRISPR-Cas9 strategy is then used for the insertion of two loxPint sequences flanking the recodonised targeted ORF segment in Cre recombinase expressing parasite line. Upon addition of rapamycin, the components of Cre recombinase heterodimerize via their rapamycin binding proteins, thereby restoring Cre activity which leads to the excision of the recodonised ORF segment located between the loxP sites. SP, signal peptide. (E) Transcriptional regulation of *P. falciparum* genes by epigenetic modification of histones at the TSS of the target genes using CRISPR/dCas9 system is depicted. It involves fusion of the enzymatic domains of PfHAT (PfGCN5) and PfHDAC (PfSir2a) to the C-terminus of deadCas9 (dCas9), a nuclease-deficient variant of Cas9 with

providing the donor DNA template, second for the expression of sgRNA using U6 promoter and third for the expression of Cas9 endonuclease (Fig. 3C). Shield mutations, also called silent mutations were introduced into the homology region near the PAM site so as to protect the modified locus from recutting with the Cas9 endonuclease (Fig. 3C). This system is highly adaptive to manipulations such as protein tagging, generating gene knockouts, or the generation of other conditional knockdowns such as destabilization domains or RNA aptamers. This system was used to regulate levels of *Pf*Hsp70x protein.[20] Furthermore, a two-plasmid approach was also successfully used for the generation of conditional knockdown mutants.[20,21]

A dimerisable Cre (DiCre) recombinase based conditional knockout system was generated using suicide-rescue based (SRB) CRISPR-Cas9 system. CRISPR-Cas9 was employed to integrate the DiCre cassette into the genomic loci of *p230p* and *pfs47* genes that are non-essential for development during the blood-stages and infection to mosquitoes (Fig. 3D). The strategy used was advantageous in the sense that it involves a Cas9 endonuclease containing plasmid with positive-negative selection cassette and a donor DNA containing rescue plasmid is linearized so as to avoid retention as an episome. Episome-based expression of DiCre often results in lower excision efficiency. The DiCre expressing parasite lines are expected

---

mutations (D10A and H840A) in its catalytic domains. Parasite is transfected with two plasmids with one carrying dCas9 fusion protein and another carrying sgRNA targeting the transcription start site (TSS) of the gene of interest. The TSS of the target gene is marked with red flag. NLS, nuclear localization signal; GS, glycine serine protein linker; FLAG, 3xFLAG tag. (F) The DiCre/loxP-inducible CRISPR-dCas9 system is depicted. The addition of rapamycin activates DiCre and removes the sequences between the loxP sites, thereby enabling the expression of dCas9-GCN5/Sir2a. (G) Blood samples undergo a sample preparation step, followed by transfer to the SHERLOCK pellet. If present, amplification of the target gene occurs via RPA followed by cleavage of the reporter molecules by the activity of Cas12a or Cas13a whichever is used for the diagnostic assay. Cas12a:crRNA complex recognizes double stranded DNA target and non-specifically cleaves ssDNA reporter molecules. Cas13a:crRNA complex recognizes single-stranded RNA target through base pairing interactions and cleaves RNA reporter molecules. The last step is the detection of end point via lateral flow strip readout or handheld fluorimeter or gold nanoparticle-based colorimetry. (H) After microinjection of a solution containing the plasmid, Cas9, Cas9 dsRNA and Ku70 dsRNA into the embryos, Cas9-gRNA complex cleaves the target site and homology directed repair facilitates the precise integration of the gene-drive cargo (anti-pathogen effector genes, DsRed, vasa-Cas9, U6A gRNA) into the *khw* genomic locus. When the transgenic mosquitoes are mated to the wild type mosquitoes, 99.5% of the offspring inherit antipathogen effector genes.

to produce efficient and rapid rapamycin-inducible gene excision, enabling the study of gene function during a single growth cycle.[22] DiCre based conditional knockout system works by the deletion of DNA sequences located between the loxP sites upon rapamycin induced dimerization of Cre recombinase.[22] The removal of DNA sequences flanked by loxP sites can be engineered not only to remove DNA sequences but also to facilitate fusion, replacement or introduction of point mutations.[22,23] A standardized small module containing a short intron with a loxP site, known as loxPint is employed that can replace native introns or be inserted into ORFs of episomes or chromosomal genes.[23] In DiCre-expressing parasite line, two loxP sites can be inserted into the targeted endogenous locus using CRISPR-Cas9 gene editing approach.[24–27]

2.6 CRISPR-Cas9 based systems to explore epigenetic regulation of essential genes of *P. falciparum*

CRISPR-Cas9 toolkit was further expanded by generation of a Cas9 null mutant, dead Cas9 (dCas9) fused at its C-terminus to the enzymatic domain of epigenetic modifiers such as *P. falciparum* histone acetyltransferase (PfGCN5) or histone deacetylase (PfSir2a) (Fig. 3E). The expression of sgRNA and dCas9 fused to epigenetic modifiers in *P. falciparum* effectively and selectively reprograms the target gene expression by histone modifications at their transcription start sites (TSS) (Fig. 3E). Examination of this system was done by epigenetic regulation of some genes involved in parasite invasion pathways, i.e. epigenetic activation of reticulocyte binding protein homologue 4 (rh4), and epigenetic knockdown of erythrocyte binding antigen 175 (eba-175) and *Pf*SET1.[28] Unregulated episomal expression of dCas9 could be detrimental to the intraerythrocytic developmental cycle (IDC) of the parasite, leading to adaptive changes in the parasite. Therefore, a leak-free inducible CRISPRi/a (CRISPR interference/activation) system was devised in which DiCre-loxP was added to regulate dCas9 fusion protein expression by transient rapamycin treatment (Fig. 3F). Up and downregulation of most of the genes selected for assessing the flexibility and robustness of this system was achieved with the help of this conditional knockdown approach.[29]

2.7 CRISPR-Cas based diagnostic systems

For malaria control and eradication, a relatively simple, ultrasensitive, point-of-care testing and field-applicable CRISPR-Cas12a based diagnostic tool was developed (Fig. 3G). This SHERLOCK (specific high-sensitivity enzymatic reporter unlocking) system has the capacity to work in resource-limited settings (RLS) to detect infections with low parasite density and differentiate between different *Plasmodium* species. SHERLOCK is a nucleic-acid based detection method that overcomes many of the constraints of previous diagnosis methods. It is an isothermal, one pot assay which involves a 10-minute sample preparation for rapid parasite extraction and the suspended sample is then transferred to lyophilized SHERLOCK pellet, incubated for 60 min, followed by measurement of fluorescence or lateral strip readout. Cas12a is guided by RNA for recognition of its target double-stranded DNA (dsDNA) with a TTTN PAM sequence but cleaves DNA sequences indiscriminately and this non-specific DNase activity of Cas12a is exploited for the degradation of a single-stranded fluorophore-quencher reporter DNA. RT-RPA (reverse transcriptase- recombinase polymerase amplification) step is added before Cas12a detection and cleavage so as to increase the amount of targeted DNA and decrease the limit of detection (LOD) (Fig. 3G). This method can detect two parasites per microliter of blood, a threshold recommended by the World Health Organization. RPA stands as a potent isothermal nucleic acid amplification technique, consisting of three key enzymes: a recombinase, an ssDNA-binding protein and a strand-displacing polymerase, which collectively facilitate DNA synthesis from target DNA paired with primers. The selection of RPA primer targets involved searching the literature for top-performing nucleic-acid amplification tests (NAATs) and identifying specific and conserved DNA sequences through alignment of species-specific strains available on the National Center for Biotechnology Information (NCBI). Future investigation of this diagnostic tool is necessary to evaluate performance in real-world environments.[30]

Another CRISPR-Cas13a based diagnostic method using SHERLOCK platform was developed to overcome the shortcomings of antigen based, microscopy or molecular diagnostic tools (Fig. 3G). It can detect all human malaria-causing *Plasmodium* species as well as specifically identity *P. vivax* and *P. falciparum*, two predominant species causing malaria globally. A prototype SHERLOCK assay was also designed that can identify the dihydropteroate synthetase (*dhps*) single nucleotide variant (SNV) A581G, which was linked to sulfadoxine resistance in *P. falciparum*. The assay includes a combination of RPA, in vitro transcription and RNA target detection using custom-designed CRISPR RNA (crRNA) and *Leptotrichia wadei* derived Cas13a (*Lw*Cas13a). The initial step involves an RPA reaction using primers containing a T7 promoter sequence to synthesize short dsDNA amplicons of *Plasmodium* 18S rRNA or *P. falciparum dhps* genes. Subsequently, in vitro transcription of dsDNA amplicons by T7 RNA polymerase produces genus-, species-, or genotype specific single-stranded RNA targets which then interacts with the LwCas13a: crRNA RNP complex through base-pairing. This interaction activates collateral RNase activity of Cas13a, causing cleavage of fluorescent or colorimetric RNA reporter molecules and results in the generation of a detectable signal. Mismatches in the target region disrupt the base-pairing interactions between the crRNA and ssRNA target, resulting in varied activation of LwCas13a depending on the presence or absence of a SNV. The reaction components may be either combined into a single or two reactions and the signal emitted is detected via a simple fluorimeter, lateral flow strip or gold nanoparticle-based colorimetry. Additional optimization and streamlining are required before the clinical implementation of this assay.[31]

Another quite similar type of diagnostic method called Rapid *P. falciparum* DNA Detection Test based on CRISPR-Cas13a (RDT-CRI) was developed for the detection of an artemisinin-resistant *P. falciparum* strain and 4 other laboratory-cultured strains of *P. falciparum* via lateral strip readout.[32]

2.8 CRISPR-Cas9 based systems of gene-drive in *Anopheles* vector for control and elimination of *P. falciparum*

Gene drive (GD) strategy for mosquito control is developed using CRISPR-Cas9 approach by employing a complex of Cas9-sgRNA to cleave a specific site within the genome of mosquito to facilitate the insertion of gene drive into the target locus and results in genetic alterations (Fig. 3H). The gene drive remains active specifically in the germline. In diploid species with two sets of chromosomes, one chromosome carries the gene drive and the other does not, a Cas9-sgRNA mediated double-stranded break occurs in the chromosome lacking the gene drive. The repair of this break can happen either through homology-directed repair (HDR) or non-homologous end joining (NHEJ). The gene drive utilizes HDR to transfer genetic information from the strand containing the gene drive to the cleaved strand, thereby promoting the proliferation of homozygous individuals with two copies of gene drive within the population. This process, known as homing, depends crucially on achieving biased inheritance of the homing endonuclease gene in reproductive cells.[33]

GD interventions are currently in the laboratory development phase and can be categorized into two main classes: (1) population suppression GDs, which aim to eliminate mosquito populations by spreading traits such as female sterility or female killing, and (2) population modification GDs, which genetically alter mosquitoes to reduce their capacity to transmit diseases while leaving the population intact. Both population suppression and modification gene drive strategies present complex ethical and ecological considerations.[34]

Autonomous CRISPR-Cas9 based highly-efficient gene drive system was designed to generate transgenic mosquitoes carrying dual anti-parasite effector genes and gene drive system for introgression of anti-parasite genes into wild vector populations, blocking transmission and thereby contributing to malaria control and eradication. This system was used to engineer *Anopheles stephensi* mosquito in such a way that it efficiently transferred a ~17 kb cargo from its insertion site to its homologous chromosome in a site-specific and precise manner. The *kynurenine hydroxylase*^{white} (*kh*^{w}) was selected as the locus for insertion of the gene drive construct. The components of gene-drive plasmid were: (1) a codon-optimised Cas9 endonuclease-encoding gene flanked by putative promoter sequences of the *vasa* gene of *An. stephensi* for expression of nuclease in both male and female germ lines; (2) a gRNA targeting the *An. stephensi kh*^{w} gene at a specific site under the control of a putative *An. stephensi* U6A gene promoter; (3) a 3xP3-DsRed gene which expresses a dominant fluorescent marker in photoreceptors of larvae and non-pigmented eyes of adults; (4) dual-antipathogen effector genes for targeting *P. falciparum*; and (5) DNA fragments homologous to *An. stepehensi kh*^{w} locus flanking the target cut-site. For generating transgenics, the embryos of *An. stephensi* were microinjected with 100 ng/uL each of the plasmid, Cas9 protein molecules, Cas9 double-stranded RNAs and Ku70 dsRNA. The inclusion of dsRNAs served the purpose of silencing the expression of Cas9 gene (dsCas9) from the gene-drive plasmid and reducing the activity of NHEJ pathway (dsKU70) thereby favoring HDR mediated insertion of the cargo. Genomic integration of the transgene was accomplished by co-injecting Cas9 protein along with gRNA encoded on the gene-drive plasmid. By crossing transgenic lines with wild-type mosquitoes, approximately 99.5% of the offspring inherit dual anti-*P. falciparum* effector genes. The findings of this system supported the system's design to be active exclusively within the germline.[35]

Gene drive systems developed using CRISPR-Cas9 technology combined with dual anti-*Plasmodium falciparum* effector genes have demonstrated potential in rapidly propagating through mosquito populations with the help of two strains that carried autonomous gene drive systems. This was shown using two single-chain variable fragment monoclonal antibodies targeting ookinetes and sporozoites of parasite in African malaria mosquito strains of *Anopheles gambiae* and *Anopheles coluzzii*. In small cage trials, these gene drive systems achieved full penetration within 3–6 months post-release. The effector molecules significantly reduced both parasite prevalence and infection intensities. Life table analyses indicated no fitness burdens affecting *A. coluzzii* gene drive dynamics whereas *A. gambiae* males showed reduced competitiveness compared to wild types. These findings supported transmission modelling of hypothetical field releases on an island, demonstrating substantial epidemiological impacts at various sporozoite threshold levels (2.5–10k) for human infection. Optimal simulations projected reductions in malaria incidence by 50–90% within 1–2 months and ≥ 90% within 3 months. However, modelling outcomes for lower sporozoite thresholds were sensitive to factors such as gene-drive system fitness burdens, gametocytemia infection intensities during parasite challenges, and the emergence of potentially drive-resistant genome target sites, thereby prolonging the predicted timeframes needed to achieve decreased malaria incidence.[36]

Slight genetic alterations to endogenous mosquito genes were explored to convert them directly into non-autonomous gene drives without disrupting their natural expression. This was achieved by utilizing the native regulatory sequences of three midgut-specific loci from the malaria vector *Anopheles gambiae* to host a prototypical antimalarial molecule and gRNAs which were encoded within artificial introns to facilitate efficient gene drive. Due to their inherent simplicity and passive drive mechanism, these traits could potentially serve as an acceptable testing pathway for gene drives aimed at malaria eradication.[37]

A midgut gene in the malaria mosquito *A. gambiae* was engineered using CRISPR-Cas9 to produce Magainin 2 and Melittin (anti-microbial peptides), additionally. This minor genetic alteration enabled an efficient system of non-autonomous gene drive, hindering development of oocyst in *P. falciparum* and *P. berghei*. It also delayed sporozoite's release and concurrently reduced the lifespan of female transgenic mosquitoes. By modelling the spread of this modification with a large-scale agent-based model of malaria epidemiology, it was found that this approach could disrupt the transmission cycle of the disease across various transmission intensities.[38]

A two-component gene drive system was designed such that it employs two different mechanisms to reduce parasite transmission through population modification of *Anopheles coluzzii*. *Lipophorin* (*Lp*) is an abundantly secreted protein in insect haemolymph that is involved in the transport of lipid essential lipid. *Lp* locus was engineered to simultaneously produce a single chain version of the mouse monoclonal antibody 2A10 that binds circumsporozoite protein of *P. falciparum*. The function of Lp protein is maintained here by utilizing a natural proteolytic site to cleave the fusion protein for the separation of its components after their secretion. The transgenic mosquitoes showed a significant decrease in the transmission of *P. berghei* that expressed the circumsporozoite protein of *P. falciparum*, to mice. Several CRISPR-Cas9 based strategies were tested to promote the dissemination of Lp::Sc2A10 transgene to mosquito populations under laboratory conditions. Initially, a suppression gene drive involving disruption of WT Lp locus revealed that the Lipophorin gene is haploinsufficient, rendering this strategy impractical. Another gene drive design which involves homing and disruption of the pro parasitic *Saglin* gene locus on the X chromosome was successful and the gRNAs expressed also targets wild-type Lp locus for homing of Lp::Sc2A10 on another chromosome. This gene drive system based on Saglin achieved high levels in caged mosquito populations and the modified mosquitoes showed a decreased ability to transmit sporozoites by depriving parasite of the *Saglin* agonist and attacking sporozoites by the Sc2A10 antibody. Multiplexes of three gRNAs were used to target both the *Saglin* and Lp loci. It enhances the likelihood that unsuccessful homing events leads to loss-of-function mutations in Lp, which could then be removed by natural selection. Loss-of-function alleles of Saglin resistant to gene drive were still expected to reduce vector competence. A functional allele resistant to homing was observed at the *Lp* locus.[34]

2.9 CRISPR-Cas9 based generation of transgenic line for assisting in *P. falciparum* research

To study the influence of environmental factors and the mechanisms driving sexual conversion, accurate measurement of the rate of sexual conversion under various conditions is crucial. *Pf*AP2G is a well-known marker that is used to evaluate sexual rings and sexually committed stages. The gexp02 gene consistently showed early responsiveness to *Pf*AP2G activation. Therefore, a two-plasmid CRISPR-Cas9 strategy was employed for the generation of fluorescent reporter lines. In these lines, gexp02 promoter controls the expression of tandem Tomato (*td*Tom) gene inserted into the lisp1 (liver specific protein 1) locus. These transgenic lines facilitate precise identification of sexual stages through flow cytometry or immunofluorescence assay from as early as the sexual ring stage. This significantly reduces the time required for assays involving sexual conversion and eliminates the requirement to add chemicals for inhibiting parasite replication.[39]

Inducible gametocyte producer (iGP) lines of NF54 strain were generated using two-plasmid CRISPR-Cas9 approach to comprehensively study the biology of gametocytes. These lines were found to consistently produce sexual commitment rates of 75% by the overexpressing the sexual commitment factor (GDV1). These lines produced viable gametocytes capable of infecting mosquitoes and the sporozoites produced retain their capacity to invade hepatocytes. Furthermore, genetic editing of these lines for molecular investigations into gametocyte stages is also feasible as was verified by the production of a tagged nuclear pore protein *Pf*NUP313.[40]

2.10 Applying CRISPR-Cas9 technology to enhance the understanding of *P. falciparum* biology

The malaria parasite extensively remodels the host cell by the synthesizing and exporting over 300 proteins outside the parasitophorous vacuole is done by the parasite to support its growth and pathogenesis. *Pf*PMV (plasmapepsin V) protein is an aspartic protease crucial for remodelling processes. It cleaves the PEXEL motif from the parasite proteins exported into host erythrocytes, altering their mechanical and adhesive properties. The essentiality of *Pf*PMV was investigated by the generation of a DiCre recombinase based conditional gene excision system using CRISPR/Cas9 gene editing tool. *Pf*PMV is indispensable for parasite's asexual growth as the lack of protein leads to early arrest of ring to trophozoite development and therefore may serve as a promising drug target.[24]

The *var* multigene family's mutually exclusive expression plays an important role in immune escape and pathogenesis of *P. falciparum* but only a limited number of factors directly associated with the *var* gene are currently known. Therefore, a CRISPR-based method was used to reveal the additional factors related with the *var* gene within their natural chromatin environment. Two-plasmid CRISPR/dCas9 approach was used to target 3HA-tagged dCas9 to *var* gene regulatory elements, followed by dCas9 immunoprecipitation and liquid chromatography- mass spectrometry to identify previously discovered as well as several novel *var* associated factors. Findings of this proteomics approach validate its effectiveness and delineate a novel var gene associated chromatin complex.[41] A gene family of non-coding RNA (ncRNA), rich in GC content, has evolved alongside *Plasmodium* species expressing *var* genes. A single member of this ncRNA gene family is predominantly transcribed when the ncRNA is positioned near and before an active *var* gene. Functional characterization of this gene family was done with the help of a CRISPR interference strategy that involves the use of sgRNA for guiding the binding of dCas9 to a conserved region common to all GC-rich genes, thereby blocking their transcription. As a result, whole family of *var* gene was downregulated in parasite's ring-stage. It also influenced the gene transcription of other clonally variant families in mature blood stage parasites. This study suggested a pivotal role of the transcription of GC-rich ncRNAs for the activation of *var* gene and identified a molecular connection between the regulation of diverse clonally variant multigene families associated with parasite infectivity.[42]

Glucosamine-phosphate N-acetyltransferase (GNA1) enzyme produces UDP-N-acetylglucosamine donor during the synthesis of glycoconjugates that are required for parasite viability and infectivity. The targeted disruption of *PfGNA1* with the help of CRISPR-Cas9 was unsuccessful, indicating its essential role for the growth of the parasite.[43] The role of *Pf*CLAMP (Claudin-like apicomplexan microneme protein) was determined with the help of a TetR-DOZI based conditional knockdown system generated using CRISPR-Cas9 gene editing approach. Tet Repressor protein (TetR) is produced as a fusion with the translational repressor (DOZI) so as to supress the aptamer-tagged *Pf*CLAMP protein expression unless anhydrotetracycline (aTc) is introduced into the medium. PfCLAMP protein is crucial for erythrocyte invasion as the knockdown of this protein inhibits the parasite's asexual growth.[44]

The relatively conserved family of actin related proteins (ARPs) are distributed between the nucleus and cytoplasm of eukaryotic cells. The nuclear ARPs are often found as part of the chromatin remodelling complex. The roles of PfARP4 and PfARP6 were investigated with the help of gene tagging and glmS ribozyme based inducible knockdown system generated via two-plasmid CRISPR/Cas9 gene editing tool. *Pf*ARP4 along with *Pf*ARP6 were found to colocalize in the nucleus of *P. falciparum*. Knockdown of *Pf*ARP4 inhibits the asexual growth of the parasite.[21]

As protein kinases are expected to have critical roles in various stages of the parasite biology, *P. falciparum* CDPKs (Calcium-dependent protein kinases) were investigated at the molecular level with the help of two-plasmid CRISPR-Cas9 approach. Efforts to knockout *Pf*CDPK1 from the asexual stages of the WT (wild-type) parasite were unsuccessful, suggesting an indispensable role of *Pf*CDPK1 in IDC of the parasite.[45] The mutant *Pf*CDPK1 (T145M) parasite was successfully generated wherein the threonine residue adjacent to the ATP binding site of *Pf*CDPK1 was mutated to methionine residue. The mutant parasite showed reduced transphosphorylation activity of this enzyme and upregulated levels of PKG (Protein Kinase G), *Pf*CDPK5 and

*Pf*CDPK6, indicating that the reduced activity of *Pf*CDPK1 was compensated by the increased levels of other kinases. This study proposes that targeting multiple kinases would be a more effective strategy so as to preserve the antimalarial potential of each one of them.[46] *Pf*CDPK1 was successfully disrupted in mutant T145M parasites which showed a slower growth as compared to WT parasite and also lost its potential to infect mosquitoes due to defect in gametogenesis.[45] *Pf*CDPK2 knockout parasites though reproduce asexually comparable to WT parasite, fail to infect mosquitoes due to defect(s) in male gametocyte exflagellation and possibly in female gametes.[47] DiCre recombinase-based conditional knockout system was generated using CRISPR-Cas9 technique to examine the function of cAMP-dependent *P. falciparum* protein kinase A (*Pf*PKA) in asexual growth of the parasite. *Pf*PKA is crucial for erythrocyte invasion by the parasite as the lack of protein leads to reduction in growth after one intraerythrocytic developmental cycle of the parasite. PfPKA knockdown also results in loss of phosphorylation at Ser610 of PfAMA1.[27]

The invasion of RBCs by the merozoites involves the discharge of proteins from rhoptries as well as micronemes of the parasite which interacts with the host cell receptors. RhopH3 is part of a three-protein complex called RhopH and has a phosphorylation site at serine 804 (S804) residue. The role of phosphorylation in regulating its activity in the parasite was elucidated by the generation of a mutant of RhopH3 (S804A) wherein S804 was mutated to alanine with the help of two-plasmid based CRISPR-Cas9 approach. The phospho-mutant parasites showed a significantly reduced growth rate.[48] In addition to merozoite invasion, its egress from host erythrocytes is also an essential step for the pathogenesis of malaria. DiCre recombinase based conditional knockout system was generated using CRISPR/Ca9 to examine the role of *Pf*MSA180 (merozoite surface antigen 180) during the asexual growth of the parasite. *Pf*MSA180 knockout causes failure of the merozoite egress and as a result, there is a significant decrease in merozoite invasion and the formation of rings. Novel medicines could be designed to interfere with the merozoite egress by a deeper understanding of the antigens involved in the process of egress.[25]

The role of catalytic subunit of *P. falciparum* casein kinase 2 (*Pf*CK2α) was investigated in the asexual as well as sexual blood stages using two-plasmid CRISPR-Cas9 approach by the generation of *Pf*CK2α-GFP expressing transgenic line, *Pf*CK2α-GFPDD expressing parasite line for the conditional regulation of destabilisation domain (DD) tagged protein levels of *Pf*CK2α with the help of stabilising ligand

Shield 1 and a DiCre/loxP system based conditional knockout parasite line. *Pf*CK2 was found to localize in both cytoplasmic and the nuclear compartment of asexual as well as sexual blood stages alike. *Pf*CK2α knockout in the early ring stages extends the time interval for the completion of schizogony, the majority of the merozoites released from the schizonts are unable to invade fresh erythrocytes and a small subset of merozoites which invade gets arrested at the trophozoite stage. *Pf*CK2α knockdown inhibited the transition of gametocytes from stage IV to stage V and *Pf*CK2α knockout in the sexual rings causes early blocking of gametogenesis. *Pf*CK2α may serve as potential drug target as is evident by its essentiality for both the asexual proliferation as well as the transmission compatible sexual blood stages.[26]

*Pf*P230p and *Pf*P48/45 belongs to 6-cysteine protein family and are essential for gamete fertility. *Pf*P230p is a paralog of *Pf*P230 which is expressed only in male gametocytes and gametes. Using CRISPR-Cas9 based system for *Pf*P230p disruption produces transgenic parasite that exhibit normal counts of male as well as female gametocytes. The exflagellation of male gametocytes produces male gametes which are incapable of attaching to erythrocytes in vitro. Consequently, characteristic exflagellation centres are absent. The *Pf*P230p disrupted parasites showed a strong reduction in ookinete, oocyst and sporozoite formation in the mosquitoes, suggesting that *Pf*P230p could serve as potential transmission blocking vaccine antigen.[49] *Plasmodium* gamete surface protein's (*Pf*s230) pro-domain and domain 1 of fourteen 6-Cysteine domains were previously reported as targets of transmission blocking antibodies and majority of these antibodies rely on human complement which is also found in the blood meal of mosquito. In a recent study, *Pf*s230 domain 7 was identified as target of murine monoclonal antibody mAb 18F25.1 but this antibody failed to show any transmission reducing activity (TRA). Therefore, a subclass-switched complement fixing variant mAb 18F25.2a was developed using CRISPR-Cas9 mediated gene editing of Hybridoma cell line. Using a guide RNA, Cas9 was instructed to create a DSB in the constant heavy region 1(CH1) of the mIgG1 locus within the 18F25.1 hybridoma cell line and a homology directed repair (HDR) resulted in the replacement of coding sequence of mIgG1 CH-CH3 with mIgG2a CH1-CH3, producing a mIgG2a switch variant. Standard membrane feeding assays (SMFA) were performed to assess the transmission reducing potential of mAB 18F25.2a and it is found to significantly decrease *P. falciparum* infection in *Anopheles stephensi* through a complement-dependent mechanism.[50]

2.10.1 Study of genes involved in apicoplast biogenesis of P. falciparum

Apicoplast is a plastid organelle originated from prokaryotes and found in *P. falciparum* but lacks a counterpart in human host and thus can offer new targets for antimalarial drugs.[51] The majority of apicoplast proteins are nuclear-encoded and transferred to the organelle through the secretory pathway but the regulatory mechanism of plastid proteome is not known. Using CRISPR-Cas9 based gene editing tool, proteostasis regulation via apicoplast localized Clp (caseinolytic protease) system was studied. The null mutant of *P. falciparum* Clp protease (*Pf*ClpP) was unable to produce viable parasites without IPP suggesting that *Pf*ClpP is indispensable for parasite survival. During the asexual blood stages, the sole essential function of apicoplast is the production of Isopentenyl pyrophosphate (IPP). *Pf*ClpP knockdown via *tetR-aptamer* conditional knockdown system revealed its robust enzymatic activity for apicoplast biogenesis. The expression of catalytically dead *Pf*ClpP demonstrated that *Pf*ClpP form oligomers as a zymogen and undergoes maturation through transautocatalysis. Both wild type (*Pf*ClpC) and mutant Clp chaperone variants showed a functional interaction with protease *Pf*ClpP. *Pf*ClpS adapter knockdown via *tetR-aptamer* conditional knockdown system highlighted its indispensable role in plastid biogenesis.[52] *P. falciparum* DNA Gyrase (*Pf*Gyr) has two subunits A and B with both having apicoplast targeting signal peptide. Two-plasmid CRISPR/Cas9 based strategy was used to disrupt *Pf*Gyr A by introducing nonsense mutations to create a stop codon in the beginning of the PfGyr gene, halting further protein translation. Disruption of *Pf*Gyr gene led to the loss of plastid acyl carrier protein (ACP) and plastid genome. A gyrase inhibitor, Ciprofloxacin has been used for malaria prevention and the mutant *Pf*Gyr A clone supplemented with IPP showed reduced sensitivity to Ciprofloxacin but high concentrations of Ciprofloxacin still inhibited IPP-rescued mutant *Pf*Gyr parasites, hinting at a possible nonapicoplast target in *P. falciparum*.[51]

2.10.2 Study of genes involved in drug resistance using CRISPR-Cas9

Previous studies indicated the association of *PfCARL* (cyclic amine resistance locus) with resistance against IZPs (imidazalopiperazines). The role of mutations in *PfCARL* that were revealed through in vitro drug evolution experiments were investigated by introducing the individual mutations into the parental line with the help of two-plasmid CRISPR-Cas9 system. 5 mutant (L830V, S1076N/I, V1103L and I1139K) lines were generated which were all sufficient to provide resistance against IZPs and some structurally unrelated antimalarials.[53]

Pantothenamide bioisosteres may serve as potent antimalarial compounds by interfering with the growth of asexual stages and by their cytocidal activity against gametocytes. The resistant parasites generated in vitro were found to have mutations in Acetyl CoA synthetase as well as Acyl CoA synthetase 11, both of which play roles in the synthesis of Acetyl CoA. By introducing these mutations into the WT parasite using a two-plasmid CRISPR-Cas9 system, it was confirmed that pantothenamides work by interfering with the Acetyl CoA biosynthesis.[54]

Polymorphisms in *kelch 13* (*k13*) are associated with artemisinin resistance phenotype.[55,56] In addition to polymorphisms in *k13*, an autophagy like mechanism was found to be associated with the resistance against the artemisinin-based combination therapy (ART). A single-nucleotide polymorphism (SNP) in autophagy-related gene (*atg18*) was investigated with the help of two-plasmid CRISPR-Cas9. Transgenic parasite with T38I mutation in the gene, *atg18* was found to grow better in nutrient deprived conditions and this mutation may also provide the additional resistance to artemisinin derivatives even in the absence of *k13* mutations.[57]

MMV020291 was reported to block invasion of erythrocytes by *P. falciparum*.[58,59] Whole genome sequencing of the resistant lines generated via in vitro selection revealed non-synonymous SNPs in the genes, *profilin* and *actin-1*. When these mutations were introduced into the WT parasite with the help of CRISPR-Cas9, the transgenic parasites harboring these mutations in *profilin* and *actin-1* became resistant towards the compound thus confirming the mechanism of resistance against MMV020291. Here, linearized donor plasmid along with recombinant Cas9 and synthetic sgRNA were electroporated into the parasite for the generation of transgenic parasites. Sulfonyl-piperazine compounds, MMV020291 and its analogues may therefore serve as novel antimalarials by interfering with the actin-1/profilin interaction.[60]

2.10.3 Study of Anopheles mosquito genes involved in infection by P. falciparum

There are many mosquito factors (agonists) which are responsible for supporting the growth of the parasite inside the female *Anopheles* mosquito. One such factor is FREP1 (Fibrinogen-related protein 1) and a modified CRISPR-Cas9 approach was used to study its role in the sexual growth of the parasite inside the mosquitoes. FREP1 knockout *A. gambiae* was generated by crossing two transgenic lines with one expressing 3 gRNAs and another expressing the Cas9 endonuclease controlled by a germline-specific promoter (*Vasa2*). The mosquitoes lacking FREP1 protein showed a prominent decrease in infection at both the oocyst and sporozoite stages, though the mutants also displayed high fitness costs.[61] The role of another mosquito factor, mosGILT (mosquito gamma interferon inducible lysosomal thiol reductase) was examined by following a very similar type of CRISPR-Cas9 based strategy of crossing two transgenic lines with one expressing 3 gRNAs and another expressing Cas9 endonuclease for the generation of mosaic mosGILT mutants of *A. gambiae*. Female mosaic mutants exhibited abnormalities in development of ovaries and resistance to *Plasmodium*.[62]

To investigate the role of *A. gambiae* C-type lectin CTL4 in safeguarding *P. falciparum*, CTL4 knockout *A. gambiae* mosquitoes were generated using CRISPR-Cas9. CTL4 knockout (KO) *A. gambiae* were generated by crossing two transgenic lines with one expressing 3 gRNAs controlled by U6 snRNA promoter and the other expressing Cas9 controlled by the promoter sequences of vasa gene. TEP1, LRIM1, and CLIPA2 are immune factors that do not while CLIPA14 does impact the melanization of ookinetes in CTL4 KO *A. gambiae*. Moreover, CTL4-associated protein, CTLMA2 provides defense to *P. falciparum* even in the absence of CTL4. CTL4 KO *A. gambiae* showed only partial melanization-based refractoriness to *P. falciparum* as there seems to be an impact of the temperature at the time of infection which is around 7 °C higher for *P. falciparum*, on the dynamics of midgut infection and the effectiveness of parasite melanization. CTL4 may serve as a promising target for developing malaria control strategies focused on transmission prevention.[63]

3. Leishmania

Human leishmaniasis is caused by *Leishmania spp*. which are members of the Trypanosomatidae family. The bite of an infected female phlebotomine sandfly transmits the protozoan parasite. Sandflies inject infectious promastigotes into human skin using during a blood meal. Following that, the promastigotes are phagocytized by macrophages and other mononuclear phagocytic cells. Amastigotes are a non-motile form that promastigotes change into once they enter the cell. Amastigotes proceed to infect additional mononuclear phagocytic cells by simply dividing and multiplying. During a blood meal, sandflies consume the contaminated cells and are infected. The amastigotes change and mature into promastigotes inside the sandfly's midgut, from whence they eventually go to the proboscis.

Thus, humans and sandflies are necessary for *Leishmania* to complete its life cycle. After malaria, leishmaniasis is thought to be the second most contagious disease in humans caused by protozoan parasites. A recent WHO report estimates that between 700,000 and 1 million new cases of leishmaniasis are reported each year. The illness presents as five distinct clinical manifestations: If treatment is not received, the most severe type of leishmaniasis, known as kala-azar, can be fatal. in visceral leishmaniasis, the parasites exhibit tropism towards visceral organs, including the liver and spleen; The most prevalent type of leishmaniasis that results in skin ulcers is cutaneous; Mouth, nose, and throat are affected by mucocutaneous leishmaniasis. Post-Kala-Azar Dermal Leishmaniasis (PKDL) is a type of cutaneous leishmaniasis that can develop after several years of successful therapy for visceral leishmaniasis (VL). Expression of the illness as Depending on the parasite, host, and other variables, the infection can cause mucocutaneous, cutaneous, or visceral leishmaniasis. Despite the rigorous study on cure of this disease over the decades there is a critical need for the discovery of new tools and strategies against Leishmaniasis. *L. major* has haploid genome that is made up of 36 chromosomes with 32 million base pairs. *L. major* contains 8272 protein encoding genes, comprising of 911 RNA genes (tRNA, rRNA, slRNA, slRNA, snRNA, snoRNA, srpRNA) and 39 pseudo-genes. *L. major* does not have broad transcription factors and its protein-coding genes are arranged into lengthy, strand-specific polycistronic clusters.[64] Disruption of a gene of interest in *Leishmania*[65] was performed by the conventional homologous recombination by replacing the target gene sequence with transgenes flanked on both the sides by DNA sequences homologous to the endogenous gene. High genomic plasticity of *Leishmania* spp. results in aneuploidy, or cells with extrachromosomal elements that causes retention of single or multiple copies of the target gene.[66] Therefore, information regarding the essentiality of the gene or its functional role in the parasite biology cannot be conclusively deduced. Another disadvantage of the conventional homologous recombination approach is that it involves cloning of long flanking sequences.[67] Furthermore, to deliver the desired gene alteration in genes that belong to a multigene family, at least two rounds of transfection are necessary, which calls for the use of multiple antibiotic selection markers.[68,69] Because most *Leishmania* species lack RNAi-related genes like dicer and argonaute, RNA interference is found to be unsuccessful against them[70] development and application of CRISPR-Cas9 technology has greatly increased the effectiveness and accessibility of gene editing in

Leishmania. Implementation of CRISPR-Cas9 for *Leishmania* parasite gene editing had a transformative effect in understanding the parasite biology and identification of molecular determinants required for host-parasite interactions. Numerous adjustments and improvements have been made to CRISPR since its initial application in the study of *Leishmania* gene editing in order to boost systematic effectiveness. In the following discussion, a few CRISPR-Cas9 systems used for gene editing in *Leishmania* along with how these systems are used to study different facets of the biology of the parasite are covered.

3.1 CRISPR-Cas9 technology in leishmaniasis

The guide RNA and the Cas9 endonuclease are the two most important element of CRISPR-Cas9 system. SpCas9, the mostly used Cas9 endonuclease, is obtained from S. pyogenes. SpCas9 is generally expressed from an episomal plasmid or may be integrated in the genome to provide constitutive expression. In a slight modification of this approach, in vitro-generated single guide RNA/Cas9 ribonucleoprotein complex (RNP) is used for gene editing. In this case, instead of the regular SpCas9 endonuclease, Cas9 isolated from *Staphylococcus aureus* (SaCas9) is used.[71] Since SaCas9 is much smaller in size compared to the SpCas9, therefore it is highly efficient for gene editing when provided as an RNP complex of SaCas9/sgRNA. For efficient transfection of the parasite the molecular size of the Cas9/sgRNA complex should be below a certain threshold that is exceeded by the bigger SpCas9. Leaving aside this strategy for gene editing that utilizes transfection with an in vitro constituted SaCas9/sgRNA complex, SpCas9 is very efficient when expressed inside the parasite from DNA sequence.[72] *Leishmania* cells have been observed to serve better tolerance to the SpCas9 despite SaCas9 showing better efficiency for genome editing.[7,71] Several CRISPR-Cas9 systems have been used for genome editing of *Leishmania* among which two stable expression systems and three transient systems known to have been employed for *Leishmania* genome editing.

3.1.1 *Stable CRISPR-Cas9 expression systems*

Stable CRISPR-Cas9 systems exhibit constitutive expression of sgRNA and Cas9 endonuclease and U6 or ribosomal RNA promoter controls the sgRNA transcription.

3.1.1.1 U6snRNA promoter for the expression of gRNA

Sollelis et al. first reported the successful CRISPR-Cas9 driven genome editing in a *Leishmania* parasite.[73] Using CRISPR-Cas9 they successfully knocked out the paraflagellar *rod-2* locus, a tandemly repeated gene family in *L. major.* Unlike the conventional approach of homologous recombination, which requires multiple rounds of transfection, the null mutants were generated in less than a month with a single round of transfection. Importantly, there were no off-target effects. The NHEJ pathway doesn't seem to exist in *Leishmania*, even though Ku proteins are present. Therefore, donor DNA containing sequences homologous to the target DNA is needed to achieve the intended gene editing. Sollelis et al. employed two plasmid systems (Fig. 4A), wherein the pTCAS9 plasmid contains SpCas9 and the pLS7 plasmid carries a sgRNA expression cassette and a puromycin resistance cassette flanked by homology regions.[73] The dihydrofolate reductase thymidylate synthase (DHFR-TS) controls the expression of Cas9 endonuclease and U6snRNA promoter drives the expression of sgRNA. *Sca*I site in the pLS7 plasmid was utilized for linearization in order to increase the system's gene editing efficiency.[73]

3.1.1.2 RNA polymerase I promoter in the CRISPR-Cas9 system for the expression of sgRNA

Zhang and colleagues used two vector systems in CRISPR-Cas9 for gene editing in *L. donovani.*[6] The pLPhyg plasmid was used to express SpCas9 endonuclease (LP = *Leishmania* promastigotes; hyg = hygromycin). The pLPhyg plasmid was transfected into *L. donovani* promastigotes in order to express Cas9 endonuclease constitutively. Since the Leishmanian RNA polymerase III U6 promoter is not well understood, an RNA polymerase I-driven ribosomal RNA promoter was employed instead. To generate the exact boundaries of sgRNA, a 68-bp hepatitis delta virus (HDV) ribozyme sequence was cloned downstream to the sgRNA sequence (Fig. 4B). Using this system one amino acid in miltefosine transporter gene LdMT was edited conferring drug resistance against miltefosine (MLF) (hexadecylphosphocholine, MIL). Repairing DSBs caused by the Cas9 endonuclease was accomplished through the use of Microhomology-mediated End joining (MMEJ). Two additional guide sequences were selected to compare the efficiencies of gene editing. The guide sequence nearest to the microhomology region was found to be the most efficient in mediating the desired gene editing. To see if the oligonucleotide donor transfection would increase the gene editing frequency of the LdMT, a 61-nucleotide

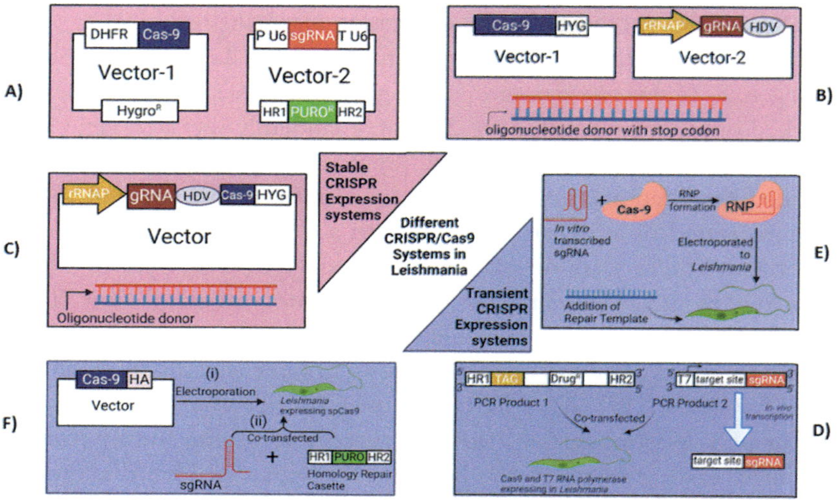

Fig. 4 Schematic diagram of different CRISPR-Cas9 strategies used for genome editing in *Leishmania*. (A) In these two plasmid systems, vector-1 expresses SpCas9 and vector-2 expresses the sgRNA under the control of *DHFR* promoter and *U6snRNA* promoter, respectively. The plasmid with the sgRNA expression cassette bears a puromycin resistance cassette flanked by homology regions. (B) SpCas9 endonuclease is expressed using vector-1. *L. donovani* promastigotes are transfected with this plasmid for constitutive expression of Cas9 nuclease. For the expression of sgRNA, an RNA polymerase I driven ribosome RNA promoter is used and to generate the exact boundaries of sgRNA, a 68-bp hepatitis delta virus (HDV) ribozyme sequence is cloned downstream to the sgRNA sequence. A 61 nt- oligonucleotide cassette comprising of two homology regions flanking the Cas9/gRNA cleavage site and with a stop codon is electroporated into the Cas9/gRNA expressing *Leishmania* cells for Homology-directed repair. (C) A single co-expression vector drives the transcription of gRNA, Cas9 and antibiotic selection marker through rRNA promoter to simplify the gene editing process. (D) A T7 RNA polymerase promoter-based system is used to enable constitutive and stable expression of Cas9 and T7 RNA polymerase that are incorporated into the *Leishmania* genome. *Leishmania* promastigotes constitutively expressing Cas9 and T7 RNA polymerase are co-transfected with two PCR products i.e. sgRNA transcribed from the linear DNA fragments and donor DNA template containing antibiotic selection markers or fluorescent protein tags. (E) RNP complex composed of *S. aureus* Cas9 (SaCas9) and in vitro-transcribed single guide RNAs (sgRNAs) is used for easy and rapid editing of *Leishmania* spp. Codelivery of an RNP complex and a repair template containing stop codons in all three reading frames flanked by short homology regions leads to the successful gene knockout. (F) *S. pyogenes* Cas9 is amplified from commercial vector and it is cloned in the *Leishmania* expression vector pSP72 αHYGα subsequently. HA tag is included at the C-terminal of Cas9 during the amplification of the ORF to track the expression of Cas9. Homology repair cassette is made by inserting a PCR amplified 600 bp puromycin N-acetyltransferase gene flanked by 30 bp sequence homologous to the target gene. Both gRNA and the homology repair cassette are co-transfected in the *Leishmania* parasite expressing Cas9-HA for gene disruption.

oligonucleotide cassette with a stop codon and two homology regions flanking the Cas9/gRNA cleavage site was electroporated into the *Leishmania* cells expressing Cas9/gRNA. In comparison to Cas9/gRNA-expressing cells without the donor homology sequence, oligonucleotide donor with homology sequences electroporated to Cas9/gRNA-expressing cells demonstrated higher MLF drug resistance. This indicates that DSBs are repaired by homology-directed repair in the presence of a donor repair template. It is because of the homology-directed repair, it is possible to introduce point mutation and endogenously tag the gene of interest in *L. donovani*. The system was also used for disruption of a non-essential gene through integration of bleomycin resistance cassette. They have also targeted *TOR1* gene (Ldbpk_366580.1) in *L. donovani* by introducing the same drug selection marker. Half of the bleomycin resistant clones they obtained died after 2 weeks of continuous culture and in the other half clones that survived, only one allele of the LdTOR1 gene was altered while the other allele remained intact indicated that CRISPR-Cas9 could partially inactivate the crucial *Leishmania* gene and demonstrated its ability to target genes with a selection marker. Additionally, the *LdMT* gene has been effectively endogenously tagged with GFP using the electroporation of a GFP donor cassette into gRNA/Cas9-expressing cells. A double-gRNA expression plasmid expressing two gRNAs simultaneously was created employing the rRNA promoter, Hammerhead and HDV ribozymes. The entire *LdMT* gene was eliminated using this vector containing two gRNAs, which also improved the inactivation efficacy of desired gene. Complete blocking of a pathway or multigene family can be achieved with vectors expressing more than one gRNAs flanked by Hammerhead and HDV ribozymes. A stable live attenuated *Leishmania* vaccine strain could be developed by entirely deleting one or more virulence genes using the vector system that co-expresses multiple gRNAs concurrently. In order to effectively modify the *L. donovani* genome, the CRISPR-Cas9 system can be used to delete a whole gene, add mutations, protein tag sequences, and antibiotic selection markers to the genome.[6] A single co-expression vector that drives the transcription of gRNA, Cas9 and antibiotic selection marker through rRNA promoter was developed to simplify the gene editing process (Fig. 4C). pLdCN and pLdCH vectors have been used to target *LdMT* genes by this strategy and successfully generated miltefosin resistant cells. This strategy has been used for *L. major*, *L. mexicana*, and *L. donovani* spp. for gene editing.[74]

3.1.2 Transient CRISPR expression systems

3.1.2.1 T7 promoter-based system

The Cas9 and T7 RNA polymerase are incorporated into the *Leishmania* genome in the T7 RNA polymerase promoter-based system to allow constitutive and stable expression of these two genes. The altered parasite's growth rate and morphology were comparable to those of the wild type. Promastigotes expressing Cas9 and T7 RNA polymerase are then co-transfected with sgRNA sequence and donor template bearing antibiotic selection markers or fluorescent protein tags. T7 RNA polymerase promoter-based system has been demonstrated by Beneke et al. to be used for in vivo transcription of PCR-amplified sgRNA templates.[75] This system was used to tag three different genes of *Leishmania* i.e. *PF16*, small myristoylated protein 1 (SMP-1) and histone H2B. *L. mex* cells expressing Cas9 and T7 RNA polymerase were co-transfected with a donor DNA flanked by 30 nt homology and a sgRNA targeting gene of interest (Fig. 4D). While the control cells did not exhibit any drug resistance, the transfected cells with both the donor DNA and the sgRNA template developed drug resistance and displayed fluorescence in the axoneme for PF16, in the nucleus for histone H2B, and in the flagellar membrane for SMP-1. This finding implied that the linear DNA segments were used to transcribe the functional sgRNA. The primary findings of this study indicated that a Leishmanian gene locus could be altered by co-transfecting two PCR products, and it also offered a quick and scalable method of gene tagging in *Leishmania*.[75]

3.1.2.2 RNP complex of CRISPR-Cas9 system

Soares Medeiros et al. have used RNP complex[71] composed of *S. aureus* Cas9 (SaCas9) and single guide RNAs (sgRNAs) transcribed in vitro- for easy and rapid editing of *Leishmania spp* (Fig. 4E). SaCas9/sgRNA complexes have been utilized to determine the role of critical genes and endogenously tag parasite proteins in *Leishmania*. This technique provides a rapid, cloning and selection-free method to edit the genome of *Leishmania* spp. SaCas9/sgRNA has been introduced to the eGFP-expressing procyclic forms of *L. major* in order to investigate the gene editing effectiveness of the RNP complex. SaCas9/sgRNA targeted the *eGFP* gene at 3 different positions in the downstream of GFP start codon. Knockout of *eGFP* gene was reported in both the presence and absence of repair template. Endogenous lipophosphoglycan (LPG) gene (*lpg2*) was also disrupted in *L. major* using RNP complex and a repair template. The reduction of phosphoglycan surface expression caused by loss of lpg2 was identified by the LPG-specific WIC79.3 monoclonal antibody.[71]

3.1.2.3 T7-pSP72-based system

In this transient system, an in vitro synthesized sgRNA and a donor template containing antibiotic selection marker gene were co-transfected in SpCas9 expressing *Leishmania* cells. To study the drug resistance genes in *Leishmania*, Fernandez-Prada et al. adopted an approach called Cos-Seq approach[76] in which intracellular amastigote of *L. infantum* complemented with cosmid library of *L. infantum* were submitted to increasing concentrations of anti-Leishmania drug compounds miltefosine, amphotericin B and pentavalent antimonials. After being separated, the cosmids from the treated amastigotes were sent for next-generation sequencing and then processed for gene enrichment analysis. This method for screening drug resistant genes identified seven cosmids and tested for drug resistance. The relevant drug resistant genes were further confirmed either by overexpression of the gene or by CRISPR-Cas mediated disruption of the gene. LinJ.06.1010 gene identified by this drug screening process coding for a putative leucine rich repeat protein (LRR) has been targeted and LRR null has been generated using CRISPR-Cas9 gene editing tool. For LRR null generation S. pyogenes Cas9 was amplified from commercial vector and subsequently cloned in the expression vector pSP72 αHYGα. The Eukaryotic Pathogen CRISPR gRNA Design Tool was used to create gRNA that targets the LRR. A 600 bp puromycin N-acetyl-transferase gene that had been PCR amplified and flanked by 30 bp of sequence homologous to the LRR region upstream and downstream to the Cas9 cleavage site was inserted to create the homology repair cassette. Both gRNA and the homology repair cassette were co-transfected in the *Leishmania* parasite expressing Cas9-HA (Fig. 4F).[76]

3.2 Applications of CRISPR-Cas9 in *Leishmania*

3.2.1 Desired editing of the target locus for functional analysis and localization studies

3.2.1.1 Incorporation of single point mutations to understand drug resistance mechanism and function of the target gene

CRISPR-Cas9 system was adapted in *Leishmania* parasites for gene editing of the 3-kb miltefosine transporter (LdMT).[6] The *LdMT* gene sequence was modified either in absence or presence of a repair DNA sequence. When no repair template was used, the double strand break was repaired by MMEJ while single amino acid substitution was only possible when only repair template is present. CRISPR-Cas9 was used to introduce the already known single point mutation (M381T) in the LdMT gene that confers miltefosine drug resistance (Fig. 5A-b). The efficiency of the gene editing

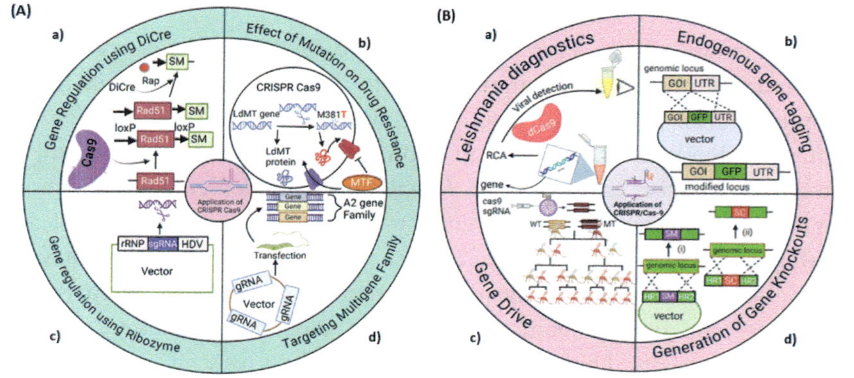

Fig. 5 Applications of CRISPR Cas9. (A) Schematic Diagram illustrating various strategies of using CRISPR-Cas9 in *Leishmania*. (a) *RAD51* gene is involved in DNA repair in *L. major*. Gene modification is done via CRISPR-Cas9 to integrate loxP sites in *RAD51*. To turn off the genes a small organic compound called rapamycin is added which activated dimerizable Cre recombinase to remove the gene (*RAD51*) floxed by the two loxP sites. Over longer periods, cells without RAD51 showed growth defects demonstrating that *RAD51* and its related genes are crucial for maintaining DNA stability and long-term growth in *L. major* (b) CRISPR-Cas9 is employed to introduce a point mutation (M381T) in the miltefosine transporter gene (*LdMT*) in *Leishmania*. This mutation results in the loss of transporter protein function responsible for miltefosine uptake, thereby contributing to miltefosine resistance in *Leishmania*. (c) A double-gRNA expression vector is introduced with Hammerhead and HDV ribozymes. Hammerhead and HDV ribozymes cleave the RNA at specific locations to ensure the guide RNAs (gRNAs) are properly processed and functional. This precise processing is essential because it enables the gRNAs to effectively bind to the Cas9 enzyme and accurately target towards the correct DNA sequence which enhance gene-targeting efficiency. (d) A triple gRNA expression vector is made, containing instructions to produce all three gRNAs simultaneously. This vector is then introduced into *L. donovani* cells. Two gRNAs are designed: one targets the *LdMT* gene, and the other targets the *A2* genes. The cells are then exposed to miltefosine, and only those with mutations in the *LdMT* gene, which confer resistance to miltefosine, survive. These surviving cells are more likely to possess active CRISPR-Cas9 machinery. (MTF – Miltefosine, SM – Selection Marker. Rap-Rapamycin). **(B) Schematic Diagram illustrating various applications of CRISPR-Cas9 in *Leishmania*.** (a) PCR analysis confirms the presence of a conserved 23-mer sequence in the blood and urine samples of VL (visceral leishmaniasis) patients, but not in healthy controls. DNA from patient blood or urine sample are employed to Recombinase Polymerase Amplification (RPA). Only the target DNA is amplified by RPA with a biotinylated primer and subsequently the target DNA is immobilized to streptavidin-coupled beads via biotin-streptavidin interactions. immobilized target DNA is recognized by CRISPR-dCas9 that has a ssDNA circle attached. The circular ssDNA is used in Rolling Circle Amplification (RCA) reaction. The resulting RCA product consists of tandem repeats of G-quadruplexes picks up a heme group and produce a colorimetric readout.

was determined by the presence of the microhomology region near the cut site of the endonuclease. NHEJ pathway was not found to be involved in the repair of the cut site as some of the components required for the pathway are missing in *Leishmania*. Furthermore, the whole 3-kilobase (kb) miltefosine transporter gene (LdMT) of *Leishmania* parasites was deleted using CRISPR-Cas9. This deletion involved designing guide RNAs (gRNAs) that targeted specific sequences flanking the *LdMT* gene, the Cas9 enzyme which is guided by these gRNAs creating double-strand breaks at these target sites. Two guide sequences driven by a single promoter were expressed independently using ribozymes. Cellular repair machinery come into action, resulting in the precise removal of the entire *LdMT* gene. This demonstrates the ability of CRISPR-Cas9 to produce targeted mutations as well as the possible drug resistance ramifications of those mutations. The M381T mutation resulted in a change from methionine (M) to threonine (T) at position 381 in the LdMT protein. This specific alteration caused the *Leishmania* parasite to become resistant to miltefosine, a key oral drug used to treat leishmaniasis. This finding highlights the applications of CRISPR-Cas9 to introduce genetic changes, enabling researchers to study that how the drug resistance and other phenotypic traits depends upon the effects of particular mutations. But by creating specific mutations like M381T, leads to resistance against essential drugs such as miltefosine.[6]

A visible colour appears on the beads in the tube in the presence of target DNA sequence. (b) Endogenous gene tagged with GFP is helpful to determine the expression level of a gene and cellular localization of its protein product. *LdMT* gene is tagged with *gfp* using CRISPR-Cas9 gene editing tool and involves insertion of a bleomycin resistance gene expression cassette into the Cas9 cleavage site for the selection of successfully modified cells. The gRNAa/Cas9-expressing cells are electroporated with a GFP donor plasmid. The endogenously GFP tagged gene enabled visual confirmation of the modifications, as the cells containing GFP emit a green glow under a fluorescence microscope. (c) *Phlebotomus papatasi* embryos are injected with Cas9 protein and four sgRNAs targeting different parts of the *rel* gene. Resulting adult sand flies are screened for the presence of mutation(s) and multiple *rel* gene mutations are observed in PCR assay. Only a small number of total injected embryos reach adulthood out of which a significant rate of mutation is accomplished indicating a reasonably high efficiency of the CRISPR-Cas9 technique in gene editing of *P. papatasi* sand flies. (d) CRISPR-Cas9 is efficient tool for precise gene knockouts in *L. donovani*, employing oligonucleotide donors in the form of plasmid with selection marker (i) or PCR amplified product containing stop codons with short homology arms (ii) to disrupt target genes effectively. (GOI – Gene of Interest, SM – Selection Marker, HR1 & HR2 – Homology Region 1 & 2, RCA- Rolling Circle Amplification).

A single mutation (A189G) in both alleles of *rad51* was introduced using the CRISPR-Cas9 system. *rad51* gene encodes a DNA recombinase enzyme that is required for homology-directed repair. Plasmids expressing Cas9 and gRNAs specific for *rad51* gene were transfected in *L. donovani* cells. Oligonucleotide donors containing stop codons and an *Eco*RI restriction site, or a conserved amino acid substitution were subsequently transfected in these *L. donovani* cells. Genomic DNA from the altered *L. donovani* clones confirmed the existence of the intended mutations and allowed identification of *rad51* function in *Leishmania*. Disruption of both the alleles of *rad51* resulted in lower survival rates indicating that *rad51* is essential for the survival of *L. donovani*.[74]

3.2.1.2 Generation of gene knockouts

Gene knockouts in *Leishmania* are effectively produced using the CRISPR-Cas9 technology, which eliminates the need for conventional selectable markers. Instead of using selection, the method involves inserting an oligonucleotide donor containing stop codons leading to disruption of the gene of interest (Fig. 5B-d). The oligonucleotide donor contains short DNA sequences homologous to the regions flanking the cut site, facilitating integration via homologous recombination. The insertion of the oligonucleotide donor disrupted the gene's sequence, preventing it from being properly transcribed and translated into a functional protein, thus achieving a gene knockout. This approach allows for efficient gene disruption without the need for additional selection markers.[6] CRISPR-Cas9 was also used to generate conditional knock-out for target genes using a dimerizable Cre recombinase strategy.[77] *Leishmania* parasites that produces Cas9, a protein that can cut DNA, and DiCre, a protein that can activate the gene knockout process. First, the original *rad51* gene and its similar genes is replaced with new versions that have special markers called loxP sites around them and a HA tag at the end for tracking (Fig. 5A-a). This replacement was done using CRISPR-Cas9 and was very efficient, with all original genes being swapped out after one attempt. They found that adding the loxP sites or the HA tag did not disturb the growth of cells with the *rad51*-like genes. However, the HA tag did interfere with RAD51, so they made another version without the tag, which grew normally. To turn off the genes, a substance called rapamycin is added. This activates DiCre to remove the gene between the loxP sites. The gene was completely removed is confirmed by the second cycle of cell growth by checking the DNA and proteins. When *rad51* and its related genes are knocked-out, but not another gene called *PIF6* (used as a control), they observed an increase in DNA damage, shown by higher levels of a marker called yH2A. This indicates that *rad51* and its related genes are involved in DNA repair. The cells grew normally shortly after the genes were knocked out, but over longer periods, cells without *rad51* or its related genes showed growth problems. These issues appeared more quickly in a different growth medium (M199). The cells also showed signs of genomic instability, meaning their DNA was more damaged or disorganized. In contrast, knocking out the *pif6* gene did not cause the same effects. The cells continued to grow normally without it, suggesting that *pif6* has a different role compared to *rad51* and its related genes. The study showed that *rad51* and its related genes are crucial for maintaining DNA stability and long-term growth in *L. major*. Removing these genes leads to DNA damage and growth problems, while the *pif6* gene does not have the same essential function.

3.2.1.3 Endogenous gene tagging

In addition to creating gene knockouts, CRISPR-Cas9 system has also been used to precisely tag endogenous genes in *L. donovani* (Fig. 5B-b). The target locus was modified by introducing a bleomycin resistance marker and a *gfp* gene sequence at the end of the GoI. The bleomycin resistance marker allowed for the selection of successfully modified cells, as only these cells could survive in the presence of bleomycin. The *gfp* gene enabled visual confirmation of the modifications, as the cells containing GFP would emit a green glow under a fluorescence microscope. This approach ensured that the surviving cells had the desired genetic modifications, which could be both selected for with bleomycin and visually confirmed with GFP fluorescence.[6,78]

3.2.2 Targeting multi-gene family and co-selection

The CRISPR-Cas9 system has also been used for efficient editing of *L. donovani* genes belonging to multigene family through a technique called coselection. This method aims to improve the deletion efficiency of *A2* genes by simultaneously targeting another gene, the miltefosine transporter gene (*LdMT*), that allows selection for miltefosine resistance. First, two guide RNAs (gRNAs) are designed, one targets the *LdMT* gene and the other targets the *A2* genes. These gRNAs are incorporated into a CRISPR-Cas9 system which uses the Cas9 enzyme to cut DNA at specific sites guided by the gRNAs. A triple gRNA expression vector is created, containing instructions to produce all three gRNAs at once, and this vector is introduced into *L. donovani* cells (Fig. 5A-d). When these cells are exposed to miltefosine, only those transfectants that contain disrupted *LdMT* gene or mutations that confer resistance to the drug survive. The surviving cells are likely to have functional CRISPR-Cas9 machinery, increasing the chances of successful *A2* gene deletions. Finally, the cells that survive miltefosine selection are analyzed to confirm the deletion of the *A2* genes. This strategy leverages the selection for miltefosine resistance to enhance the efficiency of *A2* gene deletion, offering a more effective method for *L. donovani* gene editing. This co-selection strategy substantially reduced the time to obtain *A2* deleted parasites from months without co-selection to weeks with co-targeting and selection for miltefosine resistance.[74]

3.2.3 Expression of multiple guides using ribozymes

To enhance gene-targeting efficiency and obtain the desired editing, a double- gRNA expression vector was developed, incorporating Hammerhead and Hepatitis Delta Virus (HDV) ribozymes (Fig. 5A-c). This vector likely facilitated more precise editing by providing a more efficient means of delivering the guide RNAs required for Cas9- mediated cleavage. A double-gRNA expression vector is a plasmid that carries two separate gRNA sequences. Each gRNA is designed to target a particular site in the genome. Using two gRNAs allows for more precise and efficient targeting, as it can facilitate the simultaneous editing of two closely located sites or increase the likelihood of successful editing at a single site. Ribozymes are RNA molecules with self-cleavage activity. When included in the gRNA expression vector, they enhance the maturation and functionality of the gRNAs. Hammerhead and HDV ribozymes cleave the RNA at specific sequences, ensuring that the gRNAs are precisely processed and retain their functionality of directing the Cas9 endonuclease to the desired target locus. Proper processing of gRNAs is crucial for their function, as any unprocessed or incorrectly processed gRNAs might not bind effectively to the Cas9 enzyme or might fail to correctly target the DNA sequence.[6]

3.2.4 Leishmania diagnostics

CRISPR-Cas9 has been utilized for specific diagnosis of *Leishmania* infection. The approach is reliant on identification of a conserved DNA sequence across different species of *Leishmania* parasite. The common consensus sequence is checked among various organisms such as *Trypanosoma* and *Plasmodium* species or in humans via a tool called BLAST, which compares the sequence against known genomes. Selection of DNA sequences that are unique to *Leishmania* parasites and do not match with any other sequences in the genomes of other pathogens or humans are used for CRISPR-Cas9 based detection. T-coffee is a tool that aligns multiple DNA sequences to find a consensus region, which is a sequence shared by all variants of the parasite. These unique sequences are used in the CRISPR-Cas9-based detection method to ensure accurate identification of the *Leishmania* parasite. By following these steps, detection method is specific to *Leishmania* parasite and will avoid false positives from other similar organisms or human DNA. This makes it suitable for point-of-care (POC) testing in areas with limited resources. A conserved 23-mer sequence was chosen from a 115 bp DNA sequence that contains the dCas9 protospacer adjacent motif (PAM) site. This sequence was chosen since it does not have any homology with the human genome or non-Leishmanian pathogens. This target sequence was found in the blood and urine samples of VL patients, but not in the healthy controls, according to PCR analysis. Two DNA extraction methods, M-PER for blood and chitosan-mediated extraction for urine, were employed followed by Recombinase Polymerase Amplification (RPA). Both methods successfully detected the target DNA directly from the crude samples. The sensitivity of the detection scheme was rigorously tested by adding known amounts of the VL target DNA sequence to the reaction, a process known as titration. Streptavidin-coated beads were used to capture the target DNA. The presence of the target DNA was then detected using dCas9, an enzyme that binds specifically to the target sequence. Following this, Rolling Circle Amplification (RCA) was used to amplify the signal, which allowed to observe the results through a colour change that can be seen with the naked eye (Fig. 5B-a). This method proved to be extremely sensitive, capable of detecting as little as 1 zeptomole of DNA, which is approximately 600 DNA molecules. further the complete detection process is done using blood samples that were spiked with the target DNA to simulate real patient samples. The results showed that this method could detect the target DNA at levels between 1 zeptomole and 1 attomole. This study further demonstrated that the detection system is both highly sensitive (able to detect very low amounts of DNA) and specific (accurately identifying the VL target DNA) when applied to clinical samples.[79–81]

3.2.5 CRISPR-Cas9 cytosine base editor (CBE) for large scale screen

CRISPR-Cas9 gene editing offers an alternative, simplifying methods and enabling precise gene knockouts. Tools like LeishGEdit have enhanced loss-of-function experiments by allowing researchers to delete genes in

Leishmania accurately and efficiently. LeishGEdit also facilitates bar-seq screens, where hundreds of mutants are analyzed together, aiding in genome-wide studies. The LeishGEM project aims to knockout approximately all genes in *L. mexicana* to identify key virulence factors using these barcoded mutant pools. However, sharing and maintaining these cell pools is difficult, and their use is limited. Additionally, Leishmania's reliance on homologous recombination for DNA repair, rather than non- homologous end joining (NHEJ) like in other organisms, makes CRISPR targeting inefficient. This results in low mutation rates and the need for additional steps, such as donor DNA or clone isolation, making large-scale screens impractical. Then, a CRISPR-Cas9 cytosine base editor (CBE) is developed that bypasses the need for double-strand breaks (DSBs). Instead, it directly converts cytosine to thymine, introducing STOP codons without requiring homologous recombination or donor DNA. This approach, used successfully in other large-scale screens, combines a cytidine deaminase with an impaired Cas9 to target specific DNA sites and edit them precisely.[82,83]

3.2.5.1 Gene drive

Few experiments have been performed so far, using CRISPR-Cas9 in "Gene drive" technology to genetically regulate the spread of sand flies. In the first study performed by Martin-Martin et al., a comprehensive protocol was devised to apply the CRISPR-Cas9 approach in *Lutzomyia longipalpis* sand fly, elucidating the techniques for microinjecting sand fly embryos necessary for CRISPR-Cas9 gene editing investigations.[84] Yellow genes responsible for pigmentation of sand fly were targeted to produce sgRNA and embryo microinjection protocol. A total of 775 embryos were micro-injected with a mixed composition of 6 sgRNAs targeting the exon 3 of the Yellow gene and Cas9 mRNA throughout the course of three sets of microinjections. For injected embryos, the hatching rate varied from 11.90% to 14.22%, which is in line with studies conducted on other Dipteran species like mosquitoes. However, no mutations were observed in the resultant adults.

A breakthrough was reported by Louradour et al. and colleagues, who successfully used CRISPR-Cas9 to create mutations in sand flies[85] by targeting the relish gene (*rel*), the only transcription factor in the immunological deficiency pathway. They targeted a region encoding the N-terminal domain of the *rel* gene resulted in the generation of transmissible null mutant alleles of the *rel* gene. In this work, four sgRNAs and the Cas9 protein were injected into *Phlebotomus papatasi* embryos, specifically targeting exons 1 and 2 of the *rel* gene (Fig. 5B-c). Multiple *rel* gene mutations were found in mature sand flies that were checked for the presence of mutation(s) using a particular PCR analysis; however, these were lost in later generations due to decreased fitness. Only a small percentage of all injected embryos (11/540) matured, and of those, 8/11 achieved a considerable rate of mutation, suggesting a reasonably good efficiency of the CRISPR-Cas9 approach *in P. papatasi* sand fly gene editing. The vector competence for *Leishmania* depends upon the immune response of the sand flies through the IMD pathway, as it was observed that mutant sandflies were more susceptible to bacteria and larger number of parasites when infected with *L. major*. Therefore, homozygous mutants had a very low probability of survival. The heterozygous mutants exhibited a more permissive phenotype with higher parasite burdens, while the null mutants had more bacteria in their gut than the wild type flies and they were more receptive to *L. major* growth and development. This work validates that by focusing on particular genes in the sand fly population, CRISPR-Cas9 technology may enhance gene drive applications in population reduction and replacement tactics.

4. Trypanosoma

Trypanosomes are flagellated protozoan parasites responsible for a number of important, yet neglected tropical diseases. The two most significantly studied species include Trypanosoma brucei and *T. cruzi*. *T. brucei* is the causative agent of sleeping sickness or Human African Trypanosomiasis. It is transmitted by tsetse flies of the *Glossina* genus mostly prevalent in the poorest rural areas of Africa. Pentamidine, eflornithine, nifurtimox, melarsoprol, and suramin are some of the known drugs but have their own side effects. Trypanosoma infection starts when Tsetse fly injects metacyclic trypomastigotes into the bloodstream of the host. The parasite enters the blood and lymph stream and is transformed into bloodstream trypomastigotes. Bloodstream trypomastigotes reach to other sites through blood fluids. Trypomastigotes replicate by binary fission in the blood. Bloodstream trypomastigotes are transferred to tsetse fly while taking a blood meal. In the tsetse fly midgut, the trypomastigotes are transformed into procyclic trypomastigotes that further divide by binary fission. After leaving the midgut they transform into the epimastigote form. Epimastigote invade the salivary gland where they multiply by binary fission to form metacyclic trypomastigote. It approximately takes 3 weeks to complete the life cycle in tsetse fly.[86] *T. cruzi* is the causative agent of Chagas disease. Its transmission takes place through infected triatomine insects. Its life cycle stages include intracellular amastigotes and bloodstream trypomastigotes in the host and trypomastigotes, epimastigotes, and metacyclic trypomastigotes in the triatomine insect.[86]

The genetic complexity of the parasite and being the cause of the neglected tropical disease are responsible for the least progress in understanding the parasite pathogenicity. Gene manipulation technologies are essential to study the basic parasite biology, pathogenicity mechanisms and drug resistance. The diploid nature of the parasite does not allow homologous recombination to be enough for gene manipulation and most of the genes are encoded by moderate or large families, multigene families which also restrict gene editing in Trypanosoma. Gene manipulation through RNA interference is possible in *T. brucei* while *T. cruzi* lacks the RNAi machinery.

Techniques are available to study the effect of target gene upregulation but there was a lack of a prominent gene editing tool for studying downregulation. The development of CRISPR-Cas9 has revolutionized gene editing techniques in pathogens including kinetoplastids. Here we have discussed CRISPR-Cas systems and gene editing studies which are possible in Trypanosoma due to CRISPR-Cas9.

4.1 CRISPR-Cas9 systems for gene editing in *Trypanosoma*

4.1.1 Constitutive Cas9 expression and T7 RNA polymerase-based system

CRISPR-Cas9 allows targeting of multicopy genes on supernumerary chromosomes which has not been possible hitherto through conventional gene manipulating methods. Stable Cas9 expression through its integration in the genome eliminates repeated transfections with the same plasmid. Cas9 and T7 RNAP expressing *T. brucei* parasites are generated. The engineered cells are transfected with PCR amplified sgRNA that is driven by the T7 promoter.[75] Plasmid containing donor DNA fragment and drug selection marker serve as a template for repair of the double stranded breaks. In this system rapid generation of desired genetic modification is possible without the need for gRNA cloning. This system also enables endogenous gene tagging and gene knockout through transfection of plasmid containing appropriate repair donor DNA. Genes are endogenously tagged at the 5′ and 3′ locus depending on the requirement for a N- or C- terminal fusion target proteins. The tagging allows study of the localization, identification of interacting partners through immuno-precipitation using gene specific antibodies or proximity biotinylation approach that requires fusion with a modified biotin ligase.[87]

4.1.1.1 Ribonucleoprotein system for gene editing

A ribonucleoprotein complex of SaCas9 and in vitro transcribed sgRNA provides a simple, rapid, and cloning- and selection-free gene editing system.[71] It was used for tagging endogenous genes and studying the function of essential genes. This system has two components: Cas9 derived from *S. aureus* and gRNA to guide the nuclease to the complementary target sequence. The lower mass of SaCas9 as compared to SpCas9 improves the efficacy of the RNP complex. This system enables rapid and efficient gene editing in multiple kinetoplastid parasite.[71]

4.1.1.2 Episome-based CRISPR-Cas9 system

In this approach 2 episomes are used for gene editing; one episome contains human codon-optimized *S. pyogenes* Cas9, while the second episome is used for the expression of the desired sgRNA under the control of PARP (procyclic acidic repetitive protein) promoter and a donor DNA repair template that allows incorporation of the desired modification through HDR pathway. To ensure generation of a sgRNA of correct length and with precise ends, it was flanked by a hammerhead ribozyme at its 5-end and a hepatitis delta virus ribozyme at the 3-end. This system was used for endogenous tagging of *scd6* gene with *gfp* DNA sequence, and modification of genes belonging to multigene families. Histones are encoded by multigene families due to which it is difficult to study individual histone modification. Episome-based CRISPR-Cas9 system made it possible to perform histone mutation studies in *Trypanosoma*. H3. V and H4. V were targeted for deletion and both alleles were deleted in a single transfection.[88] Additionally, Lysine 4 in histone 4, H4K4 was replaced with arginine, H4R4 to partially replicate an in-activated chromatic state. Using this approach, out of 43 copies of H4, ~26 could be successfully edited (H4R4). There is no requirement for the insertion of resistance markers in all the studies performed.

4.1.1.3 Tetracycline-induced Cas9 gene editing in *Trypanosoma*

Tet-induced Cas9 is the next approach for gene editing in *T. brucei*. Constitutive Cas9 production can be toxic and can have off-target effects. In this strategy the expression of SpCas9 and sgRNA are under the Tetracycline

(Tet)-inducible ribosomal RNA (rRNA) promoter and T7 promoter, respectively.[89] It is the first inducible Cas9 system in trypanosomatids. It provides 100% efficiency of the target gene disruption and DNA cleavage. The precise 3′ end of sgRNA was generated using HDV ribozyme cloned downstream to sgRNA sequence. The sgRNA was detected only when Cas9 was expressed showing that sgRNA by itself is not stable and requires complex with Cas9 for stability. Due to the diploid nature of Trypanosoma two rounds of electroporation and selection are required for desired gene disruption. However, using the inducible Cas9 expression system, double-allele editing was achieved without the requirement of any selectable marker. Aquaglyceroporin (AQP2) and amino acid transporter 6 (AAT6) genes were targeted for disruption using this method. Single base substitution was introduced in the AQP2 gene (AQP2L264R) without any off-target effect. This inducible system allows highly efficient, precise, and temporally regulated gene editing in trypanosomatids.

4.1.2 Transient CRISPR-Cas9 expression system

Unlike the above mentioned CRISPR-Cas9 system that requires generation of pre-established cell lines for constitutive or inducible expression of Cas9 and T7 polymerase, an all in-one-Cas9 plasmid was constructed to express T7 RNAP and Cas9 under EP procyclin promoter. This system allows transient expression of Cas9 and T7 RNAP genes for desired gene editing and can be used for endogenous tagging and gene knock-out. It requires sequential transfection with a plasmid encoding Cas9 and T7 RNAP, followed by templates for gRNAs expression and repair DNA sequences. The plasmid and sequential transfection protocol allow transient expression of the proteins and guide RNAs.[90] This strategy was used to knock out both the alleles of trypanin or subunit 8 of the GPI-anchor transamidase. The limitation with this system is that it provides fewer clones than pre-established cell lines expressing Cas9 and T7.

4.1.2.1 GFP tagged Cas9 system

In this system *T. cruzi* parasites were generated that express Cas9 endonuclease tagged with GFP protein. GFP fusion allowed selection and enrichment of Cas9-GFP-expressing cells through fluorescence-activated cell sorting. Transfection of Cas9-GFP positive parasites with in vitro transcribed sgRNA against essential genes such as GP72, α-tubulin and β-tubulin allowed phenotypic evaluation of individual knock-out parasites. The system is highly efficient and cost-effective for CRISPR-Cas9 gene editing.[91]

4.2 Applications of CRISPR-Cas9 in *Trypanosoma*

4.2.1 Disruption of multigene family

In 2014 CRISPR-Cas9 was first adapted in *T. cruzi* for the knockout of endogenous genes and members belonging to multigene family. The multiplexing activity is demonstrated by knocking down the expression of enzyme gene family of 65 members leading to a reduction of enzyme product with no apparent off-target mutations.[92] *T. cruzi* was transfected with pTrex plasmids expressing enhanced green fluorescent protein (eGFP) and nucleus localized Cas9 under G418 and blasticidin drug selection, respectively. Multicopy tubulin gene was targeted with sgRNA resulting in parasites with cytokinesis and cell shape defects. Histidine ammonia lyase (HAL) and fatty acid transporter (FATP) gene were also targeted with sgRNA resulting in a 60% decrease in HAL and a 37% decrease in the fatty acid uptake rate.

4.2.2 Functional analysis of *Trypanosoma* genes

Many genes are required for flagellum organization and attachment in *Trypanosoma*. Studying their function by conventional gene disruption strategies is cumbersome in the genetically intractable parasites such as *Trypanosoma*. CRISPR-Cas9 was used to disrupt three different genes required for flagellum organization and attachment.[93] Disruption of paraflagellar rod proteins: PFR1, and PFR2 result in flagellum detachment from the cell body and incomplete development of paraflagellar rods.[93] Similarly, disruption of GP72 caused flagellar detachment however, the paraflagellar rods remained intact. Three different strategies were used to disrupt the target genes: vectors containing single guide RNA (sgRNA) and Cas9, separately or together, or one vector containing sgRNA, Cas9, and homology region. The study shows endogenous gene disruption without cas9 toxicity and the role of PFRI, PFR2, and GP72 in flagellar attachment and motility.[93]

TcTrypanin was knocked out in *T. cruzi* using a ribonucleoprotein complex of SaCas9 and sgRNA plus donor oligonucleotide.[94] Both alleles of TcTrypanin were edited without any selectable marker. The epimastigotes show motility defects, low growth rate, and partially detached flagella with normal numbers of nuclei and kinetoplasts. There was a reduction in the parasite infectivity however, they were able of complete the life cycle under in vitro conditions.[94]

4.2.3 Endogenous tagging of genes to study cellular functions

CRISPR-Cas9 gene editing is also utilized to study the intracellular localization of parasite genes. CRISPR-Cas9 was used to reconfirm the localization of TcFCaBP (flagellar calcium binding protein) and TcVP1 (vacuolar proton pyrophosphatase) proteins while the localization of TcMCU (mitochondrial calcium uniporter) and TcIP3R (inositol 1,4,5-trisphosphate receptor) was reported for the first time using endogenous gene tagging.[95] CRSIPR-Cas9 mediated gene tagging reconfirmed the localization of TcFCaBP and TcVP1 in the flagella and acidocalcisome, respectively. TcMCU colocalized with the voltage-dependent anion channel in the mitochondria while TcIP3R show localization in acidocalcisomes.[95]

4.2.4 Improvement in the host-pathogen interaction studies

T. cruzi reporter strain expressing a fusion protein consisting of red-shifted luciferase and green fluorescent protein domains was generated.[96] Bioluminescence allows infection kinetics to be studied and fluorescence allows investigation of host-parasite interaction of individual parasite at the cellular level. The reporter parasites were made competent for the CRISPR-Cas9 mediated gene editing by incorporating T7 RNAP and Cas9.[75] CRISPR-Cas9 competent reporter parasites did not require cloning of the sgRNA or the donor DNA template for mediating desired gene modification.[96] The system allows generation of null mutants and fluorescently tagged parasites in the background for the assessment of in vivo phenotype. The engineered parasites were used for disruption of flagellar attachment protein GP72 and the endogenous tagging of DNA topoisomerase 1A with mNeonGreen.[96]

4.2.5 Conditional regulation of parasite genes

The glmS ribozyme system for conditional knockdown of target genes was introduced in *Trypanosoma*. The glmS ribozyme is activated upon binding with glucosamine that leads to its cleavage from the fused transcript resulting in destabilization and degradation of the transcript.[97,98] *T. cruzi* glycoprotein 72 (TcGP72) and vacuolar proton pyrophosphatase (TcVP1) were endogenously tagged with glmS ribozyme.[99] Endogenous glucosamine 6-phosphate produced by the parasite is enough to stimulate the glmS ribozyme activity under normal growth conditions.[99] The method is suitable enough to study the knockdown of essential genes in *T. cruzi* and to identify potential drug targets.

5. Concluding remarks and future aspects

In the field of molecular parasitology, CRISPR-Cas9 has ousted previous techniques as the preferred tool for gene editing. This technology has shortened the time required to obtain the transgenic parasite lines, which has been of tremendous help for the study of Apicomplexan parasites. Using different CRISPR-Cas9 systems and strategies, it has become simple, rapid, economical, and more feasible to create gene knockouts, tag endogenous genes, introduce point mutations, and change gene expression in Apicomplexan parasites. CRISPR-Cas9 is also being used in "Gene drive" technology to supress the target vector population or genetically modify them to prevent transmission of pathogenic diseases. With the help of CRISPR-Cas9 based genome wide screening which is dependent on the preparation of sgRNAs libraries, functionally essential genes can be identified in a short period of time and it could aid in the generation protozoan specific knockdown libraries. Although the CRISPR-Cas9 system currently has become one of the most powerful gene editing tools however reducing the off-target effects and Cas9 toxicity need more careful attention. The non-specific cleavage caused by CRISPR effectors due to similarity of the off-target DNA sequence with the guide RNA remain a challenge to solve in near future. Improvement in specificity of CRISPR-Cas9 for different cellular systems has been addressed through guide RNA engineering. Compared to protein engineering, the benefit of engineering guide RNA is that it is simple to screen for the most effective guide DNA sequence. Additionally, to minimize off target effects various database-based *in-silico* technique have been developed to design an optimal gRNA that can reduce in vivo off-target mutations. In order to reduce the off-target effects and enhance the target specificity of the traditional CRISPR-Cas9, enhanced Cas9 endonuclease with mutations in regions that are required for interaction with the target DNA sequence were generated that require strict pairing of the altered Cas9 with the target DNA without much room for mis-matches. Similarly, variants of conventional Cas9 that recognize altered PAM sequences were generated to increase the recognition landscape in the genome for incorporation of desired mutations. *De novo* discoveries of novel RNA-guided endonucleases with superior target selectivity, such as Cas12a and the OMEGA system are being made along with the identification of their DNA editing capabilities. Compared to Cas9, Cas12a has emerged as a desirable alternative in *Plasmodium,* where an abnormally high AT-rich genome is an issue. Ongoing improvement of the current CRISPR-Cas9 technology and its uses to comprehend the human disease-causing protozoan biology will undoubtedly play a major role in enhancing human health.

Acknowledgements

We express our sincere gratitude to the Department of Biotechnology (BT/PR38411/GET/119/311/2020) for providing funds to AB for carrying out studies directly or indirectly related to the chapter content. UG received fellowship as JRF/SRF from DBT. Fellowship support provided to RS by The Department of Biotechnology, Government of India is also acknowledged. The funders have no role in the design and decision to publish this chapter. The authors regret for not being able to cite the work of other colleagues in the field due to space constraints.

References

1. Barrangou R, Fremaux C, Deveau H, et al. CRISPR provides acquired resistance against viruses in prokaryotes. *Science*. 2007;315(5819):1709–1712. https://doi.org/10.1126/science.1138140
2. Jinek M, Chylinski K, Fonfara I, Hauer M, Doudna JA, Charpentier E. A programmable dual-RNA-guided DNA endonuclease in adaptive bacterial immunity. *Science*. 2012;337(6096):816–821. https://doi.org/10.1126/science.1225829
3. Doudna JA, Charpentier E. Genome editing. The new frontier of genome engineering with CRISPR-Cas9. *Science*. 2014;346(6213):1258096. https://doi.org/10.1126/science.1258096
4. Hsu PD, Lander ES, Zhang F. Development and applications of CRISPR-Cas9 for genome engineering. *Cell*. 2014;157(6):1262–1278. https://doi.org/10.1016/j.cell.2014.05.010
5. Sun W, Liu H, Yin W, Qiao J, Zhao X, Liu Y. Strategies for enhancing the homology-directed repair efficiency of CRISPR-Cas systems. *CRISPR J*. 2022;5(1):7–18. https://doi.org/10.1089/crispr.2021.0039
6. Zhang WW, Matlashewski G. CRISPR-Cas9-mediated genome editing in Leishmania donovani. *mBio*. 2015;6(4):e00861. https://doi.org/10.1128/mBio.00861-15
7. Zhang WW, Matlashewski G. Single-strand annealing plays a major role in double-strand DNA break repair following CRISPR-Cas9 cleavage in. *mSphere*. 2019;4(4) https://doi.org/10.1128/mSphere.00408-19
8. Kirkman LA, Lawrence EA, Deitsch KW. Malaria parasites utilize both homologous recombination and alternative end joining pathways to maintain genome integrity. *Nucleic Acids Res*. 2014;42(1):370–379. https://doi.org/10.1093/nar/gkt881
9. Kirti A, Sharma M, Rani K, Bansal A. CRISPRing protozoan parasites to better understand the biology of diseases. *Prog Mol Biol Transl Sci*. 2021;180:21–68. https://doi.org/10.1016/bs.pmbts.2021.01.004
10. Ghorbal M, Gorman M, Macpherson CR, Martins RM, Scherf A, Lopez-Rubio JJ. Genome editing in the human malaria parasite Plasmodium falciparum using the CRISPR-Cas9 system. *Nat Biotechnol*. 2014;32(8):819–821. https://doi.org/10.1038/nbt.2925
11. Wagner JC, Platt RJ, Goldfless SJ, Zhang F, Niles JC. Efficient CRISPR-Cas9-mediated genome editing in Plasmodium falciparum. *Nat Methods*. 2014;11(9):915–918. https://doi.org/10.1038/nmeth.3063
12. Mogollon CM, van Pul FJ, Imai T, et al. Rapid generation of marker-free *P. falciparum* fluorescent reporter lines using modified CRISPR/Cas9 constructs and selection protocol. *PLoS One*. 2016;11(12):e0168362. https://doi.org/10.1371/journal.pone.0168362
13. Lu J, Tong Y, Pan J, et al. A redesigned CRISPR/Cas9 system for marker-free genome editing in Plasmodium falciparum. *Parasit Vectors*. 2016;9:198. https://doi.org/10.1186/s13071-016-1487-4
14. Lu J, Tong Y, Dong R, et al. Large DNA fragment knock-in and sequential gene editing in Plasmodium falciparum: a preliminary study using suicide-rescue-based CRISPR/Cas9 system. *Mol Cell Biochem*. 2024;479(1):99–107. https://doi.org/10.1007/s11010-023-04711-5

15. Payungwoung T, Shinzawa N, Hino A, et al. CRISPR/Cas9 system in Plasmodium falciparum using the centromere plasmid. *Parasitol Int*. 2018;67(5):605–608. https://doi.org/10.1016/j.parint.2018.06.002
16. Zhao Y, Wang F, Wang C, et al. Optimization of CRISPR/Cas system for improving genome editing efficiency in. *Front Microbiol*. 2020;11:625862. https://doi.org/10.3389/fmicb.2020.625862
17. Nishi T, Shinzawa N, Yuda M, Iwanaga S. Highly efficient CRISPR/Cas9 system in Plasmodium falciparum using Cas9-expressing parasites and a linear donor template. *Sci Rep*. 2021;11(1):18501. https://doi.org/10.1038/s41598-021-97984-z
18. Kuang D, Qiao J, Li Z, et al. Tagging to endogenous genes of Plasmodium falciparum using CRISPR/Cas9. *Parasit Vectors*. 2017;10(1):595. https://doi.org/10.1186/s13071-017-2539-0
19. Crawford ED, Quan J, Horst JA, Ebert D, Wu W, DeRisi JL. Plasmid-free CRISPR/Cas9 genome editing in Plasmodium falciparum confirms mutations conferring resistance to the dihydroisoquinolone clinical candidate SJ733. *PLoS One*. 2017;12(5):e0178163. https://doi.org/10.1371/journal.pone.0178163
20. Kudyba HM, Cobb DW, Florentin A, Krakowiak M, Muralidharan V. CRISPR/Cas9 gene editing to make conditional mutants of human malaria parasite P. falciparum. *J Vis Exp*. 2018(139), https://doi.org/10.3791/57747
21. Liu H, Cui XY, Xu DD, et al. Actin-related protein Arp4 regulates euchromatic gene expression and development through H2A.Z deposition in blood-stage Plasmodium falciparum. *Parasit Vectors*. 2020;13(1):314. https://doi.org/10.1186/s13071-020-04139-6
22. Knuepfer E, Napiorkowska M, van Ooij C, Holder AA. Generating conditional gene knockouts in Plasmodium – a toolkit to produce stable DiCre recombinase-expressing parasite lines using CRISPR/Cas9. *Sci Rep*. 2017;7(1):3881. https://doi.org/10.1038/s41598-017-03984-3
23. Jones ML, Das S, Belda H, Collins CR, Blackman MJ, Treeck M. A versatile strategy for rapid conditional genome engineering using loxP sites in a small synthetic intron in Plasmodium falciparum. *Sci Rep*. 2016;6:21800. https://doi.org/10.1038/srep21800
24. Boonyalai N, Collins CR, Hackett F, Withers-Martinez C, Blackman MJ. Essentiality of Plasmodium falciparum plasmepsin V. *PLoS One*. 2018;13(12):e0207621. https://doi.org/10.1371/journal.pone.0207621
25. Bahl V, Chaddha K, Mian SY, Holder AA, Knuepfer E, Gaur D. Genetic disruption of Plasmodium falciparum Merozoite surface antigen 180 (PfMSA180) suggests an essential role during parasite egress from erythrocytes. *Sci Rep*. 2021;11(1):19183. https://doi.org/10.1038/s41598-021-98707-0
26. Hitz E, Grüninger O, Passecker A, et al. The catalytic subunit of Plasmodium falciparum casein kinase 2 is essential for gametocytogenesis. *Commun Biol*. 2021;4(1):336. https://doi.org/10.1038/s42003-021-01873-0
27. Wilde ML, Triglia T, Marapana D, et al. Protein kinase A is essential for invasion of Plasmodium falciparum into human erythrocytes. *mBio*. 2019;10(5) https://doi.org/10.1128/mBio.01972-19
28. Xiao B, Yin S, Hu Y, et al. Epigenetic editing by CRISPR/dCas9 in *Plasmodium falciparum*. *Proc Natl Acad Sci U S A*. 2019;116(1):255–260. https://doi.org/10.1073/pnas.1813542116
29. Liang X, Boonhok R, Siddiqui FA, et al. A leak-free inducible CRISPRi/a system for gene functional studies in Plasmodium falciparum. *Microbiol Spectr*. 2022;10(3):e0278221. https://doi.org/10.1128/spectrum.02782-21
30. Lee RA, Puig H, Nguyen PQ, et al. Ultrasensitive CRISPR-based diagnostic for field-applicable detection of. *Proc Natl Acad Sci U S A*. 2020;117(41):25722–25731. https://doi.org/10.1073/pnas.2010196117

31. Cunningham CH, Hennelly CM, Lin JT, et al. A novel CRISPR-based malaria diagnostic capable of Plasmodium detection, species differentiation, and drug-resistance genotyping. *EBioMedicine*. 2021;68:103415https://doi.org/10.1016/j.ebiom.2021.103415
32. Zheng M, Zhang M, Li H, et al. Rapid, sensitive, and convenient detection of Plasmodium falciparum infection based on CRISPR and its application in detection of asymptomatic infection. *Acta Trop*. 2024;249:107062. https://doi.org/10.1016/j.actatropica.2023.107062
33. Tajudeen YA, Oladipo HJ, Oladunjoye IO, et al. Transforming malaria prevention and control: the prospects and challenges of gene drive technology for mosquito management. *Ann Med*. 2023;55(2):2302504. https://doi.org/10.1080/07853890.2024.2302504
34. Green EI, Jaouen E, Klug D, et al. A population modification gene drive targeting both. *Elife*. 2023;12. https://doi.org/10.7554/eLife.93142
35. Gantz VM, Jasinskiene N, Tatarenkova O, et al. Highly efficient Cas9-mediated gene drive for population modification of the malaria vector mosquito Anopheles stephensi. *Proc Natl Acad Sci U S A*. 2015;112(49):E6736–E6743. https://doi.org/10.1073/pnas.1521077112
36. Carballar-Lejarazú R, Dong Y, Pham TB, et al. Dual effector population modification gene-drive strains of the African malaria mosquitoes. *Proc Natl Acad Sci U S A*. 2023;120(29):e2221118120. https://doi.org/10.1073/pnas.2221118120
37. Hoermann A, Tapanelli S, Capriotti P, et al. Converting endogenous genes of the malaria mosquito into simple non-autonomous gene drives for population replacement. *Elife*. 2021;10. https://doi.org/10.7554/eLife.58791
38. Hoermann A, Habtewold T, Selvaraj P, et al. Gene drive mosquitoes can aid malaria elimination by retarding. *Sci Adv*. 2022;8(38):eabo1733. https://doi.org/10.1126/sciadv.abo1733
39. Portugaliza HP, Llorà-Batlle O, Rosanas-Urgell A, Cortés A. Reporter lines based on the gexp02 promoter enable early quantification of sexual conversion rates in the malaria parasite Plasmodium falciparum. *Sci Rep*. 2019;9(1):14595. https://doi.org/10.1038/s41598-019-50768-y
40. Boltryk SD, Passecker A, Alder A, et al. CRISPR/Cas9-engineered inducible gametocyte producer lines as a valuable tool for Plasmodium falciparum malaria transmission research. *Nat Commun*. 2021;12(1):4806. https://doi.org/10.1038/s41467-021-24954-4
41. Bryant JM, Baumgarten S, Dingli F, et al. Exploring the virulence gene interactome with CRISPR/dCas9 in the human malaria parasite. *Mol Syst Biol*. 2020;16(8):e9569. https://doi.org/10.15252/msb.20209569
42. Barcons-Simon A, Cordon-Obras C, Guizetti J, Bryant JM, Scherf A. CRISPR interference of a clonally variant GC-rich noncoding RNA family leads to general repression of. *mBio*. 2020;11(1). https://doi.org/10.1128/mBio.03054-19
43. Cova M, López-Gutiérrez B, Artigas-Jerónimo S, et al. The apicomplexa-specific glucosamine-6-phosphate N-acetyltransferase gene family encodes a key enzyme for glycoconjugate synthesis with potential as therapeutic target. *Sci Rep*. 2018;8(1):4005. https://doi.org/10.1038/s41598-018-22441-3
44. Sidik SM, Huet D, Ganesan SM, et al. A genome-wide CRISPR screen in toxoplasma identifies essential apicomplexan genes. *Cell*. 2016;166(6):1423–1435.e12. https://doi.org/10.1016/j.cell.2016.08.019
45. Bansal A, Molina-Cruz A, Brzostowski J, et al. CDPK1 is critical for malaria parasite gametogenesis and mosquito infection. *Proc Natl Acad Sci U S A*. 2018;115(4):774–779. https://doi.org/10.1073/pnas.1715443115
46. Bansal A, Ojo KK, Mu J, Maly DJ, Van Voorhis WC, Miller LH. Reduced activity of mutant calcium-dependent protein kinase 1 is compensated in Plasmodium falciparum through the action of protein kinase G. *mBio*. 2016;7(6). https://doi.org/10.1128/mBio.02011-16
47. Bansal A, Molina-Cruz A, Brzostowski J, Mu J, Miller LH. Calcium-dependent protein kinase 2 is critical for male gametocyte exflagellation but not essential for asexual proliferation. *mBio*. 2017;8(5). https://doi.org/10.1128/mBio.01656-17

48. Ekka R, Gupta A, Bhatnagar S, Malhotra P, Sharma P. Phosphorylation of rhoptry protein RhopH3 is critical for host cell invasion by the malaria parasite. *mBio*. 2020;11(5). https://doi.org/10.1128/mBio.00166-20
49. Marin-Mogollon C, van de Vegte-Bolmer M, van Gemert GJ, et al. The Plasmodium falciparum male gametocyte protein P230p, a paralog of P230, is vital for ookinete formation and mosquito transmission. *Sci Rep*. 2018;8(1):14902. https://doi.org/10.1038/s41598-018-33236-x
50. Inklaar MR, de Jong RM, Bekkering ET, et al. Pfs230 domain 7 is targeted by a potent malaria transmission-blocking monoclonal antibody. *NPJ Vaccines*. 2023;8(1):186. https://doi.org/10.1038/s41541-023-00784-x
51. Tan S, Mudeppa DG, Kokkonda S, White J, Patrapuvich R, Rathod PK. Properties of Plasmodium falciparum with a deleted apicoplast DNA gyrase. *Antimicrob Agents Chemother*. 2021;65(9):e0058621. https://doi.org/10.1128/AAC.00586-21
52. Florentin A, Stephens DR, Brooks CF, Baptista RP, Muralidharan V. Plastid biogenesis in malaria parasites requires the interactions and catalytic activity of the Clp proteolytic system. *Proc Natl Acad Sci U S A*. 2020;117(24):13719–13729. https://doi.org/10.1073/pnas.1919501117
53. LaMonte G, Lim MY, Wree M, et al. Mutations in the Plasmodium falciparum cyclic amine resistance locus (PfCARL) confer multidrug resistance. *mBio*. 2016;7(4). https://doi.org/10.1128/mBio.00696-16
54. Schalkwijk J, Allman EL, Jansen PAM, et al. Antimalarial pantothenamide metabolites target acetyl–coenzyme A biosynthesis in *Plasmodium falciparum*. *Sci Transl Med*. 2019;11(510). https://doi.org/10.1126/scitranslmed.aas9917
55. Ashley EA, Dhorda M, Fairhurst RM, et al. Spread of artemisinin resistance in Plasmodium falciparum malaria. *N Engl J Med*. 2014;371(5):411–423. https://doi.org/10.1056/NEJMoa1314981
56. Straimer J, Gnädig NF, Witkowski B, et al. Drug resistance. K13-propeller mutations confer artemisinin resistance in Plasmodium falciparum clinical isolates. *Science*. 2015;347(6220):428–431. https://doi.org/10.1126/science.1260867
57. Breglio KF, Amato R, Eastman R, et al. A single nucleotide polymorphism in the Plasmodium falciparum atg18 gene associates with artemisinin resistance and confers enhanced parasite survival under nutrient deprivation. *Malar J*. 2018;17(1):391. https://doi.org/10.1186/s12936-018-2532-x
58. Dans MG, Weiss GE, Wilson DW, et al. Screening the Medicines for Malaria Venture Pathogen Box for invasion and egress inhibitors of the blood stage of Plasmodium falciparum reveals several inhibitory compounds. *Int J Parasitol*. 2020;50(3):235–252. https://doi.org/10.1016/j.ijpara.2020.01.002
59. Nguyen W, Dans MG, Ngo A, et al. Structure activity refinement of phenylsulfonyl piperazines as antimalarials that block erythrocytic invasion. *Eur J Med Chem*. 2021;214:113253. https://doi.org/10.1016/j.ejmech.2021.113253
60. Dans MG, Piirainen H, Nguyen W, et al. Sulfonylpiperazine compounds prevent Plasmodium falciparum invasion of red blood cells through interference with actin-1/profilin dynamics. *PLoS Biol*. 2023;21(4):e3002066. https://doi.org/10.1371/journal.pbio.3002066
61. Dong Y, Simões ML, Marois E, Dimopoulos G. CRISPR/Cas9-mediated gene knockout of Anopheles gambiae FREP1 suppresses malaria parasite infection. *PLoS Pathog*. 2018;14(3):e1006898. https://doi.org/10.1371/journal.ppat.1006898
62. Yang J, Schleicher TR, Dong Y, et al. Disruption of mosGILT in Anopheles gambiae impairs ovarian development and Plasmodium infection. *J Exp Med*. 2020;217(1) https://doi.org/10.1084/jem.20190682
63. Simões ML, Dong Y, Mlambo G, Dimopoulos G. C-type lectin 4 regulates broad-spectrum melanization-based refractoriness to malaria parasites. *PLoS Biol*. 2022;20(1):e3001515. https://doi.org/10.1371/journal.pbio.3001515

64. Ivens AC, Peacock CS, Worthey EA, et al. The genome of the kinetoplastid parasite, Leishmania major. *Science*. 2005;309(5733):436–442. https://doi.org/10.1126/science.1112680
65. Cruz A, Beverley SM. Gene replacement in parasitic protozoa. *Nature*. 1990;348(6297):171–173. https://doi.org/10.1038/348171a0
66. Cruz AK, Titus R, Beverley SM. Plasticity in chromosome number and testing of essential genes in Leishmania by targeting. *Proc Natl Acad Sci U S A*. 1993;90(4):1599–1603. https://doi.org/10.1073/pnas.90.4.1599
67. Dean S, Sunter J, Wheeler RJ, Hodkinson I, Gluenz E, Gull K. A toolkit enabling efficient, scalable and reproducible gene tagging in trypanosomatids. *Open Biol*. 2015;5(1):140197. https://doi.org/10.1098/rsob.140197
68. Papadopoulou B, Dumas C. Parameters controlling the rate of gene targeting frequency in the protozoan parasite Leishmania. *Nucleic Acids Res*. 1997;25(21):4278–4286. https://doi.org/10.1093/nar/25.21.4278
69. Sterkers Y, Lachaud L, Crobu L, Bastien P, Pagès M. FISH analysis reveals aneuploidy and continual generation of chromosomal mosaicism in Leishmania major. *Cell Microbiol*. 2011;13(2):274–283. https://doi.org/10.1111/j.1462-5822.2010.01534.x
70. Lye LF, Owens K, Shi H, et al. Retention and loss of RNA interference pathways in trypanosomatid protozoans. *PLoS Pathog*. 2010;6(10):e1001161. https://doi.org/10.1371/journal.ppat.1001161
71. Soares Medeiros LC, South L, Peng D, et al. Rapid, selection-free, high-efficiency genome editing in protozoan parasites using CRISPR-Cas9 ribonucleoproteins. *mBio*. 2017;8(6) https://doi.org/10.1128/mBio.01788-17
72. Asencio C, Hervé P, Morand P, et al. Streptococcus pyogenes Cas9 ribonucleoprotein delivery for efficient, rapid and marker-free gene editing in Trypanosoma and Leishmania. *Mol Microbiol*. 2024;121(6):1079–1094. https://doi.org/10.1111/mmi.15256
73. Sollelis L, Ghorbal M, MacPherson CR, et al. First efficient CRISPR-Cas9-mediated genome editing in Leishmania parasites. *Cell Microbiol*. 2015;17(10):1405–1412. https://doi.org/10.1111/cmi.12456
74. Zhang WW, Lypaczewski P, Matlashewski G. Optimized CRISPR-Cas9 genome editing for leishmania and its use to target a multigene family, induce chromosomal translocation, and study DNA break repair mechanisms. *mSphere*. 2017;2(1). https://doi.org/10.1128/mSphere.00340-16
75. Beneke T, Madden R, Makin L, Valli J, Sunter J, Gluenz E. A CRISPR Cas9 high-throughput genome editing toolkit for kinetoplastids. *R Soc Open Sci*. 2017;4(5):170095. https://doi.org/10.1098/rsos.170095
76. Fernandez-Prada C, Sharma M, Plourde M, et al. High-throughput Cos-Seq screen with intracellular Leishmania infantum for the discovery of novel drug-resistance mechanisms. *Int J Parasitol Drugs Drug Resist*. 2018;8(2):165–173. https://doi.org/10.1016/j.ijpddr.2018.03.004
77. Damasceno JD, Reis-Cunha J, Crouch K, et al. Conditional knockout of RAD51-related genes in Leishmania major reveals a critical role for homologous recombination during genome replication. *PLoS Genet*. 2020;16(7):e1008828. https://doi.org/10.1371/journal.pgen.1008828
78. Zhang WW, Matlashewski G. Deletion of an ATP-binding cassette protein subfamily C transporter in Leishmania donovani results in increased virulence. *Mol Biochem Parasitol*. 2012;185(2):165–169. https://doi.org/10.1016/j.molbiopara.2012.07.006
79. Bengtson M, Bharadwaj M, Franch O, et al. CRISPR-dCas9 based DNA detection scheme for diagnostics in resource-limited settings. *Nanoscale*. 2022;14(5):1885–1895. https://doi.org/10.1039/d1nr06557b
80. Daugelaite J, O'Driscoll A, Sleator RD. An overview of multiple sequence alignments and cloud computing in bioinformatics. *Int Sch Res Not*. 2013:615630.

81. Rogers WO, Wirth DF. Kinetoplast DNA minicircles: regions of extensive sequence divergence. *Proc Natl Acad Sci U S A*. 1987;84(2):565–569. https://doi.org/10.1073/pnas.84.2.565
82. Engstler M, Beneke T. Gene editing and scalable functional genomic screening in *Leishmania* species using the CRISPR/Cas9 cytosine base editor toolbox LeishBASEedit. *Elife*. 2023;12. https://doi.org/10.7554/eLife.85605
83. Huang TP, Newby GA, Liu DR. Precision genome editing using cytosine and adenine base editors in mammalian cells. *Nat Protoc*. 2021;16(2):1089–1128. https://doi.org/10.1038/s41596-020-00450-9
84. Martin-Martin I, Aryan A, Meneses C, Adelman ZN, Calvo E. Optimization of sand fly embryo microinjection for gene editing by CRISPR/Cas9. *PLoS Negl Trop Dis*. 2018;12(9):e0006769. https://doi.org/10.1371/journal.pntd.0006769
85. Louradour I, Ghosh K, Inbar E, Sacks DL. CRISPR/Cas9 mutagenesis in Phlebotomus papatasi: the immune deficiency pathway impacts vector competence for Leishmania major. *mBio*. 2019;10(4). https://doi.org/10.1128/mBio.01941-19
86. Martín-Escolano J, Marín C, Rosales MJ, Tsaousis AD, Medina-Carmona E, Martín-Escolano R. An updated view of the *Trypanosoma cruzi* life cycle: intervention points for an effective treatment. *ACS Infect Dis*. 2022;8(6):1107–1115. https://doi.org/10.1021/acsinfecdis.2c00123
87. Roux KJ, Kim DI, Raida M, Burke B. A promiscuous biotin ligase fusion protein identifies proximal and interacting proteins in mammalian cells. *J Cell Biol*. 2012;196(6):801–810. https://doi.org/10.1083/jcb.201112098
88. Vasquez JJ, Wedel C, Cosentino RO, Siegel TN. Exploiting CRISPR-Cas9 technology to investigate individual histone modifications. *Nucleic Acids Res*. 2018;46(18):e106. https://doi.org/10.1093/nar/gky517
89. Rico E, Jeacock L, Kovářová J, Horn D. Inducible high-efficiency CRISPR-Cas9-targeted gene editing and precision base editing in African trypanosomes. *Sci Rep*. 2018;8(1):7960. https://doi.org/10.1038/s41598-018-26303-w
90. Shaw S, Knüsel S, Hoenner S, Roditi I. A transient CRISPR/Cas9 expression system for genome editing in Trypanosoma brucei. *BMC Res Notes*. 2020;13(1):268. https://doi.org/10.1186/s13104-020-05089-z
91. Romagnoli BAA, Picchi GFA, Hiraiwa PM, Borges BS, Alves LR, Goldenberg S. Improvements in the CRISPR/Cas9 system for high efficiency gene disruption in Trypanosoma cruzi. *Acta Trop*. 2018;178:190–195. https://doi.org/10.1016/j.actatropica.2017.11.013
92. Peng D, Kurup SP, Yao PY, Minning TA, Tarleton RL. CRISPR-Cas9-mediated single-gene and gene family disruption in Trypanosoma cruzi. *mBio*. 2014;6(1):e02097-14. https://doi.org/10.1128/mBio.02097-14
93. Lander N, Li ZH, Niyogi S, Docampo R. CRISPR/Cas9-induced disruption of paraflagellar rod protein 1 and 2 genes in Trypanosoma cruzi reveals their role in flagellar attachment. *mBio*. 2015;6(4):e01012. https://doi.org/10.1128/mBio.01012-15
94. Saenz-Garcia JL, Borges BS, Souza-Melo N, et al. Trypanin disruption affects the motility and infectivity of the protozoan. *Front Cell Infect Microbiol*. 2021;11:807236. https://doi.org/10.3389/fcimb.2021.807236
95. Lander N, Chiurillo MA, Storey M, Vercesi AE, Docampo R. CRISPR/Cas9-mediated endogenous C-terminal tagging of Trypanosoma cruzi genes reveals the acidocalcisome localization of the inositol 1,4,5-trisphosphate receptor. *J Biol Chem*. 2016;291(49):25505–25515. https://doi.org/10.1074/jbc.M116.749655
96. Costa FC, Francisco AF, Jayawardhana S, et al. Expanding the toolbox for Trypanosoma cruzi: a parasite line incorporating a bioluminescence-fluorescence dual reporter and streamlined CRISPR/Cas9 functionality for rapid in vivo localisation and phenotyping. *PLoS Negl Trop Dis*. 2018;12(4):e0006388. https://doi.org/10.1371/journal.pntd.0006388

97. Prommana P, Uthaipibull C, Wongsombat C, et al. Inducible knockdown of Plasmodium gene expression using the glmS ribozyme. *PLoS One*. 2013;8(8):e73783. https://doi.org/10.1371/journal.pone.0073783
98. Watson PY, Fedor MJ. The glmS riboswitch integrates signals from activating and inhibitory metabolites in vivo. *Nat Struct Mol Biol*. 2011;18(3):359–363. https://doi.org/10.1038/nsmb.1989
99. Lander N, Cruz-Bustos T, Docampo R. A CRISPR/Cas9-riboswitch-based method for downregulation of gene expression in *Trypanosoma cruzi*. *Front Cell Infect Microbiol*. 2020;10:68. https://doi.org/10.3389/fcimb.2020.00068

CHAPTER SEVEN

Advances in CRISPR/Cas systems-based cell and gene therapy

Arpita Poddar[a,b,c], Farah Ahmady[a,b], Prashanth Prithviraj[a,b], Rodney B. Luwor[a,b,d,e], Ravi Shukla[c], Shakil Ahmed Polash[c], Haiyan Li[c], Suresh Ramakrishna[f], George Kannourakis[a,b], and Aparna Jayachandran[a,b,*]

[a]Fiona Elsey Cancer Research Institute, VIC, Australia
[b]Federation University, VIC, Australia
[c]RMIT University, VIC, Australia
[d]The University of Melbourne, Parkville, VIC, Australia
[e]Huagene Institute, Kecheng Science and Technology Park, Pukou, Nanjing, P.R. China
[f]Hanyang University, Seoul, South Korea
*Corresponding author. e-mail address: aparna@fecri.org.au

Contents

Progress in Molecular Biology and Translational Science, Volume 208
ISSN 1877-1173, https://doi.org/10.1016/bs.pmbts.2024.07.005

Abstract

Cell and gene therapy are innovative biomedical strategies aimed at addressing diseases at their genetic origins. CRISPR (Clustered Regularly Interspaced Short Palindromic Repeats) systems have become a groundbreaking tool in cell and gene therapy, offering unprecedented precision and versatility in genome editing. This chapter explores the role of CRISPR in gene editing, tracing its historical development and discussing biomolecular formats such as plasmid, RNA, and protein-based approaches. Next, we discuss CRISPR delivery methods, including viral and non-viral vectors, followed by examining the various engineered CRISPR variants for their potential in gene therapy. Finally, we outline emerging clinical applications, highlighting the advancements in CRISPR for breakthrough medical treatments.

1. Introduction

Cell and gene therapy are innovative fields in biomedical science with an immense potential to transform the treatment of various diseases, especially those with genetic origins. These therapies are fundamentally based on gene editing, a precise and targeted method of genetic modification that alters the DNA sequence within an organism's genome.[1,2] Therapeutic applications of gene editing often involve introducing mutations to gene-related sequences in order to replace, silence or knock down aberrant genes. Clustered Regularly Interspaced Short Palindromic Repeats (CRISPR)/CRISPR associated (Cas) nucleases is an emerging technology that has reformed the landscape of cell and gene therapy, offering unique precision and versatility in genome editing. Beginning with its discovery as a bacterial immune defence mechanism, CRISPR/Cas systems have since been re-engineered into exceptional methods for precisely modifying DNA sequences in various organisms, including humans, and revolutionised the scope of gene-editing based therapeutics.[3,4]

In this chapter, we will begin in Section 1 by introducing the history of gene therapy leading to the development of CRISPR technology, providing an overview of its journey from discovery to its current status as a transformative tool in biomedical research. In Section 2, we will then discuss the various biomolecular formats available for implementing CRISPR-mediated approaches, examining the unique characteristics of plasmid, RNA, and protein-based formats. Next, Section 3 will explore the critical considerations in CRISPR delivery methods, with viral and non-viral vectors emerging as the primary strategies. We will discuss the advantages and limitations of the available methods, guiding the selection of delivery strategies for various research and therapeutic interventions.

In Section 4, we will explore the landscape of engineered CRISPR variants, including novel Cas proteins and enhanced nucleases, highlighting their features, applications, and potential implications for advancing research on gene therapy. Subsequently, we will provide an overview of the emerging landscape of clinical applications of CRISPR gene therapy. This highlights the great potential for treating a spectrum of gene-based illnesses, cancer, and infectious diseases, and marks a crucial step towards realising the full therapeutic potential of this novel technology.

1.1 History and background

The concept of gene therapy traces back to the mid-19th to 20th centuries, emerging from the intersections of genetics, molecular biology, and clinical medicine.[5] Key milestones include Gregor Mendel's pioneering experiments on inheritance patterns in the 1850s, the connection of DNA as the genetic material in 1944, and the elucidation of DNA structure by James Watson and Francis Crick in 1953.[6,7] Further progress came with the discovery of enzymes capable of isolating and reinserting genes at specific sites along the DNA. The dawn of recombinant DNA technology in the 1970s enabled the cloning of specific genes and the exogenous production of proteins such as insulin.[8,9] These advancements demonstrated the feasibility of manipulating genes, providing engineering pathways for the growth of gene therapy.

The modern era of gene editing arguably began in the 1980s and 1990s, with the development of site-specific nucleases. These enzymes could introduce double-strand breaks (DSBs) at specific DNA locations, prompting the cell's natural repair mechanisms to fix the break, often leading to targeted mutations or gene insertions. Zinc-Finger Nucleases (ZFNs), meganucleases, and Transcription Activator-Like Effector Nucleases (TALENs) were initially investigated.[10–12] However, the most transformative breakthrough in site-specific nucleases came with the discovery of the CRISPR/Cas9 system.

Originally discovered as an adaptive immune system in prokaryotes such as bacteria and archaea, CRISPR sequences, and their associated Cas proteins, protect prokaryotic organisms against viral infections by cutting foreign DNA.[13] Subsequently, adaptation of the bacterial system for application in eukaryotic cells to mediate precise gene editing was demonstrated by Jennifer Doudna, Emmanuelle Charpentier, and their colleagues in 2012.[14]

1.2 Utility in gene therapy

CRISPR technology is bypassing formerly utilised gene editing tools by providing optimal balance between efficiency, targeting precision, versatility, flexibility and cost of use.[14–17] Because of its high adaptability and simplicity, it has been recognised as the most promising strategy for many biomedical applications, especially gene therapy. The following section summarises the unique features that highlights the benefits of the CRISPR/Cas system:

1.2.1 Precision in genome editing

The guide RNA (gRNA) component of the CRISPR system, explored in detail in subsequent sections, allows for precise targeting of specific DNA sequences within the genome. The edits can be as precise as single-base-pair changes or highly specific indels (insertions or deletions) varying from single nucleotides to full gene regulatory sites.[18]

1.2.2 Functional adaptability

CRISPR possess a remarkable functional versatility and is adaptable to an expansive range of gene therapy applications. The basis of such versatility can be attributable to exaptation, a process in which CRISPR variants or components lose primary adaptive immunity functions and gain non-defence functions. Such extensive diversifications allow for targeting of multiple genes in parallel, enabling complex genetic modifications and combinatorial therapies for multifactorial diseases.[19]

1.2.3 Efficiency of variants

The CRISPR system presents with greater efficiency and straightforwardness of implementation as compared to ZFNs and TALENs.[20] However, the first wave of engineered CRISPR variants were accompanied by off-target effects stemming from inadequate design, low specificity or prolonged expression of Cas9. Since then, significant advancements in terms of CRISPR/Cas types, variants, editing time and temperatures as well as modifications to the DSB repair pathways have streamlined the process and greatly enhanced efficiency; facilitating the development of personalised gene therapy approaches.[21]

1.3 CRISPR/Cas subtypes: prokaryotic origins and eukaryotic adaptation

The CRISPR/Cas endonuclease complex is an acquired immune system in prokaryotes for resistance to exogenous viruses or nucleic acids. The CRISPR genomic locus in bacteria comprises a CRISPR sequence array of repeats and spacers, and an upstream operon region with the genes coding for Cas proteins.[22] Based on the composition of Cas proteins and the organisation of CRISPR arrays, currently two classes and six types of

CRISPR systems are known, with multiple subtypes within each type (Table 1). Class 1 is represented by multi-subunit effector Cas protein domains and is divided into Types I, III, and IV. Class 2 consists of a single subunit effector Cas protein and is divided into Types II, V, and VI.[23]

Type I systems, characterised within Class 1, include Cas3 and utilise a multi-subunit complex for identifying the target, interference, and cleavage.[24] Type II systems are characterised by the occurrence of a single, multidomain Cas protein of Class 2 which mediates target recognition and cleavage. The most well-known subtype within Type II is the CRISPR/Cas9 system and it has been extensively studied and harnessed for genome editing applications.[25] Type III systems, including Cas10, are involved in both RNA interference (RNAi) and DNA interference.[26] Type IV systems, such as Cas5, Cas7 and Cas8, mediates RNA-guided DNA interference.[27] Type V systems are characterised by the presence of a single, multidomain Cas protein, such as Cas12, which mediates target recognition and cleavage.[28] Type VI systems, such as Cas13, mediate RNAi and is involved in targeting and cleaving of RNA molecules.[29,30]

Overall, CRISPR/Cas in prokaryotic organisms is a pathway for adaptive immunity against invading genetic elements, and their mechanisms of action vary between different types and subtypes. Since discovery, these pathways have been remodelled for genome editing applications in various eukaryotic organisms, and have been researched in multiple cell

Table 1 CRISPR/Cas classification—known classes and types.

CRISPR class	Cas component	CRISPR/ Cas type	Nuclease example	Nuclease target	References
1	Multi-subunit effector	I	Cas3	DNA	24
		III	Cas10	ssDNA and RNA	26
		IV	Cas5, Cas7, Cas8	DNA	27
2	Single-subunit effector	II	Cas9	DNA, can be modified to target RNA	25
		V	Cas12, Cas14	DNA, ssDNA, RNA (Cas12g)	28
		VI	Cas13	RNA	30

lines and tissues.[31] The most widely recognised system is the CRISPR/Cas9 complex, which is one of the best characterised for laboratory use to clinical applications.

For eukaryotic applications in target or host cells, engineered CRISPR/Cas9 consists of two key components that edit the DNA.[29] The first is the Cas9 enzyme, a nuclease which introduces DSBs in a specific target gene. As the host cell tries to repair this break, it uses the extremely error prone non-homologous end joining (NHEJ) and DSB repair systems. As a result of which the gene gets mutated to become inactive. The second component is an RNA sequence containing a hairpin loop linked to a 20-base pair guide RNA (gRNA). This RNA sequence directs Cas9 to recognise the target DNA complementary to the gRNA sequence and introduce a cut following recognition. Multiple gRNA sequences can be predesigned to match the target gene, and on simply pairing them with Cas9, efficient knockdown systems can be easily generated.[32–34]

For therapeutic gene editing, the key component of CRISPR/Cas is thus comprised of the Cas/gRNA ribonucleoprotein complex. This complex carries out the intended genomic modifications and must enter the nucleus of eukaryotic cells. To achieve this, the following section highlights the various biomolecular formats that can be chosen and utilised for intracellular delivery of CRISPR mediated gene therapy.

2. Delivery formats

There are three delivery formats for CRISPR/Cas: plasmid, mRNA, and protein/ribonucleoproteins (RNPs). Each of these formats have unique characteristics which dictate their selection and application (Table 2).

2.1 Plasmid systems

The plasmid format of CRISPR/Cas technology involves encoding both the Cas protein and the single guide RNA (sgRNA) on a single plasmid construct. This approach simplifies the gene-editing process by eliminating the need to transfect multiple separate components, thereby making it the simplest, most stable, and cost-effective option among the available formats.[35,36] In this system, the plasmid is introduced into the target cells, where it utilises the cellular machinery for transcription and translation to produce the Cas protein and sgRNA.[35,37] However, since transcription and translation occur within the cell, there can be variability in plasmid

Table 2 CRISPR/Cas delivery formats.

Format	Advantages	Disadvantages	Expression	References
Plasmid	Single component transfection Stable and easy engineering Cost efficient	Off-target effects Host genome integration	Least transient	35
mRNA	No genomic integration Lower off-target effects	Less stable and degradation prone Compromised editing efficiency	Intermediate transient	38
Protein (RNP)	Rapid Least off-target effects Minimal genomic integration	Complex transfection Repeated delivery required High cost	Most transient	48

delivery to the nucleus, potential integration into the host genome, and off-target effects.[38,39] Despite these challenges, the plasmid-based approach remains widely used and valuable in CRISPR/Cas applications due to its simplicity, cost-efficiency, and the ability to streamline transfection by eliminating the need for multiple components. Additionally, plasmids can be readily engineered and produced in large quantities.[40]

2.2 mRNA systems

The second delivery format is an mRNA (DNA)-based approach which transfers the mRNA encoding Cas and sgRNA. It is within the cell where the Cas mRNA is translated to its protein form and creates the Cas/sgRNA complex. Unlike traditional plasmid-based methods, mRNA delivery eliminates the risk of genomic integration, reducing concerns about off-target effects and immunogenicity. Advantages of this delivery format include transient expression that minimises off-target effects and the mRNA does not integrate into the host genome.[38,41] Limitations of the format include Cas mRNA being less stable and thus when producing a transient expression of Cas, makes it more prone to degradation and compromises gene editing efficiency.[42] In order to minimising these complications, chemical modifications are necessary to improve stability and translation.[43]

2.3 Protein-based systems

The third format is a protein RNP-based system which is delivered as a pre-assembled Cas/sgRNA complex into the cell, where gene editing occurs in the nucleus. A notable advantage is the direct delivery of pre-assembled Cas protein and guide RNA complexes into target cells, bypassing the need for cellular machinery to transcribe and translate CRISPR components and ensuring rapid onset of genome editing activity.[44–47] Furthermore, the protein format is the most transient and therefore has the least off-target effects, minimal host genome integration and reduced immunogenicity issues.[45,48–51] Limitations to the RNP-based method include high costs and complexity associated with large scale production, the need for repeated administration due to the transient nature and, especially in difficult-to-transfect cell types, multiple delivery or transfection attempts.[44,52] Nevertheless, ongoing research demonstrate protein-based CRISPR/Cas applications can overcome these challenges for wider use in gene editing.

Delivery of the Cas and sgRNA components are a key aspect of successful gene editing of the target cell, and therefore is required to be as efficient as possible in order to achieve favourable outcomes. The three CRISPR/Cas formats discussed above are currently the only delivery formats available. All three formats have their advantages and disadvantages, and the selection depends on the specific goals that need to be achieved. To strengthen the delivery formats and improve efficiency, there are multiple delivery methods that can be utilised, including physical delivery, viral delivery, and non-viral delivery methods further discussed in the following section.

3. Delivery methods

Several delivery methods are under investigation to efficiently deliver CRISPR/Cas components into target cells, which is essential for gene modification (Fig. 1). Both viral and non-viral CRISPR/Cas delivery approaches, each with their inherent promises and challenges, have been evaluated across multiple in vitro, *in vivo,* and *ex vivo* preclinical and clinical settings.[53,54] The following section summarises the unique strengths and weaknesses of the CRISPR/Cas9 delivery systems and emphasise areas for further improvement to optimise delivery systems for potentially clinically translatable applications.

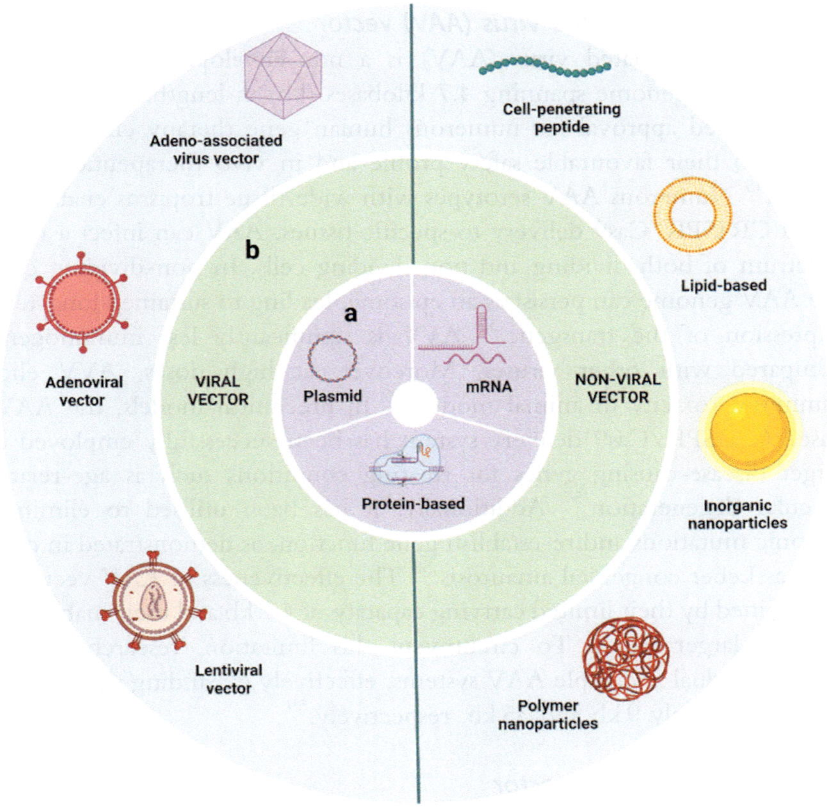

Fig. 1 ***Delivery formats and methods for CRISPR technology*****.** (A) Delivery formats include plasmid (DNA)-based, mRNA-based and protein-based formats. (B) Delivery methods consists of viral vector methods including adeno-associated virus (AAV) vector, adenoviral (AV) vector and lentiviral (LV) vector, and non-viral methods including cell-penetrating peptide (CPP), lipid-based delivery, inorganic nanoparticles such as gold, and polymer nanoparticles. *Created on Biorender.com.*

3.1 Viral vector delivery methods

Viral vectors, such as adeno-associated viruses (AAVs), adenoviruses (AVs) and lentiviruses (LVs) are widely utilised as delivery methods for introducing CRISPR/Cas9 machinery into target cells. These viral vector formats have achieved success in diverse applications, ranging from *ex vivo* transfection to the development of animal disease models.[54] Their effectiveness as delivery systems mainly stems from the inherent capacity of viruses to introduce foreign genetic material into many kinds of cells and tissues in humans with high efficiency.[53]

3.1.1 Adeno-associated virus (AAV) vector

The adeno-associated virus (AAV) is a non-enveloped, ssDNA virus, containing a genome spanning 4.7 kilobases (kb) in length.[55] AAVs have been granted approval for numerous human gene therapy clinical trials, owing to their favourable safety profile and in vivo therapeutic applicability.[55] Numerous AAV serotypes with wide tissue tropisms enable targeted CRISPR/Cas9 delivery to specific tissues. AAV can infect a broad spectrum of both dividing and non-dividing cells. In non-dividing cells, the AAV genome can persist as an episome, leading to sustained long-term expression of the transgene.[53] AVV is significantly less immunogenic compared with other viruses. Moreover, at high doses, AVV elicit minimum toxicity in animal models.[56] In preclinical models, the AAV-based CRISPR/Cas9 delivery system has been successfully employed to target disease-causing genes for treating conditions such as age-related macular degeneration.[57] Additionally, it has been utilised to eliminate intronic mutations and re-establish gene function, as demonstrated in cases such as Leber congenital amaurosis.[58] The effectiveness of AAV vectors is constrained by their limited carrying capacity of 4.7 kb and their inability to package larger genes. To circumvent this limitation, researchers have developed dual and triple AAV systems, effectively expanding the capacity to approximately 9 kb and 35 kb, respectively.[59]

3.1.2 Adenoviral (AV) vector

The adenovirus (AV) is a non-enveloped, dsDNA virus possessing an icosahedral nucleocapsid, containing a genome spanning 26–45 kb.[54] It infects both dividing and non-dividing cells. AV exhibits wide tissue tropism, reduced pathogenicity in immunocompetent individuals, sizable cargo capacity up to 38 kb, and it remains as episomal DNA without integrating into the host genome. Additionally, the transient nature of transgene expression in AV vectors limit off-target mutagenic effects.[53] AV has been utilised for targeted knock-in approach, as demonstrated in a haemophilia B (HB) mouse model, where the insertion of the coagulation factor IX (F9) gene led to phenotypic correction.[60]

3.1.3 Lentiviral (LV) vector

Lentivirus (LV) is an enveloped, ssRNA virus containing genomic RNA of 9 kb. It infects both dividing and non-dividing cells.[53] LVs are less immunogenic than AAVs or AVs because they are enveloped by a lipid bilayer instead of a viral capsid. Since LV vectors integrate non-specifically and efficiently into the host genome, they are associated with higher off-target mutagenic risks.[59] To address this constraint, non-integrating LV vectors have been devised, which have proven effective in generating disease animal models, such as mouse model of lung cancer.[61] A recent development involves the LV-based replication-incompetent virus-like particle (VLP) system, which is regarded as safer due to the absence of the viral genome and transient expression of the transgene.[53]

3.2 Non-viral vector delivery methods

Several promising synthetic non-viral vectors have been developed for delivering CRISPR/Cas9 components into target cells. These methods include cell-penetrating peptides, inorganic nanoparticles, polymer and lipid formulations. These formats are increasingly favoured for therapeutic applications over conventional viral vectors due to their tuneable properties, non-toxic nature, and in vivo efficacy.[40]

3.2.1 Cell-penetrating peptide (CPP)

Cell-penetrating peptides (CPPs) are short peptides, usually under 30 amino acids in length, that facilitate cellular membrane penetration.[53] CPPs can be linked with a target molecule including CRISPR/Cas9 components. CPPs fused with nuclear localisation signal are more beneficial for facilitating entry into the nucleus. However, CPPs are vulnerable to degradation by proteolytic enzymes and may undergo degradation before reaching the target tissue. Additionally, endosomal entrapment poses another obstacle after cellular entry.[40,53]

3.2.2 Lipid-based delivery systems

In recent years, lipid nanoparticles (LNPs) have garnered much attention. LNPs are generated by combination of negatively charged nucleic acids and positively charged liposomes via electrostatic interactions. LNPs aid in the transportation of CRISPR/Cas9 components across cellular membranes through endocytosis.[54] LNPs are comprised of four types of lipids including cationic lipid, PEG lipid, helper phospholipid, and cholesterol.[53] Due to the absence of externally exposed protein or peptide components, LNPs have low immunogenicity. A recent study reported the efficiency of LNP-mediated delivery of CRISPR/Cas9 mRNA, known as NTLA-2001. This therapeutic agent edits the transthyretin (TTR) gene in vivo to treat transthyretin amyloidosis.[62]

3.2.3 Inorganic nanoparticles

Inorganic nanoparticles, particularly gold nanoparticles (AuNPs) have been reported to deliver CRISPR/Cas9 machinery into target cells. AuNPs are inert and do not elicit an immune response.[54] AuNPs have been devised to simultaneously deliver Cas9 RNP and donor DNA to induce homology-directed repair (HDR) in the skeletal muscle of muscular dystrophy X-linked (mdx) mouse model of Duchenne muscular dystrophy (DMD).[63] In contrast to other vectors, such nanoparticles offer ease of control over their size and distribution.[53]

3.2.4 Polymer nanoparticles

Polymer nanoparticles facilitate the transport of CRISPR/Cas9 components across cell membranes via endocytosis. They offer the advantage of surface modification to enhance targeting capabilities, allowing for precise delivery of cargo to specific tissues with controlled release. Polymer nanoparticle formulations such as polyethyleneimine, polyamidoamine, and chitosan are commonly used as delivery vectors. However, the increased molecular weight of polymers often correlates with greater cytotoxicity.[54]

While promising, each of these CRISPR/Cas delivery formats faces significant challenges in transitioning to potential clinical applications. Despite their numerous advantages, the high immunogenicity of viral vectors renders repeated administration ineffective.[53] Additionally, the potential integration of the transgene can lead to unwanted genotoxicity, making viral vectors less favourable.[59] These barriers can be addressed by existing non-viral delivery formulations. Future research to increase the effectiveness of non-viral vectors to target specific tissues or cells is crucial. Already, novel delivery systems are emerging with promising potential to address these challenges.[64,65] Regardless of the vector used, the immune response against the Cas proteins presents another challenge that must be overcome for effective delivery. This issue can be resolved by encapsulating the Cas9 protein or engineering it to remove immunogenic peptides.[59] However, further optimisation to facilitate endosomal escape of non-viral vehicles is crucial for practical in vivo applications. Continued development and improvement of delivery technologies, along with engineering diverse CRISPR/Cas systems, will unlock vast possibilities for genome editing therapy.

4. Engineered systems

Various CRISPR/Cas systems are now being engineered and repurposed for a myriad of research and biomedical applications. Along with genome editing, these engineered CRISPR systems are utilised for various gene regulatory functions, including CRISPR inhibition (CRISPRi) or CRISPR activation (CRISPRa), as well as for base editing or prime editing, epigenome editing, and nucleic acid detection for diagnostics.[22] These advancements can be broadly categorised into systems that involve the double-strand break (DSB) repair pathway and require a donor template, systems that are independent of inducing DSBs and do not necessitate donor templates, and systems capable of editing RNA instead of DNA (Table 3). These innovations have significantly expanded the range of CRISPR applications in basic research, biotechnology, and therapeutic interventions.[31]

4.1 Double-strand break (DSB) dependent

CRISPR/Cas9, derived from the Type II CRISPR system, remodelled genome editing with its simplicity, efficiency, and versatility. The Cas9 protein, directed by an sgRNA complementary to a donor or target DNA template, facilitates site-specific DNA cleavage. Early efforts focused on enhancing the specificity and reducing off-target effects of CRISPR/Cas9 editing. Strategies such as the development of Cas9 nickase variants (e.g., Cas9 D10A and H840A) and high-fidelity Cas9 variants (e.g., eSpCas9, SpCas9-HF1) improved the precision of genome editing by minimising unintended DNA cleavage.[66,67] Inactivation of nuclease activity of Cas9 has led to development of a catalytically deactivated or nuclease dead Cas9 (dCas9). While the dCas9 protein does not possess DNA cleavage capability, its DNA binding activity remains unaffected. Therefore, by attaching transcriptomic activators or suppressors to dCas9, the CRISPR/dCas9 system enables the enhancement (CRISPRa) or repression (CRISPRi) of transcription for target genes.[68]

4.2 DSB independent

Applications of CRISPR technology have been engineered to bypass the need for exogenous donor templates. Base editing and prime editing are two such groundbreaking developments that carry out genome editing without inducing DSBs.[69] Base editing is achieved through fusion proteins of Cas variants (such as Cas9 or Cas12a) with deaminase enzymes. This enables the direct conversion of individual DNA bases without the need for DSBs.

Table 3 CRISPR/Cas engineered variants for gene editing.

CRISPR system examples	Editing pathway	Nuclease/enzyme	Mode of action	References
Cas9 variants, Cas14, Cas 12a, Cas12b, Cas12e	Genomic DSB dependent	DNase	Expression repressor (CRISPR inhibition)	67
Deactivated Cas (dCas) fusions	Fusion protein dependent	Transcription activator or repressor	Expression enhancer (CRISPR activation); suppressor; epigenetic regulator	75
Cas9 variants, Cas 12 variants	Genomic base editing	Deaminase	A to G or C to T single base editing	69
Cas9 variants	Genomic prime editing	Nickase and reverse transcriptase	Nucleotide conversion, insertion, deletion	71
Cas13 variants, Cas9 variants, Cas12g	Transcriptomic modulation	Adenosine deaminase acting on RNA (ADAR)	RNA editing; A to I or C to U	30

Adenine base editors (ABEs) and cytosine base editors (CBEs) target specific base pairs, enabling precise alterations from A to G or C to T, respectively.[69] Prime editing fuses a Cas9 nickase with a reverse transcriptase enzyme and a prime editing guide RNA (pegRNA). This allows for various types of nucleotide conversions, and systematic insertions (maximum of 44 bp) and deletions (maximum of 80 bp).[70] While base editing excels in single-base alterations, prime editing offers greater versatility in introducing complex modifications, albeit with slightly lower efficiency.[71]

4.3 RNA modulators

Apart from its DNA-altering capabilities, RNA modulation using CRISPR is attracting growing interest. RNA editing is achieved by engineering RNA-targeting Cas proteins that can attach to target RNA in a directed manner, while replacing their nuclease function with an adenosine deaminase acting on RNA (ADAR). Currently, Types II, III, V and VI systems possess RNA editing capacities, with Type VI CRISPR/Cas13 systems being the most recognised for this application.[30]

CRISPR/Cas13 systems are characterised by a single RNA-guided Cas13 protein with ribonuclease activity, which can target single-stranded RNA (ssRNA) and precisely cleave it. Different types of Cas13 proteins have effectively been utilised for RNA knockdown, transcriptomic labelling, splicing adaptation, and virus sensing.[30] Three RNA base editing systems have also been reported that utilise catalytically deactivated Cas13 (dCas13) with ADAR—(i) RNA Editing for Programmable A to I Replacement (REPAIR) system, (ii) RNA Editing for Specific C to U Exchange (RESCUE) system, and (iii) C to U RNA Editor (CURE) system.[72–74]

CRISPR/Cas9 systems are also leveraged for RNA editing by designing gRNAs complementary to specific RNA sequences and combing with dCas9 to target RNA molecules. This approach enables a variety of applications, including CRISPRa, CRISPRi, RNA imaging, tracking, and localisation.[25] RNA editing offers advantages over DNA editing in terms of efficiency, specificity, and the ability to make temporary, reversible genetic modifications, which helps mitigate probable risks and ethical concerns accompanying permanent genome editing. Currently, RNA editing is extensively utilised in pre-clinical investigations, opening up novel avenues in RNA-level research, diagnosis, and therapeutic interventions.[30] Extensive research is underway to develop diverse CRISPR systems suitable for various clinical applications. However, the most frequently utilised system remains the CRISPR/Cas9 complex, where it is employed in over 80% of ongoing clinical trials associated with CRISPR based gene therapy.[29]

5. Clinical applications

Cell and gene therapy has long held promise as a radical approach to treating genetic and acquired diseases. While the field has faced significant challenges, including issues with efficacy, safety, and delivery, the rise of CRISPR technology has brought renewed excitement (Fig. 2). The following section provides an overview of the preclinical and clinical applications using CRISPR, representing a pivotal step towards translating laboratory discoveries into treatments for a wide range of diseases.

5.1 Monogenic diseases

Considerable progress has been accomplished in utilising CRISPR/Cas9 in preclinical and clinical trials for gene therapy aimed at treating monogenic diseases, especially β-thalassemia, sickle cell disease (SCD) and HB.[76] Studies have demonstrated the application of CRISPR-mediated gene editing in correcting β-globin gene mutations and restoring normal haemoglobin production and improved erythropoiesis in patient-derived cells and animal models of β-thalassemia.[77,78] Patients treated with autologous CD34+ cells edited with CRISPR/Cas9 targeting the BCL11A transcription factor present with recovered levels of functional haemoglobin.

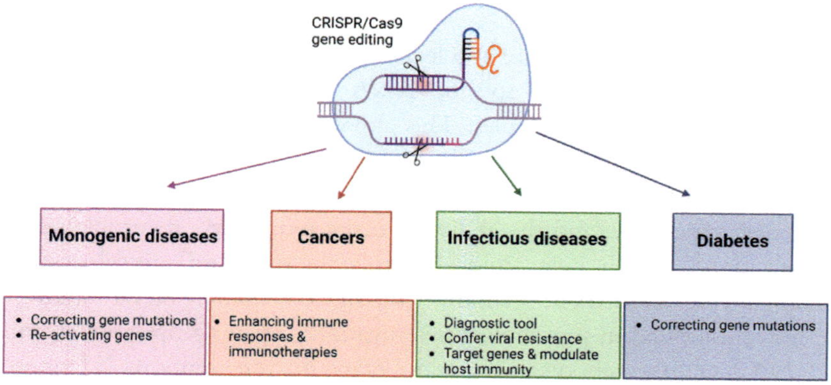

Fig. 2 ***Clinical applications of CRISPR technology in cell and gene therapy.*** Monogenic diseases, cancers, infectious disease and diabetes. *Created on Biorender.com.*

Similar to β-thalassemia, CRISPR/Cas9 system has been exploited to treat SCD by repairing the β-globin gene mutation or restarting foetal haemoglobin (HbF) expression.[76,77] Indeed, the first FDA approval came in late 2023 for gene editing for sickle cell anaemia using CRSPR/Cas9, representing the first CRISPR-based human therapy, that allows for autologous editing of the bone marrow stem cells to facilitate the production of functional haemoglobin.[79,80] For treatment of HB in mice models, Cas9/sgRNA complexes have been demonstrated to mediate somatic as well as germline gene therapy to improve the haemostatic efficiency and increase the survival of HB mice.[81] Furthermore, CRISPR/Cas9 and its advanced tools like base editing and prime editing are being examined for applications in pre-clinical and clinical investigations in DMD, Retinitis pigmentosa (RP), Hutchinson-Gilford Progeria Syndrome (HGPS), hereditary tyrosinemia (HT) and cystic fibrosis (CF).[22]

5.2 Cancers

Several trials are underway to employ CRISPR technology in order to enhance the body's immune response against various diseases.[82] Immunotherapy is highly promising for the treatment of diseases, particularly cancer.[83,84] CRISPR/Cas9 is being leveraged to enhance the functionality and safety of Chimeric Antigen Receptor (CAR)-T cell in cancer immunotherapy.[85] Through targeted genome editing, CRISPR enables precise modifications to CAR-T cells, addressing critical challenges such as immune evasion, antigen escape, and off-target effects. By disrupting immune-checkpoint inhibitory receptors, such as Programmed cell death protein-1 (PD-1) or Cytotoxic T-lymphocyte associated protein-4 (CTLA-4), CRISPR-edited CAR-T cells can overcome immunosuppressive signals within the tumour microenvironment, leading to enhanced persistence and cytotoxic activity against cancer cells.[82,86] Additionally, CRISPR-mediated knockout of endogenous T cell receptors (TCRs) mitigates the chance of graft-versus-host disease (GVHD) and alloreactivity, thereby improving the safety profile of such therapies.[86] Additionally, CRISPR is being reported to allow for engineering CAR-T cells with novel functionalities, including targeting multiple tumour antigens simultaneously, expressing bispecific CARs, or incorporating suicide genes for conditional cell elimination.[86,87]

5.3 Infectious diseases

CRISPR trials in infectious diseases represent a promising approach for combating microbial pathogens, including bacteria, viruses, and parasites.

For bacterial diseases such as tuberculosis and staphylococcal infections, CRISPR is reported as a diagnostic tool, bacteriophage-based disruptor, and for the dysregulation of virulence factors and essential metabolic pathways.[88,89] For viral diseases such as acquired immunodeficiency syndrome (AIDS) caused by human immunodeficiency virus (HIV) and hepatitis B and C, CRISPR systems utilising Cas9 or Cas13a have been explored for their potential to combat the viral pathogens through RNAi, gene editing of host cells to confer viral resistance, and development of novel antiviral therapies.[89,90] CRISPR-based assays have also been utilised for fast and sensitive detection of SARS-CoV-2 RNA, offering improved speed, specificity, and portability compared to traditional PCR-based tests.[91] In particular, CRISPR/Cas13 and Cas13-assisted restriction of viral expression and readout (CARVER) systems are used to target a wide range of ssRNA viruses.[92] Furthermore, CRISPR-based approaches are also utilised for targeting parasitic infections, such as malaria and trypanosomiasis, by disrupting essential parasitic genes or modulating host immune responses.[93]

5.4 Diabetes

CRISPR/Cas9 is harnessed for therapeutic interventions for both type 1 and type 2 diabetes. In type 1 diabetes, CRISPR is utilised to prevent autoimmune destruction of insulin-producing beta cells or regenerate functional beta cells through stem cell manipulation.[94] For type 2 diabetes, CRISPR is being employed for correcting genetic mutations associated with insulin resistance or beta cell dysfunction.[95]

Clinically, CRISPR technology thus represents a strategy for targeted intervention at the molecular level. CRISPR/Cas complexes, encompassing a diverse range of engineered Cas toolkits, offer the potential for curative therapies as well as diagnostics across a broad array of diseases. With ongoing advancements in research and clinical applications, CRISPR signifies a breakthrough approach to address major challenges in biomedical research.

6. Conclusion and future directions

The emergence of gene editing technologies based on CRISPR/Cas has steered research towards a new era for cell and gene therapy in the 21st century. Such systems offer pliable and flexible methods to modify, regulate, and visualise genomes, representing a significant milestone in the field. In this chapter, we outlined the historical evolution and current status of CRISPR technology in gene therapy. Subsequently, we looked at diverse CRISPR formats, delivery techniques, engineered variants, and clinical applications, emphasising their transformative potential in therapeutic interventions.

So far, gene therapies primarily target somatic cells over germ line cells due to technical complexity and ethical concerns. Such issues will require closer investigations and regulatory frameworks concomitant with wider application of the CRISPR/Cas system. As a result, the field of CRISPR/Cas applications may adopt more transient editing methods to mitigate long-term genetic changes or minimise genomic disruption and unwanted immune response. As research progresses, CRISPR-based therapies are anticipated to become more precise, efficient, and versatile along with advancements in delivery methods and engineered variants, thus paving the way for broader clinical applications.

References

1. Gonçalves GAR, Paiva RMA. Gene therapy: advances, challenges and perspectives. *Einstein (Sao Paulo)*. 2017;15(3):369–375.
2. Kumari S, et al. Novel gene therapy approaches for targeting neurodegenerative disorders: focusing on delivering neurotrophic genes. *Mol Neurobiol*. 2024.
3. Banda A, et al. Precision in action: the role of clustered regularly interspaced short palindromic repeats/Cas in gene therapies. *Vaccines*. 2024;12(6):636.
4. Chavez M, et al. Advances in CRISPR therapeutics. *Nat Rev Nephrol*. 2023;19(1):9–22.
5. Poddar A, Banerjee R, Shukla R. Editorial: non-viral vectors for gene therapy/nucleic acid delivery. *Front Bioeng Biotechnol*. 2023;11:1304769.
6. Malech HL, Garabedian EK, Hsieh MM. Evolution of gene therapy, historical perspective. *Hematol Oncol Clin North Am*. 2022;36(4):627–645.
7. Watson JD, Crick FHC. Molecular structure of nucleic acids: a structure for deoxyribose nucleic acid. *Nature*. 1953;171(4356):737–738.
8. Wirth T, Parker N, Ylä-Herttuala S. History of gene therapy. *Gene*. 2013;525(2):162–169.
9. Cohen SN, et al. Construction of biologically functional bacterial plasmids in vitro. *Proc Natl Acad Sci U S A*. 1973;70(11):3240–3244.
10. Carroll D. Genome engineering with zinc-finger nucleases. *Genetics*. 2011;188(4):773–782.
11. Silva G, et al. Meganucleases and other tools for targeted genome engineering: perspectives and challenges for gene therapy. *Curr Gene Ther*. 2011;11(1):11–27.
12. Joung JK, Sander JD. TALENs: a widely applicable technology for targeted genome editing. *Nat Rev Mol Cell Biol*. 2013;14(1):49–55.
13. Ishino Y, et al. Nucleotide sequence of the iap gene, responsible for alkaline phosphatase isozyme conversion in Escherichia coli, and identification of the gene product. *J Bacteriol*. 1987;169(12):5429–5433.
14. Jinek M, et al. A programmable dual-RNA-guided DNA endonuclease in adaptive bacterial immunity. *Science*. 2012;337(6096):816–821.
15. Cho SW, et al. Targeted genome engineering in human cells with the Cas9 RNA-guided endonuclease. *Nat Biotechnol*. 2013;31(3):230–232.

16. Lino CA, et al. Delivering CRISPR: a review of the challenges and approaches. *Drug Deliv.* 2018;25(1):1234–1257.
17. Pickar-Oliver A, Gersbach CA. The next generation of CRISPR–Cas technologies and applications. *Nat Rev Mol Cell Biol.* 2019;20(8):490–507.
18. Richardson C, Kelsh RN, Richardson RJ. New advances in CRISPR/Cas-mediated precise gene-editing techniques. *Dis Model Mech.* 2023;16(2).
19. Koonin EV, Makarova KS. Evolutionary plasticity and functional versatility of CRISPR systems. *PLoS Biol.* 2022;20(1):e3001481.
20. Gaj T, Gersbach CA, Barbas CF. 3rd, *ZFN, TALEN, and CRISPR/Cas-based methods for genome engineering. Trends Biotechnol.* 2013;31(7):397–405.
21. Feng S, et al. Strategies for high-efficiency mutation using the CRISPR/Cas system. *Front Cell Dev Biol.* 2021;9:803252.
22. Xu Y, Li Z. CRISPR-Cas systems: overview, innovations and applications in human disease research and gene therapy. *Comput Struct Biotechnol J.* 2020;18:2401–2415.
23. Makarova KS, et al. Evolution and classification of the CRISPR-Cas systems. *Nat Rev Microbiol.* 2011;9(6):467–477.
24. Morisaka H, et al. CRISPR-Cas3 induces broad and unidirectional genome editing in human cells. *Nat Commun.* 2019;10(1):5302.
25. O'Connell MR, et al. Programmable RNA recognition and cleavage by CRISPR/Cas9. *Nature.* 2014;516(7530):263–266.
26. Chi H, et al. Antiviral type III CRISPR signalling via conjugation of ATP and SAM. *Nature.* 2023;622(7984):826–833.
27. Pinilla-Redondo R, et al. Type IV CRISPR-Cas systems are highly diverse and involved in competition between plasmids. *Nucleic Acids Res.* 2020;48(4):2000–2012.
28. Yan WX, et al. Functionally diverse type V CRISPR-Cas systems. *Science.* 2019;363(6422):88–91.
29. Zhang S, et al. Current trends of clinical trials involving CRISPR/Cas systems. *Front Med (Lausanne).* 2023;10:1292452.
30. Gunitseva N, et al. RNA-dependent RNA targeting by CRISPR-Cas systems: characterizations and applications. *Int J Mol Sci.* 2023;24(8).
31. Asmamaw M, Zawdie B. Mechanism and applications of CRISPR/Cas-9-mediated genome editing. *Biologics.* 2021;15:353–361.
32. Cowan PJ. The use of CRISPR/Cas associated technologies for cell transplant applications. *Curr Opin Organ Transpl.* 2016;21(5):461–466.
33. Corrigan-Curay J, et al. Genome editing technologies: defining a path to clinic. *Mol Ther.* 2015;23(5):796–806.
34. Egelie KJ, et al. The emerging patent landscape of CRISPR-Cas gene editing technology. *Nat Biotech.* 2016;34(10):1025–1031.
35. Ran FA, et al. Genome engineering using the CRISPR-Cas9 system. *Nat Protoc.* 2013;8(11):2281–2308.
36. Liu C, et al. Delivery strategies of the CRISPR-Cas9 gene-editing system for therapeutic applications. *J Control Rel.* 2017;266:17–26.
37. Xu Y, Liu R, Dai Z. Key considerations in designing CRISPR/Cas9-carrying nanoparticles for therapeutic genome editing. *Nanoscale.* 2020;12(41):21001–21014.
38. Wu X, Kriz AJ, Sharp PA. Target specificity of the CRISPR-Cas9 system. *Quant Biol.* 2014;2(2):59–70.
39. Du Y, et al. CRISPR/Cas9 systems: delivery technologies and biomedical applications. *Asian J Pharm Sci.* 2023;18(6):100854.
40. Lin Y, Wagner E, Lächelt U. Non-viral delivery of the CRISPR/Cas system: DNA versus RNA versus RNP. *Biomater Sci.* 2022;10(5):1166–1192.
41. Timin AS, et al. Efficient gene editing via non-viral delivery of CRISPR-Cas9 system using polymeric and hybrid microcarriers. *Nanomedicine.* 2018;14(1):97–108.

42. Niu Y, et al. Generation of gene-modified cynomolgus monkey via Cas9/RNA-mediated gene targeting in one-cell embryos. *Cell.* 2014;156(4):836–843.
43. Yin H, et al. Structure-guided chemical modification of guide RNA enables potent non-viral in vivo genome editing. *Nat Biotechnol.* 2017;35(12):1179–1187.
44. Kim S, et al. Highly efficient RNA-guided genome editing in human cells via delivery of purified Cas9 ribonucleoproteins. *Genome Res.* 2014;24(6):1012–1019.
45. Liang X, et al. Rapid and highly efficient mammalian cell engineering via Cas9 protein transfection. *J Biotechnol.* 2015;208:44–53.
46. Zuris JA, et al. Cationic lipid-mediated delivery of proteins enables efficient protein-based genome editing in vitro and in vivo. *Nat Biotechnol.* 2015;33(1):73–80.
47. Park J, Choe S. DNA-free genome editing with preassembled CRISPR/Cas9 ribonucleoproteins in plants. *Transgenic Res.* 2019;28(Suppl 2):61–64.
48. Chen K, et al. Engineering self-deliverable ribonucleoproteins for genome editing in the brain. *Nat Commun.* 2024;15(1):1727.
49. Hendel A, et al. Chemically modified guide RNAs enhance CRISPR-Cas genome editing in human primary cells. *Nat Biotechnol.* 2015;33(9):985–989.
50. Thyme SB, et al. Internal guide RNA interactions interfere with Cas9-mediated cleavage. *Nat Commun.* 2016;7(1):11750.
51. Stahl EC, et al. Genome editing in the mouse brain with minimally immunogenic Cas9 RNPs. *Mol Ther.* 2023;31(8):2422–2438.
52. Ramakrishna S, et al. Gene disruption by cell-penetrating peptide-mediated delivery of Cas9 protein and guide RNA. *Genome Res.* 2014;24(6):1020–1027.
53. Taha EA, Lee J, Hotta A. Delivery of CRISPR-Cas tools for in vivo genome editing therapy: trends and challenges. *J Control Rel.* 2022;342:345–361.
54. Huang J, et al. CRISPR/Cas systems: delivery and application in gene therapy. *Front Bioeng Biotechnol.* 2022;10:942325.
55. Dunbar CE, et al. Gene therapy comes of age. *Science.* 2018;359(6372).
56. Xu CL, et al. Viral delivery systems for CRISPR. *Viruses.* 2019;11(1).
57. Koo T, et al. CRISPR-LbCpf1 prevents choroidal neovascularization in a mouse model of age-related macular degeneration. *Nat Commun.* 2018;9(1):1855.
58. Maeder ML, et al. Development of a gene-editing approach to restore vision loss in Leber congenital amaurosis type 10. *Nat Med.* 2019;25(2):229–233.
59. Luther DC, et al. Delivery approaches for CRISPR/Cas9 therapeutics in vivo: advances and challenges. *Expert Opin Drug Deliv.* 2018;15(9):905–913.
60. Stephens CJ, et al. Long-term correction of hemophilia B using adenoviral delivery of CRISPR/Cas9. *J Control Rel.* 2019;298:128–141.
61. Blasco RB, et al. Simple and rapid in vivo generation of chromosomal rearrangements using CRISPR/Cas9 technology. *Cell Rep.* 2014;9(4):1219–1227.
62. Gillmore JD, et al. CRISPR-Cas9 in vivo gene editing for transthyretin amyloidosis. *N Engl J Med.* 2021;385(6):493–502.
63. Lee K, et al. Nanoparticle delivery of Cas9 ribonucleoprotein and donor DNA in vivo induces homology-directed DNA repair. *Nat Biomed Eng.* 2017;1:889–901.
64. Poddar A, et al. ZIF-C for targeted RNA interference and CRISPR/Cas9 based gene editing in prostate cancer. *Chem Commun.* 2020;56(98):15406–15409.
65. Pyreddy S, et al. Targeting telomerase utilizing zeolitic imidazole frameworks as non-viral gene delivery agents across different cancer cell types. *Biomater Adv.* 2023;149:213420.
66. Gopalappa R, et al. Paired D10A Cas9 nickases are sometimes more efficient than individual nucleases for gene disruption. *Nucleic Acids Res.* 2018;46(12):e71.
67. Chen JS, et al. Enhanced proofreading governs CRISPR-Cas9 targeting accuracy. *Nature.* 2017;550(7676):407–410.
68. Whinn KS, et al. Nuclease dead Cas9 is a programmable roadblock for DNA replication. *Sci Rep.* 2019;9(1):13292.

69. Kantor A, McClements ME, MacLaren RE. CRISPR-Cas9 DNA base-editing and prime-editing. *Int J Mol Sci.* 2020;21(17).
70. Anzalone AV, et al. Search-and-replace genome editing without double-strand breaks or donor DNA. *Nature.* 2019;576(7785):149–157.
71. Scholefield J, Harrison PT. Prime editing—an update on the field. *Gene Ther.* 2021;28(7):396–401.
72. Cox DBT, et al. RNA editing with CRISPR-Cas13. *Science.* 2017;358(6366):1019–1027.
73. Abudayyeh OO, et al. A cytosine deaminase for programmable single-base RNA editing. *Science.* 2019;365(6451):382–386.
74. Huang X, et al. Programmable C-to-U RNA editing using the human APOBEC3A deaminase. *EMBO J.* 2020;39(22):e104741.
75. Moon SB, et al. Recent advances in the CRISPR genome editing tool set. *Exp Mol Med.* 2019;51(11):1–11.
76. Demirci S, et al. CRISPR-Cas9 to induce fetal hemoglobin for the treatment of sickle cell disease. *Mol Ther Methods Clin Dev.* 2021;23:276–285.
77. Frangoul H, et al. CRISPR-Cas9 gene editing for sickle cell disease and β-thalassemia. *N Engl J Med.* 2021;384(3):252–260.
78. Wilkinson AC, et al. Cas9-AAV6 gene correction of beta-globin in autologous HSCs improves sickle cell disease erythropoiesis in mice. *Nat Commun.* 2021;12(1):686.
79. Wong C. UK first to approve CRISPR treatment for diseases: what you need to know. *Nature.* 2023;623(7988):676–677.
80. Park SH, Bao G. CRISPR/Cas9 gene editing for curing sickle cell disease. *Transfus Apher Sci.* 2021;60(1):103060.
81. Huai C, et al. CRISPR/Cas9-mediated somatic and germline gene correction to restore hemostasis in hemophilia B mice. *Hum Genet.* 2017;136(7):875–883.
82. Khalaf K, et al. CRISPR/Cas9 in cancer immunotherapy: animal models and human clinical trials. *Genes (Basel).* 2020;11(8).
83. Poddar A, et al. Crosstalk between immune checkpoint modulators, metabolic reprogramming and cellular plasticity in triple-negative breast cancer. *Curr Oncol.* 2022;29(10):6847–6863.
84. Shrestha R, et al. Monitoring immune checkpoint regulators as predictive biomarkers in hepatocellular carcinoma. *Front Oncol.* 2018;8:269.
85. Wei W, Chen ZN, Wang K. CRISPR/Cas9: a powerful strategy to improve CAR-T cell persistence. *Int J Mol Sci.* 2023;24(15).
86. Allemailem KS, et al. Innovative strategies of reprogramming immune system cells by targeting CRISPR/Cas9-based genome-editing tools: a new era of cancer management. *Int J Nanomed.* 2023;18:5531–5559.
87. Liu Z, et al. Recent advances and applications of CRISPR-Cas9 in cancer immunotherapy. *Mol Cancer.* 2023;22(1):35.
88. Zein-Eddine R, et al. The future of CRISPR in Mycobacterium tuberculosis infection. *J Biomed Sci.* 2023;30(1):34.
89. Cobb LH, et al. CRISPR-Cas9 modified bacteriophage for treatment of Staphylococcus aureus induced osteomyelitis and soft tissue infection. *PLoS One.* 2019;14(11):e0220421.
90. Ashraf MU, et al. CRISPR-Cas13a mediated targeting of hepatitis C virus internal-ribosomal entry site (IRES) as an effective antiviral strategy. *Biomed Pharmacother.* 2021;136:111239.
91. Deol P, et al. CRISPR use in diagnosis and therapy for COVID-19. *Methods Microbiol.* 2022;50:123–150.
92. Huang Z, et al. CRISPR-Cas13: a new technology for the rapid detection of pathogenic microorganisms. *Front Microbiol.* 2022;13:1011399.

93. Pal S, Dam S. CRISPR-Cas9: taming protozoan parasites with bacterial scissor. *J Parasit Dis*. 2022;46(4):1204–1212.
94. Karpov DS, et al. Challenges of CRISPR/Cas-based cell therapy for type 1 diabetes: how not to engineer a "Trojan Horse". *Int J Mol Sci*. 2023;24(24).
95. Lotfi M, et al. Application of CRISPR-Cas9 technology in diabetes research. *Diabet Med*. 2024;41(1):e15240.

CHAPTER EIGHT

Advances in CRISPR-Cas systems for epigenetics

Mahnoor Ilyas[a,b], Qasim Shah[a], Alvina Gul[b], Huzaifa Ibrahim[a], Rania Fatima[a], Mustafeez Mujtaba Babar[a,c,*], and Jayakumar Rajadas[c,*]

[a]Shifa College of Pharmaceutical Sciences, Shifa Tameer-e-Millat University, Islamabad, Pakistan
[b]Atta-ur-Rahman School of Applied Biosciences, National University of Sciences and Technology, Islamabad, Pakistan
[c]Advanced Drug Delivery and Regenerative Biomaterials Lab, Stanford University School of Medicine, Stanford University, Palo Alto, CA, United States
*Corresponding authors. e-mail address: mustafeez.babar@fulbrightmail.org; jayraja@stanford.edu

Contents

Abstract

The CRISPR-Cas9 method has revolutionized the gene editing. Epigenetic changes, including DNA methylation, RNA modification, and changes in histone proteins, have been intensively studied and found to play a key role in the pathogenesis of human diseases. CRISPR-While the utility of DNA and chromatin modifications, known as epigenetics, is well understood, the functional significance of various alterations of RNA nucleotides has recently gained attention. Recent advancements in improving CRISPR-based epigenetic modifications has resulted in the availability of a powerful source that can selectively modify DNA, allowing for the maintenance of epigenetic memory over several cell divisions. Accurate identification of DNA methylation at specific locations is crucial for the prompt detection of cancer and other diseases, as DNA methylation is strongly correlated to the onset as well as the advancement of such conditions. Genetic or epigenetic perturbations can disrupt the regulation of imprinted genes, resulting in the development of diseases. When histone code editors and DNA de-/ methyltransferases are coupled with catalytically inactive Cas9 (dCas9), and CRISPRa and CRISPRi, they demonstrate excellent efficacy in editing the epigenome of eukaryotic cells. Advancing and optimizing the extracellular delivery

Progress in Molecular Biology and Translational Science, Volume 208
ISSN 1877-1173, https://doi.org/10.1016/bs.pmbts.2024.07.003

platform can, hence, further facilitate the manipulation of CRISPR-Cas9 gene editing technique in upcoming clinical studies. The current chapter focuses on how the CRISP/ Cas9 system provides an avenue for the epigenetic modifications and its employability for human benefit.

Abbreviations

Cas CRISPR-associated proteins
DNA-PKc DNA binding protein kinases – catalytic subunits
DNMT DNA cytosine-5-methyl transferase
DSB Double Strand Breaks
iPSCs induced pluripotent stem cells
KRAB Krüppel-associated box
lncRNAs Long non-coding RNAs
MECP2 methyl-CpG binding protein 2
miRNA microRNA
PAM protospacer adjacent motif
piRNAs Piwi-interacting RNA
SAM S-adenosyl methionine
sgRNA single guide RNA
siRNA small interfering RNA
Tet1CD Ten-Eleven Translocation Dioxygenase 1
tracr RNA trans-activating crRNA
VPR Viral Protein R
α-TIF Alpha-transinducing factor

1. Introduction

CRISPR is an essential part of the adaptive immunity in most prokaryotes to protect them from pathogenic components of plasmids and bacteriophages. This immune assembly consists of two genetic entities i.e., CRISPR loci (composed of spacers and repeats) and operons of Cas genes. CRISPR locus is transcribed to in long precursor CRISPR RNA which is subsequently processed and shortened into short CRISPR RNA (crRNA),. Cas genes ultimately translate into Cas proteins. These crRNAs and Cas proteins together make the nucleoprotein complexes that provide immunity against the invading pathogenic gene sequences or viruses by identifying their specific DNA sequences by complementarily binding to the crRNA.[1] In 1987, CRISPR repeats were identified and characterized using DNA from *Escherichia coli*. Early in the research, CRISPR loci's genetic data was used in medical research to genotype bacterial strains like *Mycobacterium tuberculosis* and *Streptococcus pyogenes*, even though their biological importance was unknown. Francisco Mojica hypothesized that these unusual loci containing foreign DNA segments were part of the prokaryotic immune system and his group found structural similarities between *Haloferax mediterranei's* archaeal genome and the DNA repeats in bacterial genomes. Mojica proposed the term "CRISPR" in 2002.[2] The extensive research on the CRISPR-Cas systems has diversified them quite a lot. The lack of comprehensive Cas proteins as indicators of phylogenetic history pose a great hurdle in CRISPR-Cas classification.[3,4]

According to the modified evolutionary classification, that covers the major developments on CRISPR- Cas systems, it comprises of 2 classes with 6 types and 33 subtypes, as opposed to the older classification of 2015 which included only 5 types with 16 subtypes. Discovery of numerous CRISPR– Cas systems of class II is a key progress that does require further research.[5] After repeated exposure to alien genetic entities, Cas genes evolve quicker, the genetic blueprint diversifies, and the defensive mechanisms are altered. The naturally existing anti-CRISPR proteins in phages compete for coevolution. Comparative sequence analysis, experimentation, and structural analysis suggest that all CRISPR-Cas forms share structural and functional principles and a shared ancestry despite evolutionarily divergence. Their classification depends on Cas genes, Cas operon design, and preserved Cas protein phylogenies. Studies reveal that all CRISPR-Cas systems share a common lineage, hence Class II evolved from Class I after the loss of a portion of Cas genes. Type II and V are widely utilized in bacteria due to their structural simplicity.[6] Each of these types has several subtypes associated with them. Type IV, however, is newly discovered type of this system and its subtypes are not well characterized. These systems make use of CRISPR-associated proteins acting as nucleases that attack and cleave the complementary DNA or RNA sequences, in line with the directions provided by short crRNA guides towards the targeted loci.[7] Cas protein's entire system comprises of a highly variable repeats of 25 to35 base pairs each that are disconnected by distinct spacers (typically of 30–40 base pairs each), and the nearby one or more operon of Cas genes that often contains accessory genes with them. This natural immune response of prokaryotes includes three discrete but often knotted steps: adaptation, expression of pre-crRNA and modifications, and interference.[8] CRISPR-Cas based epigenetic editing mechanisms have aided the researchers to carry out epigenetic modifications of the DNA, histones and other site specific alterations along the length of the chromatin, allowing the researchers to understand the relationship between epigenetic markers and the actual mechanisms of expression of genes.[9] However, safety and effectiveness are a major concern because these systems most commonly use viral vectors. The carcinogenic, cytotoxic and immunogenic threats most commonly occur during these delivery mechanisms.[10]

2. Manipulation of epigenetics

Epigenetics is the heritable change in the phenotype with no genotypic change. It refers to those changes that are brought about in an organism by regulation of gene expression while keeping the genetic code conserved. Despite having almost identical genetic sequences, the cells which are involved in tissue genesis of multicellular organisms are able to continue performing their individual roles throughout the life span of the organism. The attempt to determine the cause of this variance gave rise to the field of epigenetics, which examines a broad variety of variables other than DNA sequences that affect genomic function.[11] The first study initiated to define the characteristics of dCas9 coupled with the α-transinducing factor (α-TIF) or VP16 acidic activation peptide tetramer. This research served as the basis for the creation of several advanced transcriptional activators such as p300 catalytic core, S-adenosyl methionine (SAM), Viral Protein R (VPR) and Ten-Eleven Translocation Dioxygenase 1(Tet1CD) among others. Furthermore, dCas9 fusions that possess the ability to suppress gene expression have been studied and found to operate by means of direct or indirect alteration of chromatin structure by employing certain enzymes and protein domains such as methyl-CpG binding protein 2 (MECP2), Kruppel associated box (KRAB) and DNA cytosine-5-methyl transferase (DNMT) especially DNMT3a.[12] The changes include modifications in histone proteins, methylation of DNA and non-protein-coding RNAs interactions.[13] DNA methylation is carried out in CpG dinucleotides, also known as CpG sites, thus forming the CpG islands. CpG islands along with the nearby areas (shores) possess highest functional importance and their modification (methylation or demethylation) efficiently varies the expression levels of adjacent genes. This modification is catalyzed by DNA methyltransferase family i.e., DNMT1, DNMT2, and DNMT3 having further subtypes as DNMT3a, DNMT3b, and DNMT3L.[14] Regulation through non-coding RNAs also plays important role in epigenetic mechanisms. After transcription, majority of the transcripts do not undergo translation to form functional proteins, but they keep a check on important body processes. Long noncoding RNAs (lncRNAs), microRNA (miRNA), endogenous small interfering RNA (siRNA), and Piwi-interacting RNA (piRNAs) are among the significant regulatory RNAs. They are mostly considered to manage gene expression but recently, a significant number of researches have brought this to light that they are also involved in epigenetic control.[15] These regulatory mechanisms have been involved in generating and evaluating two conditional transgenic mouse strains, namely Rosa26:LSL-dCas9-p300, which induces activation of genes, and Rosa26:LSL-dCas9-KRAB, which inhibits gene expression. Target genes regulation and associated alterations to epigenetic states and corresponding traits in the brain and liver in living organisms, as well as in T-cells and fibroblasts outside of the body, may have been proven to be specifically targeted by guide RNAs to the locations where transcription begins or to distant enhancer elements. For the control of gene expression and editing of epigenome in living organisms that is easy, controlled, and specific, these mice lines are invaluable tools.[16] Two prominent epitopes of T-cell for HLA-A*02:01 have been discovered in *S. pyogenes* Cas9 (SpCas9) utilizing an advanced identification and analysis algorithm that predicts the hydrophobicity of T cell receptor contact residues and their binding to HLA molecules. These epitopes have then been assessed by using T cell assays utilizing PBMCs from healthy donors. Through a proof-of-principle study, it has been shown that it is possible to modify the Cas9 protein by targeted mutations to remove immunodominant epitopes, while still maintaining its performance and selectivity. The research emphasizes the significance of pre-existing immunity to CRISPR-associated nucleases and offers a possible approach to mitigate the T cell immune response.[17] The dCas9-SALL1-SDS3 protein, when combined with chemically modified sgRNAs, is capable of producing a higher level of target gene repression compared to both the first and second generations of CRISPRi. Furthermore, it demonstrates a high level of target selectivity. *In vitro*-transcribed dCas9-SALL1-SDS3 mRNA can be employed in conjunction with artificial small guide RNAs for the purpose of short-term transport into primary T cells and human induced pluripotent stem cells. It is also possible to employ functional gene characterization of host factors for DNA damage. This study has been carried out orthogonally to decipher the function of short interfering RNA.[18] A demethylase tool utilizing CRISPR-Cas9 technology has also been created that consists of the catalytic domain (CD) of TET1CD conjugated with a dCas9 which is a deactivated Cas9 protein. The fusion protein targets the methyl groups from specific locations within the BRCA1 promoter to remove them under the control of a particular single-guide RNAs (sgRNA), that results in the upregulation of gene transcription. The fusion constructs have a modular architecture that enables the replacement of different enzymes that change chromatin or DNA. They also help in the targeting of specific locations in the genome to keep a check on the regulatory processes of epigenome at gene-promoter regions and more targeted genomic sites.[19] To achieve maximum repressive effect at various loci, it is required to target different combinations of histones (Ezh2-dCas9 or KRAB-dCas9) and DNA methyltransferases enzymes. It is important to develop targeting tools to make epigenetic memory at a specific location in a certain type of cell.[20] Fig. 1 shows different modules of CRISPR Cas editing.

CRISPR-dCas9 is a novel form of targeted cancer treatment that acts as a tool for modulating epigenetic factors. By utilizing CRISPR-dCas9-EE complexes, for instance, it is possible to modify cancer-related epigenetic characteristics linked with various cancer traits. Due to its current developing phase, the utilization of sgRNA-dCas9 is accompanied with certain inherent obstacles. The primary obstacles in effectively targeting cancer cells using sgRNA-dCas9 are the incidence of off-target effects, the absence of well-defined guidelines for developing sgRNA, difficulties in selecting target sites, and inadequate delivery mechanisms for sgRNA-dCas9.[21] The dCas9-fusion constructs that can adjust transcriptional activity only vary in their application of transcriptional modulation. Some have an activating effect, while others have an inhibiting effect. The KRAB domain is predominantly utilized for the suppression of transcription process, yet transcriptional activation frequently uses oligomers of VP16. Since VP64 is made up of four copies of VP16, it effectively improves the assembly of the target promoter's basal transcription machinery.[22] A modular platform (pMVP) has been developed to facilitate the study of a specific gene in various experimental conditions.

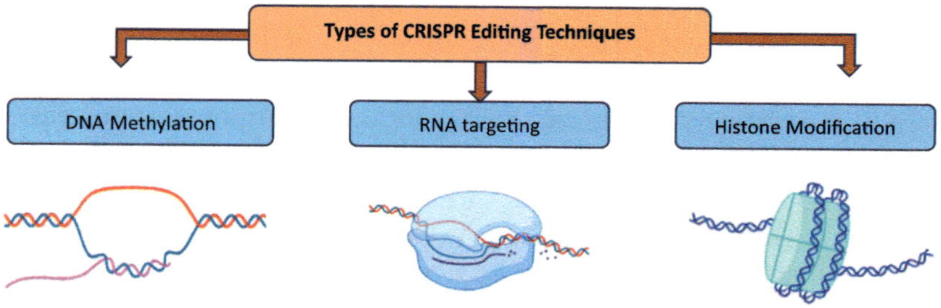

Fig. 1 CRISPR-Cas editing techniques consist of several methods, including DNA methylation, RNA targeting, and histone modification techniques. These primarily target chromatin which is an intricate macromolecule playing a crucial role in the coordination of gene expression, DNA damage repair, and DNA replication process.

This platform integrates a diverse range of promoters, epitope tags, conditional expression methods, and fluorescence reporters. Adenovirus, PiggyBac transposon, lentivirus, and Sleeping Beauty transposon are among the 35 custom destination vectors that can be obtained. In total, there are over 108,000 possible combinations of vectors reported that can be effectively employed for the epigenetic modifications.[23]

3. CRISPR-Cas protocols and strategies

The accuracy of CRISPR facilitated genome editing is compromised by alterations in the genomic structure and stability of the target DNA sequence.[24] The endonuclease activity can be eliminated by mutating Cas9 protein, as a result the Cas9 inactivates into dCas9.[21] These mutations are done at various regions of regulatory and enzymatic domains including RuvC and HNH but the dcas9 still binds and influences the DNA strand via guide RNA. The dCas9 protein are then linked with proteins leading to activation (CRISPRa) and inhibitory (CRISPRi) effects to form fused proteins complexes with transcriptional active/inhibitory states. They mediate the process of transcription by acting on the target gene functional elements i.e., promoters and enhancers.[24] Due to the easy reprogramming, excellent efficiency and specificity, compatibility and affordability, in comparison to the other editing mechanisms like zinc-funger nucleases (ZFN) or Transcription Activator Like Effector Nuclease (TALEN), today CRISPR systems are used in genetic editing more frequently.[25]

CRISPR-CRISPR- Mutant Cas9 can produce double strand breaks at specific points. Many genetic engineering applications use Cas9 protein variations created by targeting DNA recombination proteins (REC) and nuclease (nuc) activity. These variations include Cas9n, dCas9, and Light-activated Cas9. Mutations in Cas9 nickase (Cas9n) that inactivate endonuclease cleavage domains. These include a mutation called D10A that renders the endonuclease activity of the RuvC domain non-functional and a H847A mutation that disables the HNH domain. The DNA is then repaired by producing staggered DSB by Cas9n joined with two individual sgRNAs. This process improves the specificity of CRISPR-Cas9 system.[26]

Introducing several point mutations into the RuvC and HNH domains namely the D10A and H840A alterations cause Dead Cas9 (dcas9), which blocks nucleolytic activity but does not affect target binding.[27] This system (Cas9-sgRNA complex) performs two important functions of inhibition

(CRISPRi) and activation (CRISPRa). Proper selection and a complete understanding of the gRNA is important for the successful implementation of this system. For instance, utilization of a 20-nt gRNA target gene complementary sequence results in an efficient CRISPRi-based repression. Using CRISPRi, it is possible to silence numerous genes at once by delivering multiple sgRNAs. However, efforts are underway to decipher the mechanism by whichdCas9 may either increase gene expression when bound to activators like VP16 and VP64, or inhibit gene expression when bound to repressors like KRAB.[26] CRISPR-Cas9 can also introduce large site-specific transgenes into the genome using homology-directed repair (HDR). Other methods, such as gene knock-out via non-homologous end joining (NHEJ), lentiviral/gammaretroviral gene addition, and base/prime editing, are important in clinical applications but inefficient and inaccurate for treating primary immunodeficiencies or hematological disorders of genetic origin.[28] It is noteworthy, however, that knock-in requires more resources since it inserts foreign DNA sequences into the Cas9-mediated DSB that further stimulate the HDR mechanism instead of the NHEJ route. While the HDR route becomes prominent at the S/G2 stage of the cell cycle, NHEJ remains the principal pathway utilized throughout the whole cell cycle with the exception of mitosis.[29] When Ku70/Ku80 (Ku) heterodimer binds to Double Strand Breaks (DSB) ends, DNA binding protein kinases – catalytic subunits (DNA-PKc) and other scaffolding factors like nuclease Artemis, X-ray repair cross-complementing protein 4 (XRCC4), DNA ligase IV (LigIV), and Aprataxin and PNKP Like Factor (APLF) are recruited to bring the DSBs together. Nucleases sort out the incompatible ends, gaps are filled by polymerases, and finally, DNA ligases work on the ends and ligate them. This entire process eventually leads to the establishment of Insertions/Deletion mutations or INDELs.[30] HDR is initiated when the complex MRN (MRE11-RAD50-NBS1) binds to, and stabilizes the DSB ends, as a scaffold to boost Ataxia Talengiectasia mutated (ATM) signaling in response to the DSB. Long-range resections follow short-range ones. Long 3′ overhangs are used by RAD51, a DNA repair protein, to seek homology and invade strands, forming D-loops. Finally, swapped DNA strands are changed for correct editing.[31]

CRISPR-Cas may be employed to edit multiple genes in parallel. It can also be used to modify the entire genome by utilizing a library of sgRNAs. This function of gene screening used to be carried out by RNA interference (RNAi) libraries when CRISPR was not well-established. The utilization of RNAi was prevalent due to its efficacy in diminishing gene expression at the mRNA level. However, with the discovery of CRISPR technology, many drawbacks related to RNAi-based screenings were overcome and have been serving as a versatile approach since then. By using CRISPR screening, multiple screen formats are performed with much ease like, gene knockout/knockin, interference (CRISPRi), activation (CRISPRa) and epigenome screenings. Thus, CRISPR screening currently enables access to a wide variety of applications, particularly in the field of precision medicine.[32]

CRISPRi and CRISPRa have emerged as widely employed methods for controlling gene expression pattern in bacteria. CRISPRi/a techniques exert dynamic control over cellular metabolism by modulating the transcription pattern of specific genes. Targeted libraries and genome-wide libraries can both benefit from its use in constructing transcriptional regulatory circuits and clarifying genotype-phenotype associations. CRISPRi/a has primarily been developed and studied in bacterial models such as *E. coli* and *Bacillus subtilis*. However, an increasing number of research studies have shown that these tools can also be used in other non-model bacteria.[33] Due to these enormous benefits, CRISPR has rapidly become the favorite and ideal tool for genetic manipulation and also comes with incredible potential to serve as a platform for studying in vivo gene function.[34]

The CRISPR-gRNA of CRISPR-Cas9 system has demonstrated exceptional flexibility, enabling its transformation from a tool primarily used for precision genome editing to a versatile component that can be easily programmed to target and deliver various functional domains for different purposes. These objectives include manipulating specific DNA regions or sequences, conducting multicolor imaging of chromosomal regions in live cells, tracking naturally occurring mRNAs, and studying the structure and behavior of chromatin.[35] The benefits of the technique are not restrained to in vitro studies only. Mouse models are commonly employed in scientific research to study human diseases. Numerous genetically engineered models, including those for cancer, cardiomyopathy, Huntington's disease, albinism, hemophilia B, obesity, deafness and muscular dystrophy have been successfully developed. However, owing to the genetic differences between humans and rodents, it is difficult to effectively assess and follow up the long-term effectiveness of these techniques.[36]

For human diseases of clinical significance, CRISPR has been utilized to rectify the dystrophin gene in human cell culture for treating Duchenne muscular dystrophy affected patients. The repaired cells regained the expression pattern of the dystrophin protein, showing that CRISPR-Cas is a promising approach for its treatment. Similarly, a recent groundbreaking study has been presented that shows that it can modify the mutated rhodopsin gene in induced pluripotent stem cells (iPSCs) obtained from retinitis pigmentosa (RP) patients. The iPSCs were s converted into retinal organoids that regained their regular functioning, indicating the possibility of this method for treating RP.[37]

CRISPR–Cas9 approach can also be employed for investigation of, for instance, loss-of-function mutations as observed in cancer-suppressor genes by inactivating of genes like TP53. In this way, the scientists can investigate the effects of their absence and get a clear picture of the process as these alterations impact the innovation and evolution of cancer. Organoids obtained from patients are three-dimensional cell models which have resemblance to a patient's tumor and they can also be manipulated by CRISPR. Researchers can examine the performance of modified genes in a model environment which accurately shows activity as in tumor biology. These can then be employed in drug testing and formation of personalized medicine plans.[38,39] Similarly, CRISPR-Cas9 can also contribute to addressing antibiotic resistance by enhancing the sensitivity of bacteria to antibiotics.[40] The genome of *Plasmodium*, the malaria causing parasite, has been edited by CRISPR that interferes with its life cycle and decreases its ability to infect humans. Cas13, has been modified to make highly accurate and effective diagnostic tools, having the potential to detect nucleic acids in pathogenic organisms.[41]

4. CRISPR-Cas for DNA methylation

The CpG islands in the promoter region and DNA methylation are related to regulation of transcription. By directing DNA methyltransferases (DNMTs) to those specific areas, it is possible to suppress genes by inducing 5-methyl cytosine (5mC), a potent regulator of gene transcription. Following the introduction of *de novo* 5mC at specific regions, primarily promoters, the complete length of human or mouse DNMT3A fused with dCas9 (dCas9eDNMT3A and dCas9eDNMT3ACD, respectively) can be suppressed. When comparing full-length dCas9eDNMT3A with the dCas9eDNMT3ACD, the latter showed better 5mC activity and dCas9eDNMT3A induced less off-target 5mC.[42] Furthermore, apart from mammalian DNMTs, scientists have also investigated the potential of the CpG methyltransferase (M.SssI) of prokaryotes for targeted DNA methylation. The introduction of the humanized M.SssI derivative (dCas9eMQ1) has resulted in enhanced amounts of newly formed 5mC which was distributed extensively across the target region. Nevertheless, in order to utilize M.SssI for specific DNA methylation, referred to as 'Precision epigenetic editing', additional changes are necessary due to the significant off-target 5mC effects.[43]

It has also been demonstrated that manipulating the methylation of p15 in human stem/progenitor cells (HSPCs) through epigenetic editing has an impact on hematopoiesis in living organisms. The DNA methylation (DNAm) may be specifically targeted and preserved in human hematopoietic stem and progenitor cells (HSPCs). Furthermore, it highlights the functional significance of abnormal DNA methylation on the p15 gene locus. Therefore, various abnormal DNA methylation patterns linked with ageing may affect hematopoiesis in living organisms.[44] Demethylation of DNA at targeted sites has been utilized in preclinical mice models and also in cultured cells. CGG-repeats located within the promoter region of the fragile X mental retardation 1 (FMR1) gene were demethylated byintroducing dCas9-TET1CD using lentiviral expression system into post-mitotic neurons obtained from induced pluripotent stem cells (iPSCs) of patients. The demethylation procedure efficiently restored FMR1 expression and improved neuronal activity in laboratory setting. These positive effects were sustained even after the insertion of the modified cells into the brain of mice.[45,46] Methyl group from 5mC were selectively removed by dCas9 proteins fused with TET1CD. The effectiveness of demethylation process and delivery of dCas9 tool differed, due to genetic and chromatin factors. Irrespective of the partial removal of DNA methylation at the specific target sites, the reactivation of the targeted genes was not up to the mark. That might be due to the presence of a repressive microenvironment, like histones are deacetylated and contain H3K9me2, a epigenetic modification of histone H3.[4]

Similarly, upon introducing the dCas9eTET3CD systems directly into the kidneys of mice with kidney fibrosis disease using lentiviral delivery system, methyl groups were removed from its gene promoters. It reactivated two genes that have anti-fibrotic properties. This process efficiently reduced kidney fibrosis and restored normal functioning sooner and in an effective manner.[47,48]

Granulin (GRN) is a powerful multipotent mitogen and growth factor that prolongs cancer survival by maintaining the self-renewal of these cells. Its levels are found to be increased in hepatoma tissues and are connected to reduced survival rates in hepatoma patients.[49] The introduction of dCas9 epi-suppressors significantly decreases the amounts of GRN mRNA in hepatoma cells and triggered new CpG DNA methylation in the GRN promoter region. Inhibition of GRN activity through epigenetic mechanisms led to the prevention of cell proliferation, decreased formation of tumor spheres, and less cell invasion. These changes have been partially achieved through the Metalloproteinases/Tissue Inhibitors of Metalloproteinases (MMP/TIMP) pathway and by using dCas9 epi-suppressors can be used to develop specific epigenetic therapy for cancer.[50] Researchers have also utilized the dCas9/DNMT3A catalytic domain to specifically trigger MGMT methylation in CpG clusters present close to promoter, enhancer and differentially methylated regions (DMR) regions. It was noticed that there was a decrease in MGMT expression and an increase in the glioma cells sensitivity to chemotherapy in laboratory experiments with a very few adverse effects.[51,52]

Genetic or epigenetic disturbances might result in difficulties in controlling the activation of imprinted genes, thus resulting in the development of diseases. Many illnesses are caused by recurring genetic changes, including deletions, duplications and uniparental disomy (UPD). The use of CRISPR-Cas9 editing, which depends on repairing DSB, does not provide significant therapeutic benefits for most physical and mental disabilities.[53]

5. CRISPR-Cas for histone modification

Targeted modification of the epigenetic environment includes the targeting of histones as well. Therefore, the catalytic domains of enzymes responsible for preserving the histone code have been utilized for the purpose of modifying the epigenome. Histone modifications include acetylation at lysine residues among others. Histone acetyltransferases (HATs) act as its catalyst, hence, increasing the overall expression of genes. Histone deacetylases (HDACs) catalyze deacetylation while Histone methyltransferases (HMTs) cause methylation of the targeted regions. Methylation of a certain residue can either activate or suppress gene expression. Kinases control phosphorylation at the C-terminal and N-terminal, assisted by ATP while phosphatases reverse phosphorylation, while controlling gene expression by causing condensation of chromatin.[54] Histone phosphorylation is a chemical alteration that enables eukaryotic cells to quickly react to changes in their surroundings. The correlation between histone phosphorylation and changes in gene expression has been established. However, establishing a direct causal relationship between this significant epigenomic modification and specific locations within native chromatin has been difficult due to the absence of reliable and targeted deposition technologies for endogenous histone phosphorylation. In order to bridge the gap, an engineered chromatin kinase has been developed named dCas9-dMSK1. This has been achieved by directly combining a non-cutting version of the CRISPR-Cas9 with a highly active, shortened form of the human MSK1 histone kinase. Targeting dCas9-dMSK1 to human promoters has resulted in increased histone phosphorylation and activation of genes. This investigation validates the hypothesis that there is an overabundance of phosphorylation occurring at serine 28 of histone H3 (H3S28ph).[55]

Conversely, histone ubiquitination is noticeably different as opposed to other histone modifications because of its large ubiquitin moiety that contains some 76 amino acids. Ubiquitination takes place at lysine 119 and at lysine 120 on histone H2A and histone H2B respectively.[56] Three primary enzymes, E1 (activating enzyme), E2 (conjugating enzyme), and E3 (ubiquitin ligase) catalyze ubiquitination process of histones.[57] Ubiquitination governs cell-cycle regulation, DNA repair, degradation of protein, response to stress, immune responses, endocytosis, transcriptional regulation and signal transduction.[58] SUMOylation involves addition of a small SUMO proteins at specific lysine residues. SUMOylation like ubiquitination, requires three enzymes; a heterodimer activating enzyme (E1: SAE1/SAE2), a conjugating enzyme (E2: Ubc9), and a SUMO ligase (E3).[59] When a molecule of ADP-ribose, either in the form of poly ADP-ribose (pADPr) chains or mono ADP-ribose (pADPr), is transferred from co-factor NAD+ to some target molecule, catalyzed by ART (ADP-ribosyltransferase) superfamily is referred to as ADP-ribosylation. Such ribosylated histones are involved in many important functions particularly DNA damage and repair.[60]

Within the H3 histone, H3K27 can undergo acetylation in non-regulatory areas that have been epigenetically altered to resemble enhancer elements. This targeted strategy leads to precise regulation of is chenage in the structure of chromatin at specified locations exhibits a distinctive regulatory presence for the enhancer and promoter region.[61] The PRC2 enzyme from polycomb complex methylates lysine 27 on histone H3, leading to the formation of the H3K27me modification. 119 on h Genetically modified mouse embryonic stem cells (ESCS) have been developed that can specifically target the EZH1 and EZH2 subunits of the polycomb complex (PRC2). These cells have the ability to differentiate between various forms of H3K27 methylation and locate the presence of PRC2/1 on the chromatin region.[62]

A fusion protein called dCas9 HDAC3 act as an artificial histone deacetylase that can regulate gene expression. But its activity entirely depends on the factors i.e., the transcription level of the target gene, the acetylation region, the point of gRNA placement and the dosage of the gRNA.[4,63] The methyl group is removed from H3K4me1/2 and H3K9me2 through the dCas9-LSD1 complex that possesses an antagonistic effect. dCas9-LSD1 reduces the expression level of target gene as well.[64] Researchers have utilized the synergistic activation mediator (SAM) technique to activate a natural promoter. Chicken embryos have been modified by this technique, making it possible to examine gene regulatory interactions in living organisms with goodprecision.[65]

The fusion of dCas9 with HDAC3 has the ability to either repress or stimulate transcription by elimination of acetyl group on H327 which completely depends on its environment.[66] In another study, the findings of an experiment involving Cas9 fusion with PRDM9 has been found to be responsible for histone methylation indicating that the fusion protein selectively interacts with chromatin at the Cas9 cleavage site in the genome, causing a threefold increase in observed HDR effectiveness and a fivefold enhancement in the ratio of HDR to indel mutations, in comparison to the effects of Cas9. The results establish the role of chromatin maintaining the equilibrium of DNA repair mechanisms in CRISPR-Cas genome editing.[67] Of the various chemical entities affecting the histone modifications, sodium butyrate and valproic acid compounds suppress the activity of histone deacetylases. CRISPR- The impact of valproic acid on the accuracy of CRISPR-Cas9 cutting seems to differ depending on the cell type. It can improve the precision of CRISPR-Cas9 by increasing the efficiency of the HDR repair pathway, hence, indicating that these genetic and epigenetic modifications can be controlled externally through the use of different drug molecules and other chemical entities.[68]

6. CRISPR-Cas for RNA targeting

Bacteria and archaea have a similar adaptive immune system called CRISPR-Cas. The objective of this is to provide defence against viral and plasmid infection by specifically targeting and fighting particular foreign nucleic acid sequences. The DNA and histone modification techniques employing CRISPR-Cas systems have been covered in the sections above. Types III and VI CRISPR-Cas systems detect RNA originating from viral transcription and potential RNA virus invasion. The identification of viral RNA by sequence-specific methods triggers a comprehensive cellular response that usually results in widespread damage to prevent the progression of the infection.[69] Type VI CRISPR-Cas13 specifically binds the foreign RNA. Cas13 has been modified to eradicate antibiotic-resistant microorganisms and to create detection techniques that are extremely specific, sensitive, capable of multiplexing, and adaptable to field conditions, all in a lab environment that does not involve cells. Crucially, Cas13 may be reprogrammed and utilized in eukaryotes to effectively fight against harmful RNA viruses or control gene expression. This enables the suppression of mRNA, circular RNA, and noncoding RNA, hence facilitating the regulation of genetic activity. Furthermore, Cas13 has been employed for in vivo manipulation of RNA, including the ability to regulate alternative splicing, perform A-to-I and C to U editing, and induce N^6-methyladenosine (m^6A) modifications.[70] Alternatively, Class 1 CRISPR-Cas systems include type III systems. In the majority of type III systems, crRNA maturation occurs over the course of two separate periods. The initial stage entails splitting a pre-crRNA transcript into smaller pieces, with a single spacer sequence in each. Cas6 is the agent responsible for executing this phase. These methods can utilize the Cas6 variety of existing CRISPR-Cas types within the cell to process the pre-crRNA. After this cleavage, crRNAs go through further processing. The exact technique for secondary processing is still unidentified. Nevertheless, it has been demonstrated that this trimming process results in the creation of two distinct populations of crRNA molecules, which vary in terms of their length.[71]

Type III CRISPR systems consist of a distinctive characteristic such that they exclusively employ three nuclease activities, setting them apart from other known systems. The initial process involves a Cas7-mediated RNA cleavage that is unique to the sequence. The specific RNA molecule is located alongside the CRISPR RNA (crRNA) in the ribonucleoprotein complex and is broken down by Cas7. Consequently, the RNA molecule is cleaved at regular intervals of 6 nucleotides. The mechanism referred to as the 'ruler' is prevalent in all type III systems.[72]

RNA knockdown, RNA imaging and tracking, nucleic acid detection, RNA editing, resistance against RNA viruses, Splicing Alteration, induction of apoptosis, removal of repetitive sequences and gene expression regulation are some of the applications of CRISPR based RNA targeting. Programmable, specific RNA cleavage can be achieved without the need for a protospacer-adjacent motif (PAM). It is, however, noteworthy that the he DNA that corresponds to the specific RNA sequence remains unaltered. This also acts by employing three nuclease activities i.e., targeted RNA cleavage, single-stranded DNA cleavage, and RNA degradation The intricate composition of the effector complex restricts its practical use.[73] dCas9 is genetically linked to a fluorescent protein and a nuclear localization signal, allowing it to identify natural, untagged, and unaltered mRNA and PAM is used. This technique does not disrupt the normal quantities of functioning mRNA and proteins.[74] The fusion of dCas13b with the ADAR2 domain allows for the specific conversion of targeted adenosine to inosine. The tiny size of this fusion protein facilitates its distribution using adeno-associated virus (AAV).[75] The process necessitates the presence of Cas13a and a quenched fluorescent RNA reporter. The target is specifically recognized, which then induces non-specific cleavage of RNA This process is characterized by its speed and cost-effectiveness.[76]

Cas9 has been shown to have several nucleolytic activities, similar to type III systems. In type II systems, the most common nuclease activity seen is specific double-stranded DNA cleavage. The conformational shift that occurs when the crRNA/tracrRNA duplex binds to Cas9 enables it to bind to and probe DNA. The process of recognition begins by detecting the Protospacer-Associated Motif (PAM) located at the 5′ end of the protospacer sequence. Following that, a process of complementary pairing takes place between the CRISPR RNA (crRNA) and the particular DNA sequence that is being targeted. As a result, this contact stimulates the activation of the DNase activity of the Cas9 enzyme. The HNH and RuvC domains, respectively, cleave the strand that is displaced and complementary to crRNA.[77,78] RNA editing, as opposed to DNA editing, provides a reversible and manageable option for genome editing. Firstly, the identification of RNA molecules overcomes obstacles caused by DNA changes, such as chromatin accessibility. Furthermore, RNA-targeting effectors typically do not rely on PAM that reduces the likelihood of the emergence of uncontrollable mutant variants.[79] Hence, the utilization of RNA-editing technologies is currently seeing a quick expansion and holds the potential to contribute to multiple fields. CRISPR-Cas RNA-targeting technologies enable precise targeting of certain nucleic acid fragments, including RNA molecules. It enables the process of modifying RNA when a specific target is identified. The activity can also be altered by developing CRISPR-Cas RNA-targeting systems that have non-functional nuclease activity (dCas). These two features precisely determine the entirety of potential uses.[80]

Biosensing techniques are essential for monitoring biomolecules involved in both physiological and pathological processes in living organisms. They possess a tremendous capability to contribute to both diagnostic and therapeutic avenues.[81] Using precise nucleotide-by-nucleotide detection, Cas-mediated biosensing depends on the catalytic ability of Cas effectors to degrade target and collateral RNA. Thanks to the non-selective breakdown of single-stranded RNA upon target identification, researchers can measure the amount of any given nucleic acid species in a test sample. Fluorescence can be measured for this purpose or colorometric method can be used for detection, hence, providing an opportunity to detect the effectiveness of the editing mechanisms.[82]

7. Pharmacological and toxicological aspects of CRISPR-Cas

CRIPSR-Cas systems have been able to positively contribute to all fields of biological and medical sciences. The ever-increasing incidence of antimicrobial resistance infections, for instance, requires the development of innovative strategies to combat them., RNA-effector system can efficiently address these challenges by addressing bacterial resistance.[83] Widespread utilization of Cas13-based antimicrobial medications can be attributed to their efficiency to specifically target certain bacterial targets. If RNA is targeted then it lowers the chance of resistance due to DNA mutations. Based on the experimental data, Cas13-based methods are more efficient than Cas9 because they are not impacted by the location of the gene on the chromosome or the plasmid and can specifically target the microbial machinery.[84]

CRISPR-based diagnostic techniques have been extensively developed as a substitute for PCR since 2016, as the latter necessitates specialized equipment and trained personnel, resulting in a time delay. Researchers recently conducted a comprehensive review to identify the latest advancements in CRISPR-based diagnostics, with or without amplification.[85] The use of the collateral cleavage activity of single-stranded DNA in Cas effectors can aid the identification of numerous viral strains, such as the ZIKA virus, Dengue and Human Papilloma Virus among others. Moreover, these diagnostic tests are taken into account while diagnosing a microbial infection.[86] Fig. 2 represents applications of CRISPR epigenetic modifications with particular reference to human diseases.

At the interdisciplinary front, a joint system of CRISPR-Cas13a and graphene field effect transistors (GFET) biosensors has effectively enabled the rapid and sensitive sensing of SARS-CoV-2 in recent years. It can detect viral RNA quickly and efficiently in samples, as well as determine the differences among variants of the SARS-CoV-2 virus by employing the CRISPR-Cas principles mentioned above.[87]

Similarly, other related tools adopted from specific microbes like *S. pyogenes* and *Saccharomyces cerevisiae*, SpCas9 and FnCpf1, have enabled the exploration of new drugs. It utilizes the micro Cas nuclease, AsCas12f1, that has shown a much superior transformation efficiency in

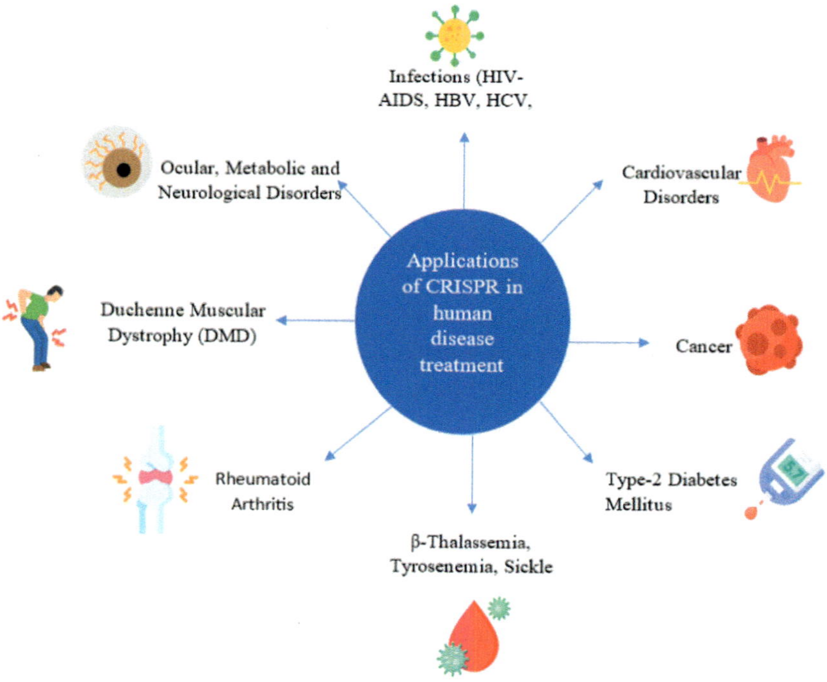

Fig. 2 Applications of CRISPR-Cas epigenetics for the treatment of human disorders. The CRISPR-Cas9 technology holds immense therapeutic promise for addressing several diseases with a known genetic basis of malfunction. It also offers opportunities for studying these diseases by generating cell or animal models. Gene editing therapy can potentially restore gene function or compensate for genomic mutations.

comparison to earlier tools. This is mainly due to theircompact size and relatively low DNA cleavage activity. Individual genes or gene clusters have been successfully deleted in *Streptomyces coelicolor* resulting in an improved efficiency of the system with a higher efficiency and low toxicity.[88]

ActivatedCas12a cuts DNA around it without any specific target, resulting into shorter DNA fragments. This helps to amplify the signal of the desired DNA and can be used for detection i.e., the target nucleic acid can trigger the initiation of cleavage by Cas12a protein. Fluorescent signals

Table 1 Characteristic features and effective therapeutic nature of CRISPR-Cas in epigenome editing.

Characteristic features	Effective therapeutic nature
1. Accurate Targeting: CRISPR-Cas systems have the ability to be accurately directed to specific sites in the genome by utilizing gRNAs.	1. Hereditary Diseases: CRISPR-Cas9 has demonstrated potential in addressing monogenic genetic illnesses by rectifying mutations that cause diseases of hereditary nature.
2. Versatility: CRISPR-Cas9, Cas12, and Cas13 possess the ability to modify genetic material in a wide range of animals, spanning from human cells to zebrafish.	2. Viral Infections: CRISPR-Cas systems have the potential to specifically target viral genomes, hence preventing their replication.
3. High Efficiency: CRISPR-Cas9 is highly effective in generating specific DNA breaks for the purpose of genetic editing.	3. Cancer Therapy: CRISPR-based approaches encompass the deactivation of oncogenes, augmentation of immune responses, and administration of cancer-eliminating compounds.
4. Multipurpose Applications: CRISPR-Cas systems can edit DNA and RNA both.	4. Immunological Diseases: CRISPR has the potential to tackle conditions related to the immune system, such as autoimmunity and immunodeficiency.
5. Potential for Clinical Application: Despite the presence of obstacles such as off-target effects, safety concerns, and delivery issues, CRISPR exhibits great potential for use in clinical treatments.	5. Personalized Medicine: CRISPR's accuracy enables customized treatments for individual patients.

are released that causes the system's single-stranded DNA fluorescent probe to cleave. This procedure can be employed for detection purposes. Therefore, the CRISPR-Cas12a system has recently become one of the most powerful tools for both gene editing and detection.[89] Table 1 summarizes the applications of CRISPR-Cas technique in therapeutics highlighting its key features.

8. Conclusion and future perspectives

CRISPR-Cas9 gene editing is a potential approach to treat genetic illnesses by utilizing inherited epigenetic changes. This innovative method provides accuracy and potential for addressing a diverse array of hereditary as well infectious diseases at their genetic origins. Scientists can use this system to precisely locate and alter diseased genes to fix mutations that can cause human diseases. The capacity to effectively modify the genome by DNA methylation, histone modification and RNA targeting has created innovative opportunities for addressing such conditions. This technology has the capacity to revolutionize the lives of many. Furthermore, it can be employed for customized therapies for precision medicine purpose. Nevertheless, it is imperative to practice caution and tackle the ethical and safety considerations related to CRISPR–Cas genome editing. Although CRISPR-Cas9 gene editing has various problems and requires attention, it is now undergoing substantial developments and has the capacity to revolutionize the therapy for many conditions. Stronger ethical and regulatory investigation and in-depth in vitro in vivo correlation studies are, however, necessary to translate the benefits of the technique for general public.

References

1. Han W, She Q. CRISPR history: discovery, characterization, and prosperity. *Prog Mol Biol Transl Sci*. 2017;152:1–21.
2. Gostimskaya I. CRISPR-Cas9: a history of its discovery and ethical considerations of its use in genome editing. *Biochem (Mosc)*. 2022;87(8):777–788. https://doi.org/10.1134/s0006297922080090
3. Chaudhuri A, Halder K, Datta A. Classification of CRISPR/Cas system and its application in tomato breeding. *Theor Appl Genet*. 2022;135(2):367–387. https://doi.org/10.1007/s00122-021-03984-y
4. Seem K, Kaur S, Kumar S, Mohapatra T. Epigenome editing for targeted DNA (de) methylation: a new perspective in modulating gene expression. *Crit Rev Biochem Mol Biol*. 2024:1–30.

5. Makarova KS, Wolf YI, Iranzo J, et al. Evolutionary classification of CRISPR–Cas systems: a burst of class 2 and derived variants. *Nat Rev Microbiol.* 2020;18(2):67–83. https://doi.org/10.1038/s41579-019-0299-x
6. Liu Z, Dong H, Cui Y, Cong L, Zhang D. Application of different types of CRISPR/Cas-based systems in bacteria. *Microb Cell Fact.* 2020;19(1):172. https://doi.org/10.1186/s12934-020-01431-z
7. Kissling L, Monfort A, Swarts DC, Wutz A, Jinek M. *Preparation and electroporation of Cas12a/Cpf1-guide RNA complexes for introducing large gene deletions in mouse embryonic stem cells. Methods Enzymol.* 616. Elsevier; 2019:241–263.
8. Koonin EV, Makarova KS. Origins and evolution of CRISPR-Cas systems. *Philos Trans R Soc B: Biol Sci.* 2019;374(1772):20180087. https://doi.org/10.1098/rstb.2018.0087
9. Goell JH, Hilton IB. CRISPR/Cas-based epigenome editing: advances, applications, and clinical utility. *Trends Biotechnol.* 2021;39(7):678–691. https://doi.org/10.1016/j.tibtech.2020.10.012
10. Wang S-W, Gao C, Zheng Y-M, et al. Current applications and future perspective of CRISPR/Cas9 gene editing in cancer. *Mol Cancer.* 2022;21(1):57. https://doi.org/10.1186/s12943-022-01518-8
11. Howie H, Rijal CM, Ressler KJ. A review of epigenetic contributions to post-traumatic stress disorder. *Dialogues ClNeurosci.* 2019;21(4):417–428. https://doi.org/10.31887/DCNS.2019.21.4/kressler
12. Yeo NC, Chavez A, Lance-Byrne A, et al. An enhanced CRISPR repressor for targeted mammalian gene regulation. *Nat methods.* 2018;15(8):611–616.
13. Sar P, Dalai S. CRISPR/Cas9 in epigenetics studies of health and disease. *Prog Mol Biol Transl Sci.* 2021;181:309–343. https://doi.org/10.1016/bs.pmbts.2021.01.022
14. Kiselev IS, Kulakova OG, Boyko AN, Favorova OO. DNA methylation as an epigenetic mechanism in the development of multiple sclerosis. *Acta Naturae.* 2021;13(2):45–57. https://doi.org/10.32607/actanaturae.11043
15. Pathania AS, Prathipati P, Pandey MK, et al. The emerging role of non-coding RNAs in the epigenetic regulation of pediatric cancers. *SemCancer Biol.* 2022;83:227–241. https://doi.org/10.1016/j.semcancer.2021.04.015
16. Gemberling MP, Siklenka K, Rodriguez E, et al. Transgenic mice for in vivo epigenome editing with CRISPR-based systems. *Nat Methods.* 2021;18(8):965–974.
17. Ferdosi SR, Ewaisha R, Moghadam F, et al. Multifunctional CRISPR-Cas9 with engineered immunosilenced human T cell epitopes. *Nat Commun.* 2019;10(1):1842.
18. Mills C, Riching A, Keller A, et al. A novel CRISPR interference effector enabling functional gene characterization with synthetic guide RNAs. *CRISPR J.* 2022;5(6):769–786.
19. Choudhury SR, Cui Y, Lubecka K, Stefanska B, Irudayaraj J. CRISPR-dCas9 mediated TET1 targeting for selective DNA demethylation at BRCA1 promoter. *Oncotarget.* 2016;7(29):46545.
20. O'Geen H, Bates SL, Carter SS, et al. Ezh2-dCas9 and KRAB-dCas9 enable engineering of epigenetic memory in a context-dependent manner. *Epigenetics Chromatin.* 2019;12:1–20.
21. Rahman MM, Tollefsbol TO. Targeting cancer epigenetics with CRISPR-dCAS9: principles and prospects. *Methods.* 2021;187:77–91. https://doi.org/10.1016/j.ymeth.2020.04.006
22. Jusiak B, Cleto S, Perez-Piñera P, Lu TK. Engineering synthetic gene circuits in living cells with CRISPR technology. *Trends Biotechnol.* 2016;34(7):535–547.
23. Haldeman JM, Conway AE, Arlotto ME, et al. Creation of versatile cloning platforms for transgene expression and dCas9-based epigenome editing. *Nucl Acids Res.* 2019;47(4):e23.

24. Cai R, Lv R, Shi XE, Yang G, Jin J. CRISPR/dCas9 tools: epigenetic mechanism and application in gene transcriptional regulation. *Int J Mol Sci.* 2023;24(19):14865.
25. Puthumana J, Chandrababu A, Sarasan M, Joseph V, Singh ISB. Genetic improvement in edible fish: status, constraints, and prospects on CRISPR-based genome engineering. *3 Biotech.* 2024;14(2):44. https://doi.org/10.1007/s13205-023-03891-7
26. Yue M, Wang Y, Chen H, Sun Z, Ju X-D. Recent progress in CRISPR/Cas9 technology. *J Genet Genom.* 2016;43. https://doi.org/10.1016/j.jgg.2016.01.001
27. Brezgin S, Kostyusheva A, Kostyushev D, Chulanov V. Dead Cas systems: types, principles, and applications. 2019;20(23) https://doi.org/10.3390/ijms20236041
28. Allen D, Kalter N, Rosenberg M, Hendel A. Homology-directed-repair-based genome editing in HSPCs for the treatment of inborn errors of immunity and blood disorders. *Pharmaceutics.* 2023;15(5) https://doi.org/10.3390/pharmaceutics15051329
29. Leal AF, Herreno-Pachón AM, Benincore-Flórez E, Karunathilaka A, Tomatsu S. Current strategies for increasing knock-in efficiency in CRISPR/Cas9-based approaches. *Int J Mol Sci.* 2024;25(5):2456.
30. Xue C, Greene EC. DNA repair pathway choices in CRISPR-Cas9-mediated genome editing. *Trends Genet.* 2021;37(7):639–656. https://doi.org/10.1016/j.tig.2021.02.008
31. Yang H, Ren S, Yu S, et al. Methods favoring homology-directed repair choice in response to CRISPR/Cas9 induced-double strand breaks. *Int J Mol Sci.* 2020;21(18):6461.
32. Ancos-Pintado R, Bragado-García I, Morales ML, et al. High-throughput CRISPR screening in hematological neoplasms. *Cancers.* 2022;14(15) https://doi.org/10.3390/cancers14153612
33. Call SN, Andrews LB. CRISPR-based approaches for gene regulation in non-model bacteria. *Front Genome Ed.* 2022;4. https://doi.org/10.3389/fgeed.2022.892304
34. Dow LE. Modeling disease in vivo with CRISPR/Cas9. *Trends Mol Med.* 2015;21(10):609–621. https://doi.org/10.1016/j.molmed.2015.07.006
35. Wang H, La Russa M, Qi LS. CRISPR/Cas9 in genome editing and beyond. *Annu Rev Biochem.* 2016;85:227–264.
36. Xu Y, Li Z. CRISPR-Cas systems: overview, innovations and applications in human disease research and gene therapy. *Comput Struct Biotechnol J.* 2020;18:2401–2415. https://doi.org/10.1016/j.csbj.2020.08.031
37. Kolanu ND. CRISPR–Cas9 gene editing: curing genetic diseases by inherited epigenetic modifications. *Glob Med Genet.* 2024;11(01):113–122.
38. Awwad SW, Serrano-Benitez A, Thomas JC, Gupta V, Jackson SP. Revolutionizing DNA repair research and cancer therapy with CRISPR–Cas screens. *Nat Rev Mol Cell Biol.* 2023;24(7):477–494.
39. Sachs N, De Ligt J, Kopper O, et al. A living biobank of breast cancer organoids captures disease heterogeneity. *Cell.* 2018;172(1):373–386.e310.
40. Ates A, Tastan C, Ermertcan S. Precision genome editing unveils a breakthrough in reversing antibiotic resistance: CRISPR/Cas9 targeting of multi-drug resistance genes in methicillin-resistant Staphylococcus aureus. *bioRxiv.* 2024 2023.2012. 2031.573511.
41. Zhang Y, Li S, Li R, et al. Advances in application of CRISPR-Cas13a system. *Front Cell Infect Microbiol.* 2024;14:1291557.
42. Gjaltema RA, Rots MG. Advances of epigenetic editing. *Curr OpChem Biol.* 2020;57:75–81.
43. Subramanian AT, Roy P, Aravind B, Kumar AP, Mohannath G. Epigenome editing strategies for plants: a mini review. *Nucl.* 2024:1–13.
44. Saunderson EA, Encabo HH, Devis J, et al. CRISPR/dCas9 DNA methylation editing is heritable during human hematopoiesis and shapes immune progeny. *Proc Natl Acad Sci.* 2023;120(34):e2300224120.

45. Qian J, Guan X, Xie B, et al. Multiplex epigenome editing of MECP2 to rescue Rett syndrome neurons. *Sci Transl Med*. 2023;15(679):eadd4666.
46. Qian J, Liu SX. *CRISPR/dCas9-Tet1-mediated DNA methylation editing bio-protocol*. 2024;14:8.
47. Xiao X, Wang W, Guo C, et al. Hypermethylation leads to the loss of HOXA5, resulting in JAG1 expression and NOTCH signaling contributing to kidney fibrosis. *Kidney Int*. 2024.
48. Xu X, Tan X, Tampe B, et al. High-fidelity CRISPR/Cas9-based gene-specific hydroxymethylation rescues gene expression and attenuates renal fibrosis. *Nat Commun*. 2018;9(1):3509.
49. Zhang L, Saito H, Higashimoto T, et al. Regulation of muscle hypertrophy through granulin: relayed communication among mesenchymal progenitors, macrophages, and satellite cells. *Cell Rep*. 2024;43(4).
50. Wang H, Guo R, Du Z, et al. Epigenetic targeting of granulin in hepatoma cells by synthetic CRISPR dCas9 epi-suppressors. *Mol Ther Nucl Acids*. 2018;11:23–33.
51. Stepper P, Kungulovski G, Jurkowska RZ, et al. Efficient targeted DNA methylation with chimeric dCas9–Dnmt3a–Dnmt3L methyltransferase. *Nucl Acids Res*. 2017;45(4):1703–1713.
52. Zapanta Rinonos S, Li T, Pianka ST, et al. dCas9/CRISPR-based methylation of O-6-methylguanine-DNA methyltransferase enhances chemosensitivity to temozolomide in malignant glioma. *J Neuro-oncol*. 2024;166(1):129–142.
53. Syding LA, Nickl P, Kasparek P, Sedlacek R. CRISPR/Cas9 epigenome editing potential for rare imprinting diseases: a review. *Cells*. 2020;9(4):993.
54. Lee HT, Oh S, Ro DH, Yoo H, Kwon YW. The key role of DNA methylation and histone acetylation in epigenetics of atherosclerosis. 2020;9(3):419–434. https://doi.org/10.12997/jla.2020.9.3.419
55. Li J, Mahata B, Escobar M, et al. Programmable human histone phosphorylation and gene activation using a CRISPR/Cas9-based chromatin kinase. *Nat Commun*. 2021;12(1):896.
56. Hosseini A, Minucci S. Chapter 6 - alterations of histone modifications in cancer. In: Tollefsbol TO, ed. 2nd ed. Academic Press; 2018:141–217. Epigenetics in Human Disease; 6.
57. Kim HS, Shi J. Chapter 9 - Epigenetics in precision medicine of pancreatic cancer. In: García-Giménez JL, ed. Academic Press; 2022:257–279. Epigenetics in Precision Medicine; 30.
58. Peng H, Zong D, Zhou Z, Chen P. Chapter 13 - Pulmonary diseases and epigenetics. In: Tollefsbol TO, ed. *Medical Epigenetics*. Boston: Academic Press; 2016:221–242.
59. Simonet NG, Rasti G, Vaquero A. Chapter 21 - The histone code and disease: posttranslational modifications as potential prognostic factors for clinical diagnosis. In: García-Giménez JL, ed. *Epigenetic Biomarkers and Diagnostics*. Boston: Academic Press; 2016:417–445.
60. Zha J-J, Tang Y, Wang Y-L. Role of mono-ADP-ribosylation histone modification (review). *Exp Ther Med*. 2021;21(6):577. https://doi.org/10.3892/etm.2021.10009
61. Eisenhut P, Marx N, Borsi G, et al. Manipulating gene expression levels in mammalian cell factories: an outline of synthetic molecular toolboxes to achieve multiplexed control. *N Biotechnol*. 2023.
62. Lavarone E, Barbieri CM, Pasini D. Dissecting the role of H3K27 acetylation and methylation in PRC2 mediated control of cellular identity. *Nat Commun*. 2019;10(1):1679.
63. Kwon DY, Zhao Y-T, Lamonica JM, Zhou Z. Locus-specific histone deacetylation using a synthetic CRISPR-Cas9-based HDAC. *Nat Commun*. 2017;8(1):15315.
64. Cortés-Fernández de Lara J, Núñez-Martínez HN, Tapia-Urzúa G, et al. A novel cis-regulatory element regulates αD and αA-globin gene expression in chicken erythroid cells. *Front Genet*. 2024;15:1384167.

65. Williams RM, Senanayake U, Artibani M, et al. Genome and epigenome engineering CRISPR toolkit for in vivo modulation of cis-regulatory interactions and gene expression in the chicken embryo. *Development*. 2018;145(4):dev160333.
66. Neja S, Dashwood WM, Dashwood RH, Rajendran P. Histone acyl code in precision oncology: mechanistic insights from dietary and metabolic factors. *Nutrients*. 2024;16(3):396.
67. Chen E, Lin-Shiao E, Trinidad M, Saffari Doost M, Colognori D, Doudna JA. Decorating chromatin for enhanced genome editing using CRISPR-Cas9. *Proc Natl Acad Sci*. 2022;119(49):e2204259119.
68. Björnson Y, Huang CY, Rollins JL, et al. The effect of histone deacetylase inhibitors on the efficiency of the CRISPR/Cas9 system. *Biochem Biophysics Rep*. 2023;35:101513.
69. van Beljouw SP, Sanders J, Rodríguez-Molina A, Brouns SJ. RNA-targeting CRISPR–Cas systems. *Nat Rev Microbiol*. 2023;21(1):21–34.
70. Kordyś M, Sen R, Warkocki Z. Applications of the versatile CRISPR-Cas13 RNA targeting system. *Wiley Interdiscip Rev RNA*. 2022;13(3):e1694.
71. Burmistrz M, Krakowski K, Krawczyk-Balska A. RNA-targeting CRISPR–Cas systems and their applications. *Int J Mol Sci*. 2020;21(3):1122.
72. Nickel L, Ulbricht A, Alkhnbashi OS, et al. Cross-cleavage activity of Cas6b in crRNA processing of two different CRISPR-Cas systems in Methanosarcina mazei Gö1. *RNA Biol*. 2019;16(4):492–503.
73. Ko SC, Woo HM. CRISPR–dCas13a system for programmable small RNAs and polycistronic mRNA repression in bacteria. *Nucl Acids Res*. 2024;52(1):492–506.
74. Hao M, Ling X, Sun Y, et al. Tracking endogenous proteins based on RNA editing-mediated genetic code expansion. *Nat Chem Biol*. 2024:1–11.
75. Montagud-Martínez R, Márquez-Costa R, Heras-Hernández M, Dolcemascolo R, Rodrigo G. On the ever-growing functional versatility of the CRISPR-Cas13 system. *Microbial Biotechnol*. 2024;17(2):e14418.
76. Yang H, Zhang Y, Teng X, Hou H, Deng R, Li J. CRISPR-based nucleic acid diagnostics for pathogens. *TrAC Trends Anal Chem*. 2023;160:116980.
77. Chi H, Hoikkala V, Grüschow S, Graham S, Shirran S, White MF. Antiviral type III CRISPR signalling via conjugation of ATP and SAM. *Nature*. 2023;622(7984):826–833.
78. Jinek M, Jiang F, Taylor DW, et al. Structures of Cas9 endonucleases reveal RNA-mediated conformational activation. *Science*. 2014;343(6176):1247997.
79. Khanzadi MN, Khan AA. CRISPR/Cas9: nature's gift to prokaryotes and an auspicious tool in genome editing. *J Basic Microbiol*. 2020;60(2):91–102.
80. Gunitseva N, Evteeva M, Borisova A, Patrushev M, Subach F. RNA-dependent RNA targeting by CRISPR-cas systems: characterizations and applications. *Int J Mol Sci*. 2023;24(8):6894.
81. Tong B, Dong H, Cui Y, Jiang P, Jin Z, Zhang D. The versatile type V CRISPR effectors and their application prospects. *Front Cell Dev Biol*. 2021;8:622103.
82. Liu X, Hussain M, Dai J, et al. Programmable biosensors based on RNA-guided CRISPR/Cas endonuclease. *Biol Proced Online*. 2022;24(1):2.
83. Shen H, Nourmohammadi S, Zhou Y, et al. Validating the target functions and synergistic multi-target, multi-pathway action mode of compound kushen injection using CRISPR/CAS. *bioRxiv*. 2024 2024.2005. 2003.592304.
84. Collias D, Beisel CL. CRISPR technologies and the search for the PAM-free nuclease. *Nat Commun*. 2021;12(1):555.
85. Liu FX, Cui JQ, Wu Z, Yao S. Recent progress in nucleic acid detection with CRISPR. *Lab Chip*. 2023;23(6):1467–1492.
86. Hang X-M, Liu P-F, Tian S, Wang H-Y, Zhao K-R, Wang L. Rapid and sensitive detection of Ebola RNA in an unamplified sample based on CRISPR-Cas13a and DNA roller machine. *Biosens Bioelectron*. 2022;211:114393.

87. Yin W, Li L, Yang Y, et al. Ultra-sensitive detection of the SARS-CoV-2 nucleocapsid protein via a clustered regularly interspaced short palindromic repeat/Cas12a-mediated immunoassay. *ACS Sens*. 2024.
88. Hua H-M, Xu J-F, Huang X-S, Zimin AA, Wang W-F, Lu Y-H. Low-toxicity and high-efficiency streptomyces genome editing tool based on the miniature type V–F CRISPR/Cas nuclease AsCas12f1. *J Agric Food Chem*. 2024.
89. Li Y, Zhou S, Wu Q, Gong C. CRISPR/Cas gene editing and delivery systems for cancer therapy. *Wiley Interdiscip Rev Nanomed Nanobiotechnol*. 2024;16(1):e1938.

CHAPTER NINE

Current progress in CRISPR–Cas systems for cancer

Hunaiza Fatima[a,b,1], Hajra Ali Raja[a,c,1], Rabia Amir[b], Alvina Gul[b], Mustafeez Mujtaba Babar[a,d,*], and Jayakumar Rajadas[d,*]
[a]Shifa College of Pharmaceutical Sciences, Shifa Tameer-e-Millat University, Islamabad, Pakistan
[b]Atta-ur-Rahman School of Applied Biosciences, National University of Sciences and Technology, Islamabad, Pakistan
[c]Health Services Academy, Ministry of Health, Islamabad, Pakistan
[d]Advanced Drug Delivery and Regenerative Biomaterials Lab, Stanford University School of Medicine, Stanford University, Palo Alto, CA, United States
*Corresponding authors. e-mail address: mustafeez.babar@fulbrightmail.org; jayraja@stanford.edu

Contents

Abstract

Cancer has been a primary contributor to morbidity and mortality worldwide. With an increasing trend of incidence and prevalence of cancer, progress has also been made in its treatment, starting from radiation and chemotherapy to immunotherapy and gene therapy. CRISPR-Cas technique, a promising gene editing tool, has been employed in cancer research for novel treatment regimens, identification of therapeutic targets, and unraveling the genetic mechanisms behind oncogenesis. CRISPR-

[1] Authors contributed equally.

Progress in Molecular Biology and Translational Science, Volume 208
ISSN 1877-1173, https://doi.org/10.1016/bs.pmbts.2024.07.007

based genome editing helped in identifying the roles of specific genetic factors linked to treatment resistance, metastasis, and cancer development. CRISPR allows the discovery of genes and treatment options through specifically interrupting tumor activators or activating tumor suppressor genes in cancer cells. Advancements in CRISPR technology, especially the use of immune cells like chimeric antigen receptor (CAR) T cells, has the potential to revolutionize personalized cancer treatment by precisely targeting and killing cancer cells. Furthermore, reactivating tumor suppressor genes makes cancer cells more susceptible to chemotherapy or immunotherapy. CRISPR-mediated genome editing can, hence, help to overcome resistance to traditional cancer treatments. The current manuscript covers that how is the CRISPR technology propelling revolutionary development in the field of cancer research, providing advance perspectives on the molecular causes of the disease and creating new lines for the development of more precise and potent cancer therapies.

Abbreviations

CAR	Chimeric Antigen Receptor
CRISPRa	CRISPR Activation
CRISPRi	CRISPR Interference
crRNA	Crispr RNA
dCas9	Dead Cas9
DSB	Double Strand Break
FDA	Food and Drug Administration
gRNA	Guide RNA
HDR	Homology Directed Repair
IARC	International Agency for Research on Cancer
KO	Knockout
NGS	Next Generation Sequencing
NHEJ	Non-Homologous End Joining
pegRNA	Prime Editing Guide RNA
PAM	Protospacer Adjacent Motif
ROS	Reactive Oxygen Species
TALENs	Transcription Activator Like Effector Nucleases
TNBC	Triple-Negative Breast Cancer
WES	Wole Exome Sequencing
WHO	World Health Organization
ZFNs	Zinc Finger nucleases

1. Introduction

Cancer has been documented in humans and other animals throughout recorded history. The earliest accounts of illness date back to Egyptian texts and describe it as an incurable condition. However, cancer cases were uncommon then. Since 1990, early onset cancer incidents and mortality have increased globally, with an approximate 75% increase from

1990 to 2019.[1] In 2019, there were almost 10 million deaths due to cancer. However, that number was fewer than 6 million in 1990. Recent statistics from International Agency for Research on Cancer (IARC) documented 20 million new cases and 9.7 million deaths because of cancer in 2022.[2] World Health Organization (WHO) predicted 35 million additional instances of cancer worldwide by 2050, a 77% increase from current levels.[3] Rise in disease burden has prompted an enhanced research towards finding cancer therapy. Over the last few decades, several conventional and contemporary options have been used to combat cancer. Traditional remedies include use of medicinal plants that promote apoptosis and inhibit tumor cell growth, invasion, angiogenesis, and metastasis. These medicines reduce pain, strengthen the patient's immune system, or manage adverse drug reactions associated with the treatment. Standard alternative treatments include radiation therapy, chemotherapy, and surgery either alone or in combination. These techniques possess the inherent problems of poor specificity, toxicity, chemoresistance and patient compliance.[4] Improved understanding of disease pathogenesis has contributed to the development of targeted drugs and immunotherapy. Because of the superior efficacy and high targeting, molecular-targeted medicines are progressively replacing the conventional chemotherapeutic options as they act by intruding specific molecular aberrations driving cancer. Two major categories of molecular targeted therapy are monoclonal antibodies (mAbs) and small molecule kinase inhibitors (SMKIs). SMKIs *e.g.*, trastuzumab (Herceptin), Ceritinib, Gefitinib, Afatinib work by blocking the enzymatic activity of kinases in signaling pathways while monoclonal antibodies (mAbs) including cetuximab and panitumumab recruit immune system against cancerous cells by making the latter more visible to immune system. Nevertheless, during clinical application, transient tumor regressions restrict their advantages.[5,6] The rapid advancements of contemporary immunotherapeutics have resulted in a reduction of malignant cancers and enhanced patient resilience. A number of advanced immunotherapies are being licensed for use in healthcare facilities to treat the patients. Few examples of immunotherapies are the sipuleucel-T, ipilimumab, pembrolizumab and nivolumab and talimogene laherparepvec (T-Vec). Despite of remarkable specificity, a decreased patient tolerance is a major obstacle to these therapeutic approaches. Immune therapy works only for a small set of cancers and patient response rates vary greatly amongst cancer subtypes.[7] To overcome the limitation of immune evasion a range of gene therapy techniques are used. Gene therapy is an approach to cure a disease by genetic modification of abnormal cells and include the likes of transcription activator-like effector nucleases (TALENs), Zinc Finger nucleases (ZFNs) and clustered regularly interspaced short palindromic repeats (CRISPR)—Cas9 as they have demonstrated promising results by inserting, deleting or mutating target genes. Over the TALENs and ZFNs, CRISPR—Cas system drawn the interest of researchers because of more flexible and effective editing options.[8] Though the CRISPR—Cas system has immense promise, research is continuing to improve the CRISPR/Cas therapy in terms of editing efficiency, delivery methods and minimizing the off-target effects.

2. Basics of cancer biology

Cancer is a complex genetic and epigenetic disorder characterized by uncontrolled cell division. It has been the most feared of human diseases. Till early 20th century, little was known about its pathogenesis. However, advances in sequencing techniques such as next-generation sequencing (NGS), whole exome sequencing (WES), RNA sequencing (RNA-Seq) and whole genome sequencing (WGS) enabled the comprehensive analysis of most cancer's abnormal genes *i.e.*, the genes that regulate the cell activities specially their growth and division.[9] Any intrinsic or extrinsic damage to genome leads to mutations in genes that can transform normal cells into the malignant ones. Intrinsic damage can be a result of replication errors, chemical instability of DNA bases or attack by reactive oxygen species (ROS). Some extrinsic factors such as ionizing radiation, chemical toxins or UV light can also damage DNA. Though the cells possess mechanisms to correct these sorts of damages, errors can still occur and the genome can become permanently altered, a process known as mutation. Genes that are in charge of preserving genomic integrity can experience certain inactivating mutations, as well, which makes it easier for new mutations to arise.[10] Mutational changes in the genes responsible for cell proliferation, cell death/apoptosis and genes that protect DNA from mutational processes make the normal cell to evolve into malignant one. Mutations in normal homeostatic genes result in the activation of growth promoting genes or oncogenes and inactivation of tumor suppressor genes. This alters the balance between birth and death rates of cells resulting in increased cell multiplication and formation of a new body or neoplasia that when spreads to other body parts is called a malignancy.[10]

For cancer cells to attain unlimited cell division, firstly the mutated genes must have to activate cell proliferation pathways and inactivate the cascades controlling the cell death. Cell proliferation is controlled by signals/growth factors that are produced locally and bind to a receptor. This binding initiates a signaling cascade that proliferates the cell and terminates upon signal removal. To avoid controls on cell proliferation cancerous cells become irresponsive to external signals causing a constant activation of proliferative pathways. In conjugation to oncogenes, some tumor suppressors genes are inactivated that further suppress tumorigenicity *e.g.*, retinoblastoma gene RB. In normal or untransformed cells, RB in conjugation with transcription factor E2F restricts the cell's entry into a newcell cycle. While in a transformed cell RB is inactivated/suppressed through phosphorylation by cyclin D1, disrupting RB/E2F that activates some genes necessary for entrance into cell cycle leading to cancerous cell division.[11] Another example of tumor suppressor gene is p53, an apoptosis inducer gene. Malignant cells evade apoptosis caused by Internal or external factors by inactivating the expression of p53.[11] Similarly, cancer cells morphologically develop mechanisms to enhance their growth, for instance by modifying connexins. Connexins are proteins that surround the gap junctions and allow communication between cells. Cancer cells break these connections with neighbors to assist abnormal proliferation.[10] Inactivation of tumor suppressors and destroying these connections with neighbors are the mechanisms of cancer cells to avoid the cell death. Moreover, tumor cells attain immortality by reactivating telomerase that maintains telomere length in a transformed cell. A telomer is a repetitive sequence at the end of chromosome that protects DNA from degradation. Telomer length reduces with expanding number of divisions, hence the need to be reconstructed to avoid senescence. Telomerases are the enzymes that rebuild the telomer and cancer cells reactivate telomerase to achieve immortality.[12] Another challenge for a cancerous cell is a continuous supply of nutrients that is ensured through fast-paced angiogenesis. To achieve this, tumor cells turn on many pro-angiogenic signals and suppress its negative regulators. The complex interplay between these cascades allow to intervene the process through CRISPR mediated gene therapy.[10] These mechanisms adopted by the cancer cells for their survival have been summarized in Fig. 1.

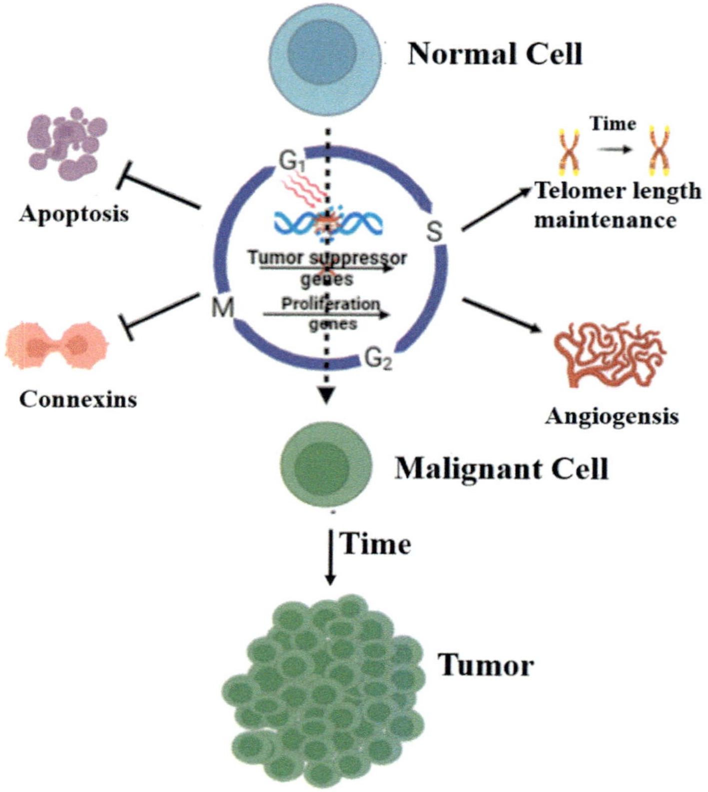

Fig. 1 Conversion of normal cell into tumorigenic cell. Alteration in expression levels of cell cycle regulators make the cell resistant to pro-apoptotic signals. Continuous proliferation and telomerase renewal mechanisms make cell resistant to senescence, hence, achieving malignancy. Over time these dividing malignant cells make a tumor.

3. CRISPR-Cas techniques for cancer genome editing

Cluster Regulatory Interspaced Palindromic Repeats (CRISPR) associated protein (Cas) is a technique adopted from prokaryotic defense mechanism. Many bacteria and archaea display an adaptive immunity by incorporating the invader's genome termed "spacer" as a memory at specific location into their genome. These spacers are separated by short palindromic sequences, hence the locus is known as Clustered regularly interspaced short palindromic repeats or CRISPR. When the infection recurs, the spacers between repeats are produced as tiny CRISPR RNAs (crRNAs), which Cas proteins use to target invader's genetic sequence-specifically. This capacity of CRISPR-Cas system of targeting specific DNA sequence has created novel ways to modify and maniputlate genome in a wide variety of cells. A synthetic guide RNA (gRNA) complementary to target is used instead of crRNA that guides the Cas nuclease to cleave the sequence. The cleaved ends are joined back either through homology directed repair (HDR) or non-homologous end joining (NHEJ). During NHEJ, manipulations occur naturally and hence, can be ensure the addition or deletion of sequence. Cancerous cells have incorporated genetic or epigenetic mutations in genes that alter their expression, thus driving t carcinogenesis. These genes can be targeted to regulate their expression for precise cancer therapy.[13,14] Identified oncogene is used to synthesize gRNA that binds to gene and makes the recruited nuclease to break the strand. Now, the gene can inactivate through frame shift mutation *e.g.*, MYC gene inactivation in lymphoma animal models presented reduction in tumor growth.[15] Similarly, CRISPR/Cas mediated inactivation of interleukin 30 has shown to inhibit tumorigenesis in colon cancer cells.[16] Alternatively, reactivation of tumor suppressor genes is achieved to attenuate malignancy *e.g.*, DKK2 gene expression inhibits prostate cancer by modulating TGF -β signaling in PC3 cell lines.[17] CRISPR/Cas9 mediated manipulation of genes involved in drug resistance can be achieved to overcome chemoresistance. Recent studies have shown that the technique can be used to target important genes involved in these mechanisms, thereby increasing the sensitivity of cancer cells for the drugs. For instance, stable knockout of drug transporter gene *ABCB1* increased the therapeutic efficacy of [^{3}H]-paclitaxel in MDR SW620/Ad300 cells.[18]

3.1 Cas protein variants improved CRISPR specificity

Despite the extraordinary capability of CRISPR/Cas system to edit genes, it continues to evolve over time to improve its specificity and efficacy. A major constraint in CRISPR-Cas practicability isthe unintended off targeting. A number of Cas variants have been identified and engineered to improve specificity of cleavage. Cas9 nuclease is the earliest and commonly used protein that recognizes a protospacer adjacent motif (PAM) sequence adjacent to the target site. Cas9 can cleave target sites identical to the DNA bearing some mismatches between guide RNA (gRNA) and target DNA. Fig. 2 represents the mechanisms CRISP-Cas system employs for genetic modifications. To limit this issue, several modifications in CRISPR/Cas9 systems have been incorporated to achieve specificity. Firstly, the gRNA

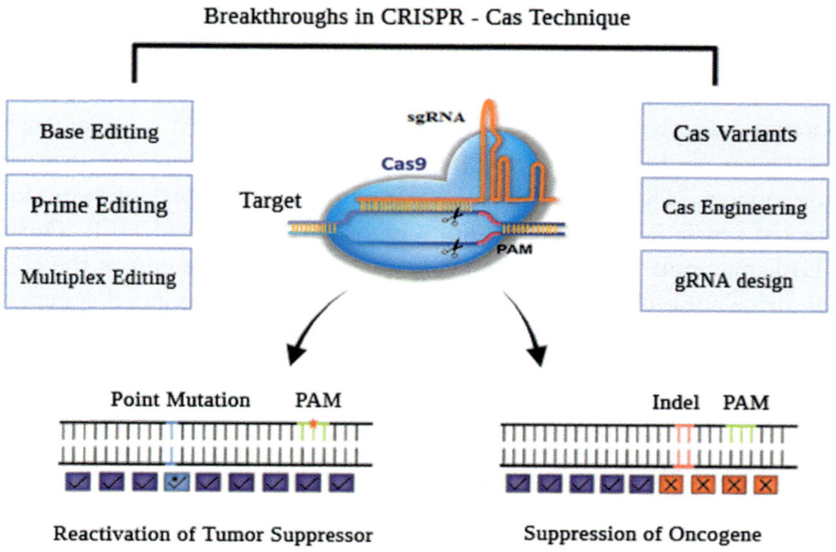

Fig. 2 CRISPR—Cas editing and major breakthroughs in CRISPR—Cas technique: Target site is guided by small guide RNA (sgRNA) which edit tumor suppress genes to active them and mutate oncogenes to suppress their expression. Advancements in CRISPR Cas design and execution are made to enhance specificity and efficiency.

are designed for sites with fewer off-target effects and a truncated gRNA at 5' end expected to bear less mismatches is used to reduce non specificity.[19] Secondly, Cas9 nickase mutants are employed to create single strand breaks (SSB). A pair of Cas9 nickase generates two single strand breaks close to each other equivalent to a double strand break (DSB). A paired Cas9 nickase increases specificity because an off-target single nick should be corrected immediately, in order to prevent unwanted mutations. Moreover, the likelihood of two nicks occurring simultaneously in the same configuration as target sites in other regions of the genome and producing double-strand breaks is minimum. The double nicking approach can greatly boost specificity but at the expense of low efficiency.[20] Hence, an alternative approach is the fusion of catalytically inactive Cas9 (dCas9) with Fok1 nuclease, collectively called fCas9. Fok1 is a restriction enzyme that cleaves away from the site of binding. Targeting of this site is achieved through gRNA and dead Cas9 (dCas9) while the nuclease activity is achieved through Fok1. As Fok1 is active only when it makes a dimer, two fCas9 monomers are engaged to simultaneously target sequence with greater specificity. The fCas9 is reported to cleave with a specificity 140 fold greater than wild type Cas9 with no compromise in efficiency.[21] Besides the synthetic variants of Cas9 multiple other Cas proteins have been identified that work with less off-target effects. Cas12a and their variants are employed in genetic engineering as they bind to a T rich PAM. EGFR and BRAF genes are disrupted by using CRISPR—Cas12a system to suppress cancer.[22,23] Successful engineering of CART cells by using Cas12a is reported by.[24]

3.2 CRISPR—Cas techniques for enhanced efficiency

Several breakthroughs in the field of CRISPR—Cas system have been incorporated to improve the efficiency of its editing. Prime editing is a very specific and efficient way to insert, delete or replace a base in target gene. This editing methodology needs one RNA called prime editing guide RNA (pegRNA) as a scaffold for Cas9 nickase and a reverse transcriptase enzyme. pegRNA contains the sequence to be edited upon binding Cas9 nickase that creates single strand break and RT repair according to complementary RNA. Hence, it is also termed as "Find and Replace" technique. A recent study conducted to correct KRAS mutations guided by pegRNA in human cancer cells has been found to successfully correct G13D KRAS.[25] Another programable method of CRISPR—Cas editing in cancer is base editing. Base editing is a precise method of creating point mutations. This system has a Cas9 nickase attached with a base converting enzymes. Base coupling between the gRNA and target DNA sequence causes a small length of ssDNA to be shifted into a "R-loop". The deaminase enzyme changes the nucleotide bases in ssDNA. Opposite strand is nicked by Cas and the mismatch of R loop is repaired by cellular mechanisms.[26] Base editing has been employed to treat point mutations in genes leading to cancer. For example, base editing of TERT promoter mutation to reduce the TERT gene expression has been carried out recently. TERT gene gets activated because of various mutations in glioblastoma cells. One of which is a point mutation in its promoter that allows the binding of E26 transcription factor family to promoter. Adenine base editing of mutated promotor reduces the expression levels of TERT leading to cell senescence and, hence, proliferation arrest.[27]

Another advancement in CRISPR—Cas called multiplex editing is the concurrent engineering of many loci by using a single CRISPR-Cas complex. The idea behind delivering multiple gRNAs is that a single CRISPR complex might potentially address many intended or unexpected genomic regions at once when it is introduced into a cell. Multiplex editing is widely tested in cancer immunotherapy. Ren and colleagues generated CART cells resistant to PD1 inhibition through multiplex genome editing.[28] These advancements in CRISPR—Cas system have lead to a new era of genome engineering for cancer therapy and immunotherapy.

4. CRISPR-based screening for cancer therapeutics

Genome editing has become a viable treatment approach for several disorders.[29] Targeting DNA domain-binding proteins and various other genomic engineering techniques, including ZFNs and TALENs, have been designed for use in cancer therapy. However, their limited capacity to target epigenetic alterations originating in cancer cells have restricted their application.[30] Recently, effective, long-term safe cancer treatment has been proposed by CRISPRs associated with the HNH domain protein Cas9 which is a more versatile genome editing approach.[31] Identifying the underlying genetic pathways of cancer growth and discovering possible therapeutic targets have been made possible by CRISPR-based screening, which has proven to be an effective tool in cancer investigation.

Treatments for a number of diseases, including leukemia, multiple myeloma, glioblastoma, lung cancer and breast cancer, have been studied using CRISPR Cas9 technique. The CRISPR-Cas and next generation sequencing (NGS) have been combined for effective cancer treatment.[32] Instead of using protein-DNA interactions to facilitate target recognition, CRISPR-Cas9 system mediates binding through a RNA molecule. This contrasts with earlier genome editing techniques as CRISPR-associated (Cas) proteins and alternating repeat-spacer units, or CRISPR loci, are components of a prokaryotic host defense mechanism that blocks plasmids and viral genomes.[33]

The Zhang Lab was the very first group to use CRISPR-Cas9 screens on large scale to investigate drug resistance in cancers. Thanks to the ease of usage and dependability of CRISPR-Cas9 technology, which is crucial for precision medicine, new therapeutic targets, drug resistance gene and biomarkers have been identified and targeted.[34] New advancements in CRISPR technology and its fusion with additional omics methodologies exhibit the potential to revolutionize cancer therapy and ease patient consequences. Hence, finding targetable genes linked to the onset, progression, and drug resistance in cancers is a necessary step in creating CRISPR libraries specifically designed for cancer screening. gRNAs that target these genes are chosen using bioinformatics techniques covering the whole genome. Optimizing factors including gRNA design, library complexity, and gene delivery techniques can enhance the screening process while reducing off-target impacts.

5. Types of CRISPR-based screening for cancer therapeutics

The three primary categories of CRISPR screens termed as activation or CRISPR activation (CRISPRa), interference (CRISPRi) and knockout (CRISPRKO), share a common process.[35] To make sure that only one copy of the established single-guide RNAs (sgRNAs) is integrated per cell, the sgRNAs are cloned in a lentivirus to prepare a library and then transduced into Cas expressing cells or dCas expression lines at low multiplication rates. Second, screening based on biological assays is performed on cells transduced with CRISPR libraries. Cells carrying the sgRNA will either disappear from the population or become more abundant if the target sequence modifies cell viability in response to selection pressure. Finally, following a phenotypic selection, CRISPR screens use next-generation sequencing (NGS) and distinct sgRNA sequences to detect changes in sgRNA frequency[36]. Table 1 summarizes the different approaches employed using the CRISPR screening methods for cancer.

5.1 CRISPR knockout (CRISPRKO) screening

A commonly used method for negative selection screening is CRISPRKO screening, which effectively introduces precisely targeted loss-of-function mutations at particular genomic sites. It finds decreased levels of sgRNAs in a group of cells that produce Cas9. DSBs are formed at the sgRNA target sites, impairing the targeted gene's functionality by point mutation or deletion of DNA fragments through NHEJ repair.[37] An *in vivo* study conducted by Wang and colleagues, pooled multiple CRISPRKO screens in a triple negative breast cancer (TNBC) mouse model. They discovered that Cop1 (an E3 ubiquitin ligase) indirectly controls the macrophage infiltration in cancer cells. Eliminating the Cop1 reduces the levels of cytokines released from cancer cells for macrophages. Decreased macrophage infiltration in tumor cells enhances their sensitivity to antitumor immune therapy and hence increase the survival of TNBC mouse.[38]

Table 1 CRISPR based screening methods and their applications in cancer treatment.

Screening method	Description/application in cancer treatment
CRISPR Knockout Screens	Cancer cells' genes are disrupted extensively to find those that are crucial for survival, growth, or resistance to medication.
CRISPR Interference Screens	Targeted gene expression stimulation highlights the genes involved in cancer spread or make cancer cells more susceptible towards the treatment.
CRISPR Activation Screens	Identifies genes whose down regulation prevents treatment resistance, metastasis or growth of cancer cells leading to gene repression.
CRISPR Epigenetic Screens	Specific alteration of epigenetic regulators to specify the genes and pathways connected to the development, spread, or resistance to treatment of cancer.
CRISPR Synthetic Lethality	Finds gene pairs whose simultaneous disruption results in the death of cancer cells, providing precision medicine treatment targets.
CRISPR Functional Genomics	Thorough evaluation of gene function in cancer cells to comprehend the hereditary causes of the disease and pinpoint the targets for treatment.

5.2 CRISPR interference (CRISPRi) screening

An inactive or dead Cas9 protein (dCas9) with no endonuclease activity is used in | CRISPRi screening. Cas9 binds to target sequence and prevents the transcription initiation or elongation without changing the genome's sequence.[39] This technique effectively silences the target sequence by inducing loss of function without causing lethality.[40]

5.3 CRISPR activation (CRISPRa) screening

CRISPR activation screens enable the positive screening by gain of function mutation in promoter region of target genes. This is employed to increase the gene expression levels. This aids in enhancing our knowledge of how genes are upregulated in cancer.[41] The benefits of CRISPRKO technology including high specificity, efficiency and low off targeting, are carried over to the CRISPRa screening method. Although there are many different molecular designs, the most important change is to the sgRNA's 5-end while targeting different genes. This economical method works well for building extensive gene expression libraries.[41]

6. Modes of CRISPR-based screening for cancer therapeutics

6.1 *In vitro* screening

Tumor models can replicate various molecular processes that promote tumor growth in a cell and are, hence, considered effective in cancer research. In this situation CRISPR-Cas systems, \ can help create practical and approachable *in vitro* mammalian cell models that can be used to discover the genes and signaling pathways that underlie the genesis and recurrence of cancer.[42]

To conduct *in vivo* CRISPR screening, cloning of the proposed sgRNA library in an expression vector is a prerequisite. Because lentiviral vectors may integrate into the host genome, they are frequently used. The sgRNA cassettes are amplified from genomic DNA and sequenced to determine the best candidate genes.[43] *In vitro* cell lines have made it much easier to study the complex mechanisms that underlie the development, growth, and signal transduction of tumors in cancer research. These *in vitro* models are more avid owing to the use of the CRISPR/Cas9 technology.[41] Loss-of-function (CRISPR), CRISPR/Cas interference (CRISPRi), and CRISPR/Cas activation (CRISPRa) screens are used in *in vitro* CRISPR/Cas9 screening. For instance, it has been reported that loss-of-function screens (CRISPR) targeting about 180,00 genes in melanoma cell lines aided in identifying genes that lead to RAF inhibitor resistance.[37]

6.2 *In vivo* screening

The CRISPR/Cas9 system can be used for *in vivo* screening as well, using both an indirect and direct screening method, in addition to *in vitro* screening. Transduced with guide RNA libraries, immortalized cancer cell lines that constitutively express the Cas9 nuclease are injected into animals in an indirect screen to induce tumor growth and metastasis.[44] In a directscreen, adeno-associated viruses (AAVs) are used to directly transport guide RNA libraries and Cas9 to animal models, creating tissue-specific cancer models, such as those in the brain, liver, and lung.[43] Fig. 3 summarizes the applications of these *in vivo* screens.

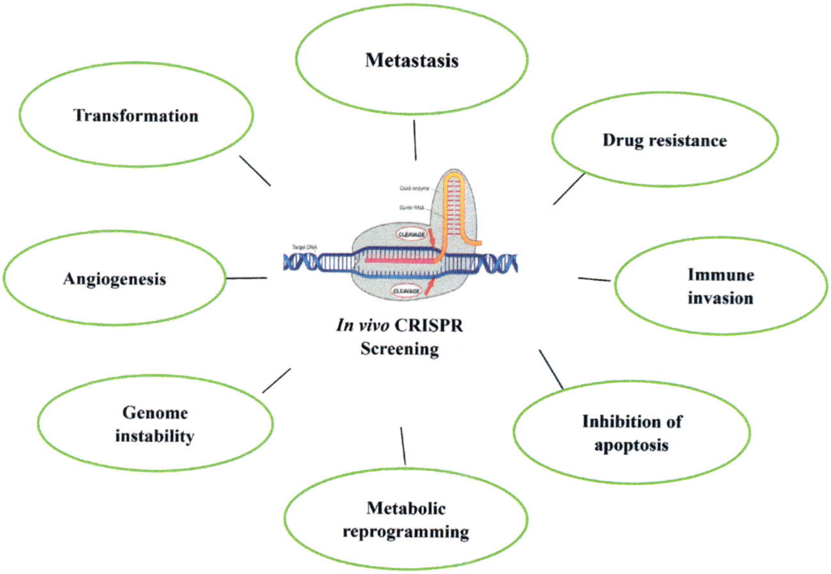

Fig. 3 CRISPR based *in vivo* screening of cancer hallmarks and potential targets for therapeutic interventions.

6.3 CRISPR for cancer immunotherapy

Immunotherapy supports the immune system's ability to combat illnesses and infections. Immuno-modulatory medicines, T-cell transfer treatment, immune checkpoint inhibitors and monoclonal antibodies are the key components involved.[45] Unfortunately, not every patient cohort responds to these medications or therapies in the same way. Additionally, their efficacy has been demonstrated in combination with chemotherapy and other conventional methods including radiation therapy and surgery.[42] In 2018, Nobel Prize in Physiology was awarded for the discovery of immune checkpoint blockers and their use as bispecific and trispecific monoclonal antibodies and it marked a turning point in its recognition.[46]

Additionally, the CRISPR/Cas9 technology opened up new lines for clinicians and researchers to investigate combinatorial approaches to gene therapy.[47] In 2016, the first experiment conducted in humans using CRISPR-Cas9 system provided more options to patients with many long-term advantages. Patients also produced more therapeutic immune cells such as the development of CAR-T cells and protein knockouts of programmed cell death proteins *e.g.*, PD1. Therefore, CRISPR-Cas9 may provide option of immunotherapy for cancer as a much-needed therapeutic option.[42]

6.4 CRISPR/Cas9 in CAR-T-cell therapies

CAR-T cells, a type of immunotherapy, represent a significant breakthrough in the individualized treatment of cancers. They have become an effective treatment option for people suffering from cancers, including multiple myeloma (MM), acute lymphoblastic leukemia (ALL), chronic lymphocytic leukemia (CLL) and lymphoma.[48] Multiple CAR-T cells based products have been licensed by US Food and Drug Administration (FDA) against follicular lymphoma, B cell acute lymphoblastic leukemia, large B-cell lymphoma (LBCL) and mantle cell lymphoma.[49] Additionally, studies on CAR-T cells have the potential in treating solid tumors like non-small-cell lung cancer, gliomas, stomach cancer, and liver cancer.[50]

7. Future perspectives for CRISPR-Cas system in cancer

CRISPR-Cas systems hold a great promise and potential for the treatment of cancer in the future. Enhancing the effectiveness and specificity of CRISPR-mediated genome editing *in vivo* would require advancements in delivery technologies, such as viral vectors and nanoparticle-based delivery systems.[51] Furthermore, by combining CRISPR with other modern techniques like organoid models and single-cell sequencing, researchers can learn more about the molecular mechanisms behind the development of therapeutic resistance and cancer progression.[52]

In addition, the possibility of genetic changes is there with the development of base editing technologies and multiplexed CRISPR systems, enabling more accurate and thorough targeting of mutations linked to cancer.[53] Furthermore, the continuous enhancement of CRISPR-based screening methodologies expedite the identification of innovative therapeutic targets and the creation of customized combination therapies for particular patients.[54]

Regarding cancer immunotherapy, CRISPR-mediated modification of immune cells presents enormous promise for increasing their anti-tumor efficacy and surpassing immune evasion strategies utilized by cancerous cells.[55] Furthermore, the creation of readily available and reasonably priced therapy alternatives for patients through the use of CRISPR-edited immune cells in off-the-shelf allogeneic cell treatments has the potential to completely transform the industry.[56]

Overall, despite the obstacles still present, there is a hope for a future in which precision medicine becomes the norm and the incidence of cancer is greatly decreased thanks to the ongoing innovation and integration of CRISPR-Cas systems into cancer therapy. It is certain that CRISPR technology will continue to influence the field of cancer treatment and enhance patient outcomes for people all over the world with continued investment in research and development and interdisciplinary collaboration.

8. Conclusion

In conclusion, the application of CRISPR-Cas systems in cancer therapy is an important development in the oncology industry that brings novel alternatives for therapeutic intervention, precise genome editing, and individualized treatment plans. Through the effective, adequate and acceptable use of CRISPR technology, scientists have made great progress in underpinning the molecular causes of cancer and creating advance approaches to treat it. The potential of CRISPR-based techniques to transform cancer therapy paradigms and improve patient outcomes is enormous. However, there are still many issues that need to be resolved, such as improving the clinical translation of CRISPR-based therapies and streamlining delivery strategies to reduce off-target effects. Nevertheless, CRISPR-Cas technology and its advancements are set to play important role in the future of precision oncology, opening up new possibilities for tailored and precise cancer therapy as the research and innovation continues.

References

1. Aljazeera. Cancer cases in people below 50 up nearly 80 percent in last three decades; 2024. Retrieved from https://www.aljazeera.com/news/2023/9/7/cancer-cases-in-people-below-50-up-nearly-80-percent-in-last-three-decades.
2. World Health Organization. Global cancer burden growing, amidst mounting need for services; 2024. Retrieved from https://www.who.int/news/item/01-02-2024-global-cancer-burden-growing--amidst-mounting-need-for-services.
3. MedicalNewsToday. An estimated 35 million new cancer cases to occur in 2050, WHO warns; 2024. Retrieved from https://www.medicalnewstoday.com/articles/an-estimated-35-million-new-cancer-cases-to-occur-in-2050-who-warns.
4. Mun EJ, Babiker HM, Weinberg U, Kirson ED, Von Hoff DD. Tumor-treating fields: a fourth modality in cancer treatment. *Clin Cancer Res*. 2018;24(2):266–275.
5. Huang M, Shen A, Ding J, Geng M. Molecularly targeted cancer therapy: some lessons from the past decade. *Trends Pharmacol Sci*. 2014;35(1):41–50.

6. Min H-Y, Lee H-Y. Molecular targeted therapy for anticancer treatment. *Exp Mol Med*. 2022;54(10):1670–1694.
7. van Elsas MJ, van Hall T, van der Burg SH. Future challenges in cancer resistance to immunotherapy. *Cancers*. 2020;12(4):935.
8. Chen M, Mao A, Xu M, Weng Q, Mao J, Ji J. CRISPR-Cas9 for cancer therapy: opportunities and challenges. *Cancer Lett*. 2019;447:48–55.
9. Nakagawa H, Fujita M. Whole genome sequencing analysis for cancer genomics and precision medicine. *Cancer Sci*. 2018;109(3):513–522.
10. Bertram JS. The molecular biology of cancer. *Mol Asp Med*. 2000;21(6):167–223.
11. Weinberg RA. Tumor suppressor genes. *Science*. 1991;254(5035):1138–1146.
12. Saretzki G. *Role of Telomeres and Telomerase in Cancer and Aging*. vol. 24. MDPI; 2023:9932.
13. Hille F, Charpentier E. CRISPR-Cas: biology, mechanisms and relevance. *Philos Trans R Soc B: Biol Sci*. 2016;371(1707):20150496.
14. Yang Y, Xu J, Ge S, Lai L. CRISPR/Cas: advances, limitations, and applications for precision cancer research. *Front Med*. 2021;8:649896.
15. Li Y, Casey SC, Felsher DW. Inactivation of MYC reverses tumorigenesis. *J Intern Med*. 2014;276(1):52–60.
16. D'Antonio L, Fieni C, Ciummo SL, et al. Inactivation of interleukin-30 in colon cancer stem cells via CRISPR/Cas9 genome editing inhibits their oncogenicity and improves host survival. *J Immunother Cancer*. 2023;11:3.
17. Kardooni H, Gonzalez-Gualda E, Stylianakis E, Saffaran S, Waxman J, Kypta RM. CRISPR-mediated reactivation of DKK3 expression attenuates TGF-β signaling in prostate cancer. *Cancers*. 2018;10(6):165.
18. Lei ZN, Teng QX, Wu ZX, et al. Overcoming multidrug resistance by knockout of ABCB1 gene using CRISPR/Cas9 system in SW620/Ad300 colorectal cancer cells. *MedComm*. 2021;2(4):765–777.
19. Fu Y, Sander JD, Reyon D, Cascio VM, Joung JK. Improving CRISPR-Cas nuclease specificity using truncated guide RNAs. *Nat Biotechnol*. 2014;32(3):279–284.
20. Mohamad Zamberi NN, Abuhamad AY, Low TY, Mohtar MA, Syafruddin SE. dCas9 tells tales: probing gene function and transcription regulation in cancer. *CRISPR J*. 2024;7(2):73–87.
21. Guilinger JP, Thompson DB, Liu DR. Fusion of catalytically inactive Cas9 to *Fok*I nuclease improves the specificity of genome modification. *Nat Biotechnol*. 2014;32(6):577–582.
22. Huang H, Huang G, Tan Z, et al. Engineered Cas12a-Plus nuclease enables gene editing with enhanced activity and specificity. *BMC Biol*. 2022;20(1):91. https://doi.org/10.1186/s12915-022-01296-1
23. Yoon A-R, Jung B-K, Choi E, et al. CRISPR-Cas12a with an oAd induces precise and cancer-specific genomic reprogramming of EGFR and efficient tumor regression. *Mol Ther*. 2020;28(10):2286–2296.
24. Siegler EL, Simone BW, Sakemura R, et al. Efficient gene editing of CART cells with CRISPR-Cas12a for enhanced antitumor efficacy. *Blood*. 2020;136:6–7. https://doi.org/10.1182/blood-2020-141115
25. Jang G, Kweon J, Kim Y. CRISPR prime editing for unconstrained correction of oncogenic KRAS variants. *Commun Biol*. 2023;6(1):681. https://doi.org/10.1038/s42003-023-05052-1
26. Pal M, Herold MJ. CRISPR base editing applications for identifying cancer-driving mutations. *Biochem Soc Trans*. 2021;49(1):269–280.
27. Li X, Qian X, Wang B, et al. Programmable base editing of mutated TERT promoter inhibits brain tumour growth. *Nat Cell Biol*. 2020;22(3):282–288.

28. Ren J, Liu X, Fang C, Jiang S, June CH, Zhao Y. Multiplex genome editing to generate universal CAR T cells resistant to PD1 inhibition. *Clin Cancer Res.* 2017;23(9):2255–2266.
29. Martinez-Lage M, Puig-Serra P, Menendez P, Torres-Ruiz R, Rodriguez-Perales S. CRISPR/Cas9 for cancer therapy: hopes and challenges. *Biomedicines.* 2018;6(4):105.
30. Sachdeva M, Sachdeva N, Pal M, et al. CRISPR/Cas9: molecular tool for gene therapy to target genome and epigenome in the treatment of lung cancer. *Cancer Gene Ther.* 2015;22(11):509–517.
31. Deng H-X, Zhai H, Shi Y, et al. Efficacy and long-term safety of CRISPR/Cas9 genome editing in the SOD1-linked mouse models of ALS. *Commun Biol.* 2021;4(1):396.
32. Salsman J, Dellaire G. Precision genome editing in the CRISPR era. *Biochem Cell Biol.* 2017;95(2):187–201.
33. Li Y, Peng N. Endogenous CRISPR-Cas system-based genome editing and antimicrobials: review and prospects. *Front Microbiol.* 2019;10:491504.
34. Xing H, Meng L hua. CRISPR-cas9: a powerful tool towards precision medicine in cancer treatment. *Acta Pharmacol Sin.* 2020;41(5):583–587. https://doi.org/10.1038/s41401-019-0322-9
35. Cong L, Ran FA, Cox D, et al. Multiplex genome engineering using CRISPR/Cas systems. *Science.* 2013;339(6121):819–823.
36. He C, Han S, Chang Y, et al. CRISPR screen in cancer: status quo and future perspectives. *Am. J. Cancer Res.* 2021;11(4):1031–1050.
37. Shalem O, Sanjana NE, Hartenian E, et al. Genome-scale CRISPR-Cas9 knockout screening in human cells. *Science.* 2014;343(6166):84–87.
38. Wang X, Tokheim C, Gu SS, et al. In vivo CRISPR screens identify the E3 ligase Cop1 as a modulator of macrophage infiltration and cancer immunotherapy target. *Cell.* 2021;184(21):5357–5374.
39. Qi LS, Larson MH, Gilbert LA, et al. Repurposing CRISPR as an RNA-guided platform for sequence-specific control of gene expression. *Cell.* 2021;184(3):844.
40. Gilbert LA, Horlbeck MA, Adamson B, et al. Genome-scale CRISPR-mediated control of gene repression and activation. *Cell.* 2014;159(3):647–661.
41. Ding S, Liu J, Han X, Tang M. CRISPR/Cas9-mediated genome editing in cancer therapy. *Int J Mol Sci.* 2023;24(22). https://doi.org/10.3390/ijms242216325
42. Bhat AA, Nisar S, Mukherjee S, et al. Integration of CRISPR/Cas9 with artificial intelligence for improved cancer therapeutics. *J Transl Med.* 2022;20(1):1–18. https://doi.org/10.1186/s12967-022-03765-1
43. Chow RD, Chen S. Cancer CRISPR screens in vivo. *Trends Cancer.* 2018;4(5):349–358. https://doi.org/10.1016/j.trecan.2018.03.002
44. Chen S, Sanjana NE, Zheng K, et al. Genome-wide CRISPR screen in a mouse model of tumor growth and metastasis. *Cell.* 2015;160(6):1246–1260.
45. Kirkwood JM, Butterfield LH, Tarhini AA, Zarour H, Kalinski P, Ferrone S. Immunotherapy of cancer in 2012. *CA: A Cancer J Clin.* 2012;62(5):309–335.
46. Huang P-W, Chang JW-C. Immune checkpoint inhibitors win the 2018 Nobel Prize. *Biomed J.* 2019;42(5):299–306.
47. Xia A-L, He Q-F, Wang J-C, et al. Applications and advances of CRISPR-Cas9 in cancer immunotherapy. *J Med Genet.* 2019;56(1):4–9. https://doi.org/10.1136/jmedgenet-2018-105422
48. Han D, Xu Z, Zhuang Y, Ye Z, Qian Q. Current progress in CAR-T cell therapy for hematological malignancies. *J Cancer.* 2021;12(2):326.
49. Labanieh L, Mackall CL. CAR immune cells: design principles, resistance and the next generation. *Nature.* 2023;614(7949):635–648.

50. Cao B, Liu M, Wang L, et al. Remodelling of tumour microenvironment by microwave ablation potentiates immunotherapy of AXL-specific CAR T cells against non-small cell lung cancer. *Nat Commun.* 2022;13(1):6203.
51. Akinc A, Querbes W, De S, et al. Targeted delivery of RNAi therapeutics with endogenous and exogenous ligand-based mechanisms. *Mol Ther.* 2010;18(7): 1357–1364.
52. Sachs N, Clevers H. Organoid cultures for the analysis of cancer phenotypes. *Curr OpGenet Dev.* 2014;24:68–73.
53. Anzalone AV, Randolph PB, Davis JR, et al. Search-and-replace genome editing without double-strand breaks or donor DNA. *Nature.* 2019;576(7785):149–157.
54. Doench JG. Am I ready for CRISPR? A user's guide to genetic screens. *Nat Rev Genet.* 2018;19(2):67–80.
55. Stadtmauer EA, Fraietta JA, Davis MM, et al. CRISPR-engineered T cells in patients with refractory cancer. *Science.* 2020;367(6481):eaba7365.
56. Xu L, Wang J, Liu Y, et al. CRISPR-edited stem cells in a patient with HIV and acute lymphocytic leukemia. *N Engl J Med.* 2019;381(13):1240–1247.

CHAPTER TEN

Current progress in CRISPR-*Cas* systems for autoimmune diseases

Juveriya Israr[a] **and Ajay Kumar**[b,*]
[a]Institute of Biosciences and Technology, Shri Ramswaroop Memorial University, Lucknow, Barabanki, Uttar Pradesh, India
[b]Department of Biotechnology, Faculty of Engineering and Technology, Rama University, Mandhana, Kanpur, Uttar Pradesh, India
*Corresponding author. e-mail address: ajaymtech@gmail.com

Contents

Progress in Molecular Biology and Translational Science, Volume 208
ISSN 1877-1173, https://doi.org/10.1016/bs.pmbts.2024.07.011

Abstract

A body develops an autoimmune illness when its immune system mistakenly targets healthy cells and organs. Eight million people are affected by more than 80 autoimmune diseases. The public's and individuals' well-being is put at risk. Type 1 diabetes, lupus, rheumatoid arthritis, and multiple sclerosisare autoimmune diseases. Tissue injury, nociceptive responses, and persistent inflammation are the results of these stresses. Concerns about healthcare costs, health, and physical limitations contribute to these issues. Given their prevalence, it is crucial to enhance our knowledge, conduct thorough research, and provide all-encompassing support to women dealing with autoimmune diseases. This will lead to better public health and better patient outcomes. Most bacteria's immune systems employ CRISPR-*Cas*, a state-of-the-art technique for editing genes. For *Cas* to break DNA with pinpoint accuracy, a guide RNA employs a predetermined enzymatic pathway. Genetic modifications started. After it was developed, this method was subjected to much research on autoimmune diseases. By modifying immune pathways, CRISPR gene editing can alleviate symptoms, promote immune system tolerance, and decrease autoimmune reactivity. The autoimmune diseases that CRISPR-*Cas9* targets now have no treatment or cure. Results from early clinical trials and preclinical studies of autoimmune medicines engineered using CRISPR showed promise. Modern treatments for rheumatoid arthritis,multiple sclerosis, and type 1 diabetes aim to alter specific genetic or immune mechanisms. Accurate CRISPR editing can fix autoimmune genetic disorders. Modifying effector cells with CRISPR can decrease autoimmune reactions. These cells include cytotoxic T and B lymphocytes. Because of improvements in delivery techniques and kits, CRISPR medications are now safer, more effective, and more accurately targeted. It all comes down to intricate immunological reactions and unexpected side consequences. Revolutionary cures for autoimmune problems and highly personalized medical therapies have been made possible by recent advancements in CRISPR.

Abbreviations

AAV	Adeno-associated viral
BAFF	B cell-activating factor
CRISPRa	CRISPR activation
CRISPRi	CRISPR interference
CRISPRko	CRISPR knockout
crRNAs	CRISPR RNAs
DMARDs	Disease-modifying antirheumatic medications
DSB	Double-strand break
HDR	Homology-directed repair
IL-17RA	IL-17 receptor A
IBD	Inflammatory Bowel Disease
MS	Multiple sclerosis
NHEJ	Non-homologous end joining
NSAIDs	Non-steroidal anti-inflammatory drugs
PD-1	Programmed cell death 1
RA	Rheumatoid arthritis
SLE	Systemic lupus erythematosus
T1D	Type 1 Diabetes

1. Introduction

Autoimmune diseases arise from immune system dysregulation, leading to the destruction of normal cells, tissues, and organs. An abnormal immune system response that does not detect self-antigens causes inflammation and tissue damage. Autoimmune illnesses can damage joints, skin, muscles, blood vessels, and organs. This category includes rheumatoid arthritis, lupus, MS, type 1 diabetes, and Hashimoto's thyroiditis.[1] variable conditions and regions have variable autoimmune disease prevalence. About 5–10% of affluent countries' populations may have autoimmune disorders, which afflict millions worldwide.[2] Autoimmune diseases like Multiple sclerosis and Rheumatoid arthritis are widespread, whereas autoimmune hepatitis and scleroderma are rarer. Many autoimmune diseases affect women more than men. Accurate autoimmune disease prevalence data is essential for public health planning, resource allocation, and patient treatment.

Several types of autoimmune disorders cause tissue damage and inflammation due to abnormal immune responses to self-antigens. Systemic lupus erythematosus, multiple sclerosis, and RA are among the most common autoimmune disorders. RA is a persistent inflammatory condition that results in pain, swelling, and rigidity of the joints. It causes joint abnormalities and functional impairment due to synovial membrane inflammation, cartilage deterioration, and bone tissue erosion.[3]

Demyelination and axonal destruction are hallmarks of multiple sclerosis, a central nervous system neuroinflammatory disease. The condition causes neurological symptoms such as sensory, motor control, and cognition problems, often causing impairment.[4] The accumulation of inflammatory and immunological complexes in systemic lupus erythematosus can impact many organs and tissues. This illness can include rash, arthritis, nephritis, and mental symptoms. These symptoms can affect several organs and systems including the skin, joints, kidneys, heart, lungs, and neurological system.[5] Identifying, diagnosing, and managing these common autoimmune illnesses is difficult. Complex clinical presentations and diverse outcomes sometimes require individualized therapy strategies and interdisciplinary care.

Current treatment for autoimmune illnesses involves drugs that modulate the aberrant immune response, reduce inflammation, and manage symptoms. Non-steroidal anti-inflammatory drugs (NSAIDs), corticosteroids, Disease-modifying antirheumatic pharmaceuticals (DMARDs), biologic therapies, and immune suppressants are employed for the treatment of this condition.[6] For pain and inflammation, NSAIDs like naproxen and ibuprofen are used to treat autoimmune diseases like lupus and rheumatoid arthritis. They merely reduce symptoms, not the illness process. Long-term usage of NSAIDs increases the risk of cardiovascular disease, kidney damage, and gastrointestinal ulcers. Corticosteroids, strong anti-inflammatory medications, are routinely prescribed for autoimmune illnesses. These therapies include prednisone and methylprednisolone. Despite its symptom-management benefits, long-term corticosteroid use can cause weight gain, osteoporosis, diabetes, and infection susceptibility. DMARDs such as methotrexate, hydroxychloroquine, and sulfasalazine slow autoimmune disorders and protect joints. Not all patients respond well to DMARDs, which may take weeks or months to work.

B-cell-targeted treatments, IL-6 inhibitors, and TNF inhibitors have transformed the treatment of autoimmune diseases by targeting the immune system components that cause them. Biologics may be beneficial, but they cost, require regular injections or infusions, and may heighten the likelihood of infections and other adverse effects. In autoimmune illnesses, immune suppressants reduce inflammation and immunological response. These include azathioprine, cyclosporine, and mycophenolate mofetil. However, extended immunosuppressant usage can increase infections, cancers, and other serious consequences. Many patients encounter partial responses, relapses, and treatment-related side effects, making autoimmune disorder management difficult. These drugs can alleviate symptoms and decrease autoimmune disease progression, but they do not cure them. To improve long-term outcomes, autoimmune disease treatments must be less intrusive, more targeted, and less likely to cause side effects.

CRISPR technology shows potential in treating autoimmune illnesses as a substitute for conventional drugs by providing precise gene editing to regulate immune responses. This precise method has the potential to decrease the necessity for prolonged pharmaceutical usage and limit any adverse effects. This research highlights the ability of CRISPR technology to provide effective and customized treatments for autoimmune illnesses.

2. Explanation of the fundamentals of CRISPR-*Cas* systems and their method of action

As an adaptive defense mechanism, the CRISPR-*Cas* system protects bacteria and archaea from genetic invaders such as viruses and plasmids.[7] There are essentially three steps to how the CRISPR-*Cas* systems work-adaptation, expression, and interference.When a bacterium or archaeon is adapting, its CRISPR-*Cas* system takes in short pieces of DNA from outside sources and inserts them into its genome as "spacer" sequences. By alternately organizing repeat repetitions and spacer sequences, the characteristic CRISPR array is generated.During the process of transcription, the CRISPR array undergoes conversion into pre-CRISPR RNA, which is a lengthy precursor molecule of RNA. Following processing, the pre-crRNA is transformed into truncated CRISPR RNAs (crRNAs) that possess complementary target DNA sequences on each of their unique spacers.The third and last stage of Interference involves the creation of a ribonucleoprotein complex by combining the crRNA with a CRISPR-associated (*Cas*) protein. Following its crRNA-guided path to its specific target, the Cas protein causes a targeted DNA double-strand break (DSB). *Cas* proteins, also called *Cas9* or *Cas12a*, are endonucleases, meaning they may cut the target DNA strands in half, creating a double-strand break (DSB). While repairing double-strand breaks (DSBs), the DNA repair machinery in cells often uses the error-prone non-homologous end joining (NHEJ) repair method, which causes numerous insertions or deletions (indels) at the precise spot. Another option is to utilize homology-directed repair (HDR) to target a specific area of DNA and alter its sequence there.

According to what Doudna, Charpentier[8] explained in 2014, this is accomplished by making use of a donor DNA template. By precisely recognizing and cleaving the DNA sequences of foreign genetic material, CRISPR-*Cas* systems enable bacteria and archaea to build adaptive immunity. Several applications have made use of this method, including genome editing, gene control, and nucleic acid detection.

3. Overview of CRISPR components (e.g., *Cas* proteins, guide RNA)

CRISPR-*Cas* systems have two primary constituents: *Cas* proteins and gRNA. *Cas* proteins, also known as CRISPR-associated proteins, have a crucial function in the process of genome editing. Proteins are accountable for the fragmentation of certain DNA sequences. Genome editing applications mostly employ *Cas9*, a *Cas* protein obtained from the *Streptococcus pyogenes* bacterium. In addition to *Cas13* and *Cas12a*, several additional *Cas* proteins have been modified for the aim of genome editing and other purposes.[9] Guide RNA (gRNA), a diminutive RNA molecule, employs base pairing to accurately direct the *Cas* protein to the specific DNA region for cleavage. The gRNA is composed of two primary elements—a scaffold sequence that interacts with the *Cas* protein and a targeting sequence that complements the DNA area of interest. The CRISPR-*Cas* method employs a targeting sequence, usually composed of 20 nucleotides, to guide the *Cas* protein to the precise genomic region that needs to be broken or altered. The scaffold sequence is essential for increasing the stability of the structure and promoting the interaction between the gRNA and *Cas* protein.[10] A complex consisting of a *Cas* protein and guide RNA may be designed to accurately modify targeted regions of DNA inside the genome. The adaptable genome editing approach has greatly advanced the field of molecular biology, providing extensive opportunities for its application in biotechnology, gene therapy, and functional genomics.

4. Historical background and key milestones in CRISPR research

The beginnings of CRISPR can be traced back to the late 1980s when Japanese scientists made the discovery of repetitive DNA sequences in bacterial and archaeal genomes.[11] The potential role of these sequences in bacterial immunity was not recognized by researchers until the early 2000s, so their significance was previously unknown. Francisco Mojica introduced the term "CRISPR" in 2005 to describe the repetitive DNA sequences and the corresponding genes.[12] In 2007, a study conducted by researchers at Danisco (now DuPont) and the University of Alicante demonstrated that CRISPR sequences confer acquired immunity to bacteria against viruses by specifically targeting and cutting viral DNA. This discovery represented a significant advancement in our knowledge of the role of CRISPR. The subsequent study offered more clarification on the function of CRISPR-associated (*Cas*) proteins, namely the *Cas9* protein, which is an RNA-guided endonuclease capable of precisely cutting specific DNA regions.[10,13] In 2012, scientists such as Emmanuelle Charpentier and Jennifer Doudna provided evidence that the *Cas9* protein may be directed to identify and cut particular DNA regions in a regulated setting by utilizing artificial RNA molecules. The discovery described in the paper by Jinek et al.[10] established the basis for the advancement of CRISPR, an influential technique used to modify genomes. This groundbreaking study established the basis for the future utilization of CRISPR-*Cas9* in the accurate and efficient manipulation of genetic material in different organisms. Subsequently, there have been a multitude of significant developments and discoveries in the field of CRISPR research. Significant progress in this area includes the use of CRISPR-*Cas9* to alter the genetic material of eukaryotic cells, the creation of new CRISPR systems with unique capabilities (such as base editing and prime editing), and the application of CRISPR technology in various fields such as agriculture, biotechnology, and biomedicine. The efficiency, adaptability, and user-friendly nature of CRISPR-*Cas* systems have brought about a revolution in molecular biology. These technologies have immense potential for advancing science and enhancing health in the future.When the immune system mistakenly attacks healthy cells and tissues, it results in autoimmune diseases. There are a multitude of autoimmune disorders, with a few prevalent ones being (Fig. 1).

5. CRISPR/*Cas9* and rheumatoid arthritis

Inflammation of the joints is a hallmark of rheumatoid arthritis (RA), a chronic inflammatory disorder. Pain, swelling, and degeneration of joints

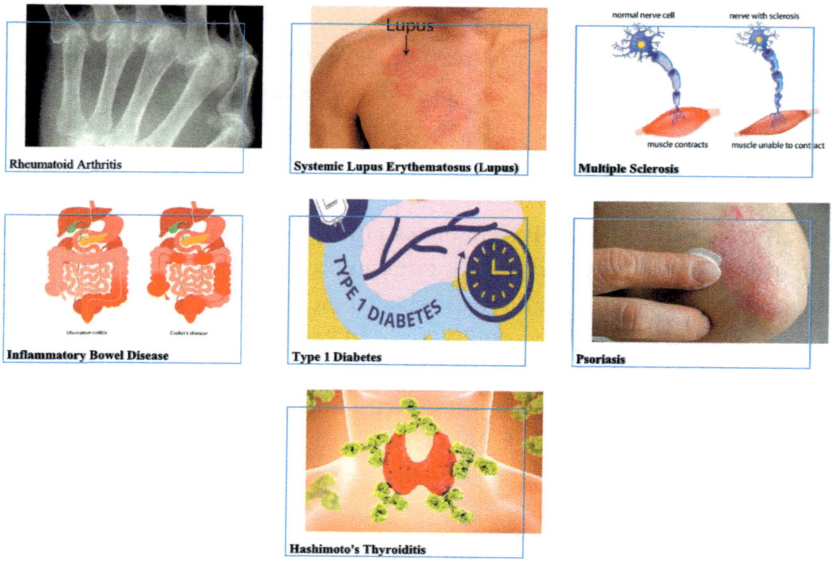

Fig. 1 Some common examples of autoimmune diseases.

can all result from inflammation. An important step in the onset of RA is immune system dysregulation. By focusing on key genes implicated in RA progression, CRISPR/*Cas9* technology has arisen as a promising tool for RA research and, maybe, treatment.

5.1 Taking action against TNF-alpha

The pro-inflammatory cytokine tumor necrosis factor-alpha (TNF-alpha) plays a pivotal role in the onset of RA. Patients with RA develop chronic inflammation and joint damage as a result of excessive production of TNF-alpha. As a potential treatment method for RA, the CRISPR/*Cas9* technology has been investigated for its ability to target the TNF-alpha gene in preclinical research.[14]

5.2 Modulating immune responses

It is feasible to target immune cells, such as T cells and B cells, that are involved in the development of RA using CRISPR/*Cas9*. This allows for the modification of immune responses. The researchers hope that they may be able to control immune responses and reduce inflammation in RA joints by altering certain genes that are found in these immune cells themselves.[15]

5.3 Enhancing drug efficacy

Improving Medications' Efficacy Another area that might benefit from the CRISPR/*Cas9* technology is the study of RA drug targets and action mechanisms. Researchers can uncover novel therapeutic targets and optimize drug efficacy in patients with RA by altering genes that are associated with drug resistance or their responsiveness to treatment. The CRISPR/*Cas9* technique, in general, has a great deal of potential for expanding our understanding of the molecular pathways that are responsible for RA and generating tailored therapeutics for this complicated inflammatory illness. Investigations using CRISPR/*Cas9* in preclinical RA models may lead to therapeutic uses in the future for the treatment of afflicted individuals.[16,17]

6. CRISPR/*Cas9* and systemic lupus erythematosus (lupus)

Lupus, also known as systemic lupus erythematosus (SLE), is an autoimmune disorder that can have a multiplicity of effects on different parts of the body. Autoantibodies and persistent inflammation are only two of the many symptoms that can result from an immunological imbalance. Lupus research and treatment might benefit greatly from the application of CRISPR/*Cas9* technology, which could lead to a better knowledge of the illness and its origins.

6.1 Targeting B cells and autoantibody production

Lupus development relies on the stimulation of B cells and the generation of autoantibodies; CRISPR/*Cas9* can target these genes precisely. The researchers intend to manipulate the production of autoantibodies and decrease disease activity in lupus patients by modifying particular genes related to B cell function.[15]

6.2 Altering immune cell signaling pathways

Lupus disrupts the type I interferon system, which may be targeted using CRISPR/*Cas9* technology to alter genes involved in immune cell signaling pathways. By focusing on these pathways, lupus patients may be able to reduce inflammation and aberrant immune responses.[18]

6.3 Exploring genetic risk factors

CRISPR/*Cas9* enables the examination of genetic risk factors linked to the vulnerability to lupus. Through the modification of certain genetic loci associated with lupus, scientists can obtain valuable knowledge about the molecular processes that drive the disease and pinpoint possible targets for treatment. In summary, CRISPR/*Cas9* technology shows great potential in enhancing our comprehension of the molecular pathways that drive lupus and could potentially lead to the creation of precise treatments for this complex autoimmune disorder. Current research utilizing CRISPR/*Cas9* in preclinical lupus models has the potential to discover new therapeutic targets and perhaps apply these discoveries to clinical treatments for lupus patients.[19]

7. CRISPR/*Cas9* and multiple sclerosis (MS)

Deterioration of nerve cells in the CNS, inflammation, and the loss of myelin (demyelination) characterize multiple sclerosis (MS), an inflammatory disease that persists over time. A critical component in the onset of multiple sclerosis is the impairment of the immune system. We can only hope that CRISPR/*Cas9* technology will lead to a better understanding of multiple sclerosis and, maybe, new ways to treat this complex neurological disease.

7.1 Regulating immune responses

T cells and B cells, two types of immune cells, are involved in MS progression; CRISPR/*Cas9* can be used to target these cells precisely. Through genetic manipulation, the researchers hope to control immune responses, reduce inflammation, and shield the central nervous system's myelin from immune system assault.[15]

7.2 Restoring myelin integrity

The CRISPR/*Cas9* technology may be engaged to modify genes associated with the restoration and growth of myelin. Researchers are focusing on specific genes related to remyelination to stimulate the restoration of damaged myelin sheaths in individuals with MS. This approach has the potential to slow down the advancement of the illness and enhance neurological function.[20]

7.3 Exploring genetic risk factors

The utilization of CRISPR/*Cas9* technology enables researchers to investigate genetic risk factors linked to susceptibility to MS. Through the manipulation of certain genetic loci associated with MS, scientists can uncover the precise molecular pathways that contribute to the illness and pinpoint prospective targets for therapeutic intervention.

8. CRISPR/*Cas9* and type 1 diabetes

Several hereditary disorders, including Type 1 Diabetes (T1D), may be amenable to treatment in the future thanks to CRISPR/*Cas9* technology. In type 1 diabetes (T1D), the pancreatic beta cells responsible for making insulin are attacked and killed by the immune system. CRISPR/*Cas9* offers hope for precise genome editing to modify immune responses that play a role in the onset of type 1 diabetes or to correct genetic mutations associated with the disease. Gene editing, immunological regulation, and beta cell transplantation are three areas that have been explored about the use of CRISPR/*Cas9* in type 1 diabetes research.

8.1 Gene editing for T1D

Researchers found that the CRISPR/*Cas9* method successfully corrected a T1D-related mutation in the INS gene.[21] The use of pluripotent stem cells obtained from patients allowed this to happen.

8.2 Immune regulation in T1D

Researchers in recent research used CRISPR/*Cas9* technology to target immune cells and control the immunological response in type 1 diabetics.[16,17] This study's results provide hope for the future of immunotherapy methods that use this strategy.

8.3 Beta cell transplantation

To improve the effectiveness and survival of beta cell transplants for the treatment of type 1 diabetes, scientists have looked at CRISPR/*Cas9*-mediated gene editing.[22]

8.4 Clinical applications

Current clinical studies are in the first phases of employing CRISPR/*Cas9* for the treatment of T1D. Ongoing research is primarily concerned with ensuring safety and evaluating the effectiveness of this approach.[23] This-research emphasizes the promise of CRISPR/*Cas9* technology in enhancing research and therapy strategies for T1D.

9. CRISPR/*Cas9* and psoriasis

Psoriasis is a long-lasting skin disorder caused by the immune system attacking the skin cells, resulting in the growth of excessive and aberrant skin cells. This leads to the development of red and scaly patches on the skin. The utilization of CRISPR/*Cas9* technology has demonstrated promise in the examination and management of psoriasis by specifically targeting crucial genes implicated in the development of the illness.

9.1 Gene editing in psoriasis

The technique of CRISPR/*Cas9* has been used to specifically target genes linked to psoriasis, such as IL-17 and TNF-α, to regulate immune responses and reduce inflammation in the affected skin of psoriasis patients.[24]

9.2 Targeting immune cells

The use of CRISPR/*Cas9* has been examined in modifying immune cells, namely T cells and dendritic cells, to control inflammatory reactions in psoriasis.[25]

9.3 Personalized medicine approach

CRISPR/*Cas9* technology enables accurate modification of the genome, allowing tailored treatment plans for psoriasis that take into account individual genetic differences.[26]

9.4 Future treatment potential

CRISPR/*Cas9* has the potential to create innovative treatment approaches for psoriasis, such as modifying keratinocytes by gene editing and targeting cytokine pathways.[27–29]

10. CRISPR/*Cas9* and inflammatory bowel disease (IBD)

Inflammatory bowel disease (IBD) is a catch-all word for a variety of gastrointestinal disorders, including ulcerative colitis and Crohn's disease.

CRISPR/*Cas9* technology has recently gained a lot of attention for its potential as a tool to study the genetic factors that contribute to inflammatory bowel disease (IBD) and to find therapies for it.

10.1 Genetic studies in IBD

The utilization of CRISPR/*Cas9* has facilitated the examination of certain genetic variations linked to IBD susceptibility, such as NOD2 and IL23R. This has resulted in a better understanding of the processes underlying the illness.[30]

10.2 Gut microbiome modulation

The CRISPR/*Cas9* technology has been utilized to genetically modify probiotic bacteria to specifically transport therapeutic compounds to the gut. This has the potential to alter the gut microbiome in individuals with IBD.[31]

10.3 Therapeutic applications

The use of CRISPR/*Cas9*-mediated gene editing has demonstrated potential in creating individualized therapies for IBD, by specifically targeting important pathways involved in inflammation and immunological responses.[32]

10.4 Future directions

Current investigations are examining the application of CRISPR/*Cas9* in precision medicine for IBD, specifically targeting the modification of immune cells, cytokine signaling, and gut barrier function.[27–29]

11. CRISPR/*Cas9* and Hashimoto's Thyroiditis

Hypothyroidism can be caused by an inflammatory disorder known as Hashimoto's Thyroiditis, which affects the thyroid gland. Research and treatment methods for Hashimoto's Thyroiditis stand to benefit greatly from the application of CRISPR/*Cas9* technology.

11.1 Genetic research on Hashimoto's Thyroiditis

It has employed the use of CRISPR/*Cas9* to examine the genetic elements that contribute to the condition. This includes studying the involvement of certain genes, such as HLA-DR and CTLA-4, in autoimmune thyroid illness.[33]

11.2 Immune cell engineering

Researchers have investigated the application of CRISPR/*Cas9* to alter immune cells implicated in the development of Hashimoto's Thyroiditis. The objective is to control autoimmune reactions and reduce inflammation in the thyroid gland.[22]

11.3 Therapeutic applications

The use of CRISPR/*Cas9* technology shows potential in the development of precise treatments for Hashimoto's Thyroiditis. This includes the ability to edit genes in thyroid cells to regulate immune responses and restore normal thyroid function.[25]

11.4 Future directions

Current research is investigating the use of CRISPR/*Cas9* in Hashimoto's Thyroiditis to develop customized medical methods. This involves employing precise genome editing to target specific genetic variants and immune dysregulation in particular patients.[34]

Recent progress in CRISPR-*Cas* systems has led to significant developments in treating autoimmune diseases. Here are the corresponding papers that have contributed to these advancements (Table 1).

12. Integrative research on autoimmune diseases using CRISPR-*Cas* systems

Recent research suggests that CRISPR-*Cas* systems may be able to effectively treat autoimmune diseases by focusing on genes that cause the disease or dysregulated immune pathways. Thus, the IL-17 receptor A (IL-17RA) gene was selectively targeted utilizing CRISPR-*Cas9* technology in a mouse model of rheumatoid arthritis.[40] By lowering RA-associated joint inflammation and bone deterioration, the researchers proved the therapeutic potential of CRISPR-mediated gene editing in autoimmune arthritis by effectively deleting the IL-17RA gene. In their study on systemic lupus erythematosus (SLE) therapy,[33] showed that CRISPR-*Cas9* might alter immune cell function. To combat autoimmune diseases, the researchers devised a novel immunomodulatory approach. In a mouse model of systemic lupus erythematosus (SLE), they were able to increase immunological tolerance and reduce autoimmune responses by concentrating on the programmed cell death 1 (PD-1) gene in T cells. In

Table 1 The table represents the autoimmune disease and its applications of the CRISPR-Cas system.

S. no.	Research title	Autoimmune disease	Applications of the CRISPR-Cas system	References
1	Gene editing using CRISPR/Cas9 in the context of Rheumatoid Arthritis	Rheumatoid arthritis	Manipulating genes to regulate immune cells and control inflammation	35
2	Gene therapy for Systemic Lupus Erythematosus using CRISPR/Cas9 technology	Systemic Lupus Erythematosus	Altering the functioning of immune cells and the generation of autoantibodies	36
3	Utilizing CRISPR-Cas9 for precise genome editing in Multiple Sclerosis	Multiple Sclerosis	Addressing genetic alterations and immune dysfunction	37
4	Applications of CRISPR/Cas9 in the context of Inflammatory Bowel Disease	Inflammatory Bowel Disease	Modifying the composition of gut microbiota and its impact on immune responses	38
5	The use of CRISPR/Cas13a to specifically target microRNA in autoimmune diseases	A range of autoimmune disorders	Control of microRNA expression in autoimmune disorders	39

light of these recent discoveries, CRISPR-*Cas* systems are being considered more seriously as novel therapeutic techniques for identifying the genetic bases of autoimmune diseases and developing targeted treatment plans.

Even though there has been a lot of advancement in CRISPR-based treatments for autoimmune illnesses, preclinical models have shown that certain possible uses are effective. An animal model of psoriasis was used by Wang et al.[41] to demonstrate how CRISPR-*Cas9* technology may be used to specifically target the IL-17 receptor A (IL-17RA) gene. As a treatment option for autoimmune skin diseases, this work highlights the promise of CRISPR-mediated gene editing. This is accomplished by restoring normal IL-17RA expression in skin cells and reducing the severity of inflammation similar to psoriasis. By targeting the B cell-activating factor (BAFF) gene, the researchers used the CRISPR-*Cas9* technology to generate a mouse model of systemic lupus erythematosus (SLE).[42] By lowering inflammation in the kidneys and decreasing the generation of self-targeting antibodies by suppressing B-cell activating factor (BAFF) expression, the researchers found a practical way to treat systemic lupus erythematosus (SLE). Important information on the safety and effectiveness of CRISPR-based treatments for autoimmune illnesses can be uncovered through animal research, which provides the groundwork for future clinical trials. By focusing on disease-related pathways, CRISPR technology can alter immune responses in autoimmune diseases. One strategy involves going after the dysregulated immune pathways or the genes that cause autoimmune diseases. As an example, a study conducted by Xu et al.[40] showed that a reduction in joint inflammation and bone disintegration was achieved by employing CRISPR-*Cas9* to alter the *IL-17 receptor A* (IL-17RA) gene in a mouse model of rheumatoid arthritis (RA). In a similar vein, a mouse model of systemic lupus erythematosus (SLE) was used by Wu et al.[33] to reduce autoimmune responses and increase immunological tolerance through the application of CRISPR-*Cas9* technology. The PD-1 gene, which is located in T cells, was the focus of this precise targeting, and the result was successful.

Tolerance and immunological balance restoration areother goals of immune cell function modulation. As an example, a study conducted by Lei et al.[42] used CRISPR-Cas9 to restrict the expression of the B cell-activating factor (BAFF) gene in a mouse model of systemic lupus erythematosus (SLE), which reduced inflammation and autoantibody production in the kidneys of the mice. It is also possible to control gene expression and immune response levels using CRISPR-based approaches like CRISPR activation (CRISPRa) and CRISPR interference (CRISPRi) that do not alter the genome in any way whatsoever.[43] Sanjana et al.[44] observed that CRISPR knockout (CRISPRko) and CRISPR activation (CRISPRa) screens are two types of genome-wide screening methods that use CRISPR technology. These methods can reveal new therapeutic targets and disease-specific pathways related to autoimmune illnesses. These screening procedures involve systematically altering gene expression and assessing resulting changes in immune function. This can help identify important regulators of immune function and potential targets for therapy. CRISPR offers new perspectives on the processes and therapy options for autoimmune diseases by utilizing its versatile tools and strategies to modify immune responses and target particular pathways associated with the diseases.

13. Recent advancements in CRISPR-based therapies

CRISPR is increasingly being explored as a potential tool for addressing autoimmune diseases, because of recent advancements that have expanded possibilities for studying and treating these conditions. Novel genome editing techniques utilizing CRISPR have been devised, exhibiting enhanced precision, efficiency, and specificity compared to previous methods. Recent advancements in base editing and prime editing techniques have enabled the precise insertion of certain sequences or modification of individual nucleotides, without the occurrence of double-strand breaks.[45,46] These advancements enhance the safety and versatility of CRISPR-mediated genome editing, a technique that can be employed to correct mutations responsible for autoimmune disorders. Advancements in CRISPR delivery systems for *in vivo* applications represent a significant breakthrough. This progress will provide accurate gene editing within afflicted organs or tissues. Recent research has demonstrated that various delivery techniques can efficiently transport CRISPR components to specific cell types or tissues affected by autoimmune disorders. The strategies encompass adeno-associated viral (AAV) vectors, lipid nanoparticles, and cell-based delivery systems.[27–29,47] These advancements facilitate the expedited transition of CRISPR-based therapies for autoimmune disorders from the laboratory to clinical settings. One advantage of utilizing CRISPR in conjunction with high-throughput screening techniques is the capacity to discover novel pathways and therapeutic targets for autoimmune diseases. To discover key regulators of immune responses and autoimmune illness, researchers have created genome-wide CRISPR activation (CRISPRa) and CRISPR knockout (CRISPRko) screens.[48,49] These screening procedures considerably expand our understanding of the etiology of illnesses and potential targets for therapeutic action. Recent discoveries and enhancements have significantly enhanced the effectiveness and potential of CRISPR technology in treating autoimmune diseases. Individuals suffering from autoimmune disorders might derive advantages from the novel genome editing tools, effective delivery methods, and high-throughput screening platforms offered by CRISPR technology. These advancements significantly contribute to the creation of personalized treatments and targeted pharmaceuticals. The progress in CRISPR-based technologies and delivery systems has created opportunities for precise genome editing in immune cells, enabling focused treatments in autoimmune diseases.[46] Emphasize base editing as an innovative technique that utilizes CRISPR to accurately modify individual nucleotides without causing double-strand breaks. Base editors have the potential to be used for correcting disease-causing mutations associated with autoimmune disorders in immune cells by converting one DNA base pair to another. Cytosine base editors (CBEs) and adenine base editors (ABEs) are two examples of such editors. Prime editing is an advanced technique that utilizes CRISPR technology to achieve accurate genome editing. Prime editing involves the utilization of a reverse transcriptase enzyme along with the CRISPR-*Cas9* system to accurately add or delete particular nucleotides from the genome. This process does not require the usage of donor DNA templates or the creation of double-strand breaks.[45] The prospect of prime editing to effectively and accurately rectify genetic abnormalities in immune cells associated with the disease is highly promising. Techniques for transporting CRISPR components into immune cells have progressed, with the development of both viral and non-viral vectors to ensure effective delivery. To achieve precise modification of genetic material both inside and outside of living organisms, CRISPR-*Cas* systems have been improved for transportation into immune cells through the use of lentiviral, adenoviral, and adeno-associated viral (AAV) vectors.[41] Furthermore, researchers have conducted inquiries into alternative strategies for delivering CRISPR components into immune cells, such as lipid nanoparticles, cell-penetrating peptides, and electroporation-based techniques.[27–29,47] Overall, the progress in developing customized therapies for autoimmune diseases is facilitated by the emergence of CRISPR-based technologies and delivery systems. These advancements offer encouraging possibilities for precise genome editing in immune cells. State-of-the-art research and clinical trials are actively investigating the safety and effectiveness of CRISPR-based therapies for autoimmune disorders. An illustrative instance is the application of CRISPR-*Cas9*-mediated genome editing in both preclinical models and clinical trials to specifically target genes or immune pathways that are dysregulated or responsible for causing disease. An illustrative instance is a study conducted by Xu et al.,[40] which demonstrated the potential efficacy of inhibiting the IL-17 receptor A (IL-17RA) gene by CRISPR-*Cas9* as a therapeutic approach for mice with rheumatoid arthritis (RA). This would lead to reduced inflammation in the joints and less bone deterioration. A separate investigation carried out by Wang et al.[41] employed CRISPR-*Cas9* to diminish psoriasis-like skin inflammation in a mouse model of the ailment by suppressing IL-17RA expression. These initial findings provide an opportunity for further application in clinical settings and offer insights into the feasibility and efficacy of CRISPR-based treatments for autoimmune conditions. Research is now being conducted on the application of CRISPR-based immunomodulatory techniques in the treatment of autoimmune diseases, in conjunction with gene editing technologies. A clinical experiment, led by Carl June at the University of Pennsylvania (ClinicalTrials.gov Identifier: NCT04008365), is currently investigating the use of CRISPR-*Cas9*-modified T cells to specifically target and suppress autoreactive immune cells in individuals with myasthenia gravis. Similarly, the clinical trial now being conducted at the University of California, San Francisco, seeks to evaluate the efficacy and safety of CRISPR-*Cas9*-modified regulatory T cells in treating type 1 diabetes. The trial is registered under the identifier NCT04582275 on ClinicalTrials.gov. These innovative clinical trials offer hope for patients to have access to novel therapeutic options, representing considerable advancement in the development of CRISPR-based therapies for autoimmune diseases. The growing enthusiasm and potential of CRISPR technology to revolutionize autoimmune therapy is backed by cutting-edge research and clinical investigations that evaluate CRISPR-based interventions for autoimmune diseases. Scientists are diligently striving to overcome challenges such as immunological reactions, delivery issues, and unintended effects to develop CRISPR-based medications for autoimmune diseases that are both safe and efficient, while also being customized to meet the specific requirements of individual patients.

14. Challenges and limitations

Several significant obstacles must be surmounted before the implementation of CRISPR-based therapies for autoimmune diseases in clinical settings. A significant challenge in using CRISPR-*Cas* systems is the potential for off-target effects, where genomic regions other than the intended target are modified, leading to unforeseen consequences.[50] To guarantee the safety and effectiveness of CRISPR-based therapies in clinical environments, additional efforts are needed to reduce off-target effects and enhance the specificity of CRISPR-*Cas* systems. A significant obstacle is in delivering CRISPR components to targeted cells or organs in a live organism. To obtain therapeutic effects while minimizing systemic toxicity and immunological reactions, it is necessary to employ effective delivery techniques.[41] An important challenge in bringing CRISPR-based medicines into clinical use is the creation of delivery systems that are both safe and effective. These vehicles can be either viral vectors, nanoparticles, or cell-based approaches. Furthermore, the utilization of CRISPR for long-term therapeutic objectives may encounter challenges due to immune responses that specifically target *Cas* proteins and guide RNAs.[41] To overcome the limitations posed by immune recognition and clearance, it is necessary to develop methods that can decrease immune responses and prolong the duration of therapeutic benefits for CRISPR components in living organisms. The clinical development and application of CRISPR-based therapies for autoimmune illnesses face significant challenges arising from ethical and regulatory considerations. When developing and using CRISPR-based interventions, it is important to consider ethical factors like as ensuring patient safety, obtaining informed permission, and providing fair access to innovative medicines.[51] To successfully transition CRISPR-based treatments from the laboratory to clinical applications for treating autoimmune illnesses, it is crucial to address and overcome significant concerns and obstacles such as off-target effects, limitations in delivery methods, immune system reactions, and ethical considerations.

CRISPR-*Cas* systems encounter hurdles like as off-target effects, immunological reactions, and delivery limits, which hinder their therapeutic potential, particularly in the treatment of autoimmune illnesses. An off-target effect refers to the occurrence of genetic modifications at genomic loci other than the intended target region. Recent advancements in CRISPR-Cas technology have raised ongoing concerns regarding potential unintended effects on non-targeted areas.[50] To minimize unintended consequences and enhance precision in CRISPR-*Cas* systems, various techniques have been developed, including improving the design of guide RNA and utilizing high-accuracy *Cas* nucleases. Another hindrance to the therapeutic application of CRISPR is the occurrence of immunological responses against CRISPR components, including Cas proteins and guide RNAs. As per the findings of Wang et al.,[41] the immune system can recognize these foreign elements and can either kill the modified cells or make the therapeutic effects useless. To enhance the safety and efficacy of CRISPR-based therapies, scientists are investigating methods to reduce immune reactions, such as employing immune-evading delivery mechanisms or modifying CRISPR components to minimize their immunogenicity. One of the major challenges in utilizing CRISPR-*Cas* systems in clinical applications is the delivery mechanism. The efficacy of CRISPR therapeutics relies on their precise and efficient delivery to specific cells or organs within living beings.[41] To optimize the targeting of CRISPR-*Cas* systems to their desired destinations, scientists are investigating several delivery techniques, such as lipid nanoparticles, cell-based delivery systems, and viral vectors.

To surmount these challenges, collaboration among researchers from diverse disciplines is imperative, and progress in CRISPR technology, delivery systems, and immunomodulation approaches must be relentlessly pursued. To guarantee the safety and effectiveness of CRISPR-*Cas* systems in clinical environments, researchers must tackle issues related to off-target effects, immunological reactions, and limitations in delivery. This will enable the complete realization of the potential of these systems in the treatment of autoimmune diseases. To ensure responsible and efficient clinical use, it is imperative to carefully consider and resolve the ethical, regulatory, and safety issues associated with the development of CRISPR-based therapies. The ethical implications of genome editing arise from its uncertain consequences, especially when modifying germline cells that can be inherited by future generations.[51] Issues of consent, equity, and justice arise when considering patients' capacity to make informed decisions and the fairness of access to CRISPR-based treatments. When considering regulatory factors, it is crucial to remember the multitude of regulations and authorities established to guarantee the safety and efficacy of new medicines.[52] Evaluating clinical and preclinical data, assessing risks and benefits, and granting approval for clinical studies and, ultimately, market authorization are essential responsibilities of regulatory bodies such as the FDA and the EMA. Given the risks associated with off-target effects, unexpected outcomes of genome editing, and immune reactions to CRISPR components, it is crucial to prioritize safety while developing CRISPR-based treatments.[53] Approaches to enhance safety encompass monitoring immune responses in clinical trials, advancing more accurate CRISPR-*Cas* systems, and optimizing delivery systems to minimize off-target effects. Ensuring the responsible research and implementation of CRISPR-based therapies necessitates addressing ethical, regulatory, and safety concerns. Transparency, collaboration among stakeholders, and adherence to rigorous scientific and ethical norms are crucial for the ethical implementation and effective translation of CRISPR-based therapies into clinical practice.

15. Future directions and opportunities

To advance in the field of CRISPR-based therapies for autoimmune diseases, it is feasible to explore several strategies to overcome current challenges and improve the efficacy and safety of these treatments. Enhancing the specificity and minimizing non-target impacts of CRISPR-*Cas* systems is a viable approach. Enhancing guide RNA design strategies and developing more precise *Cas* nucleases can achieve this objective.[50] To minimize unintended effects, recent advancements in CRISPR technology have introduced novel variations such as CRISPR-*Cas* base editors and prime editors. These variants enable precise modifications to the DNA without causing any breaks.[45,46] Another objective of new delivery systems is to enhance the precise targeting of CRISPR components to specific cells or tissues affected by autoimmune diseases. The studies conducted by[27–29,41,47] propose the need for developing new viral vectors, lipid nanoparticles, or cell-based delivery methods that are specifically optimized for precise intracellular distribution. An approach to enhance the safety and efficacy of CRISPR-based medications involves developing delivery mechanisms that can evade the immune system and selectively target particular organs. To ensure lasting therapeutic efficacy, it is crucial to tackle immune responses directed towards CRISPR components. To prolong the duration of therapeutic benefits, it is possible to employ tactics that restrict the immune system from identifying and removing CRISPR components. One possible approach is to modify CRISPR proteins to reduce their immunogenicity or utilize delivery vehicles that can evade the immune system.[41] Prudent advancement and execution of CRISPR-based medications likewise necessitate meticulous consideration of ethical and regulatory issues. To establish public confidence and promote the ethical execution of clinical trials, it is essential to prioritize transparency, involve patients actively, and comply with ethical principles and regulatory guidelines.[51] Researchers can expedite the development of safe, efficient, and personalized CRISPR-based therapeutics for autoimmune diseases by employing these strategies and capitalizing on ongoing advancements in CRISPR technology, delivery methods, and regulatory frameworks.

The CRISPR research field is currently undergoing tremendous advancements, driven by emerging trends, interdisciplinary collaborations, and innovative approaches. These developments are facilitating the creation of new medications and technology. CRISPR is being applied in various disciplines. The integration of CRISPR with other cutting-edge techniques and disciplines is becoming more prevalent. These disciplines encompass synthetic biology, machine learning, and single-cell sequencing. By employing a multidisciplinary approach, researchers can get a deeper understanding of complex biological systems and develop more efficient CRISPR-based tools and strategies.[54,55] According to Jaitin et al.,[56] the combination of CRISPR-*Cas* systems and single-cell sequencing techniques has provided new insights into cell fate determination, lineage tracing, and disease causes. This powerful combination allows for precise manipulation and characterization of individual cells. Another emerging trend is the broadening application of CRISPR-based technologies, which now extend beyond genome editing to include epigenome editing, RNA targeting, and synthetic biology. The utilization of CRISPR-*Cas* systems has facilitated the effective and accurate manipulation of cellular pathways, the control of gene expression, and the exploration of gene function.[43,46] These novel techniques demonstrate promise in enhancing our understanding of fundamental biological mechanisms and addressing a diverse array of biological challenges. Furthermore, collaborative endeavors among academic, industrial, and clinical institutions are driving advancements in CRISPR research, hence facilitating the practical application of scientific discoveries. Interdisciplinary collaborations facilitate the exchange of knowledge, tools, and resources, hence expediting the advancement of CRISPR-based diagnostics, technologies, and treatments.[57] The Innovative Genomics Institute and the Broad Institute's Center for CRISPR Research are collaborative efforts that unite researchers from diverse backgrounds to address significant challenges and further our understanding of CRISPR. In summary, CRISPR research is experiencing a significant transformation as a result of novel approaches, collaborations across many disciplines, and innovative techniques. These factors are facilitating advancements in genome editing, functional genomics, and other interconnected areas. By harnessing the collective expertise and creativity of scientists from several fields, CRISPR technology is propelling progress in the fields of biology, medicine, and biotechnology, paving the way for revolutionary applications and breakthroughs.

Future research on enhancing CRISPR-*Cas* systems for the treatment of autoimmune illnesses encompasses various areas, aiming to enhance the specificity, efficacy, and safety of therapeutic interventions. Developing more accurate CRISPR-*Cas* systems to minimize off-target consequences represents a promising advancement in the correct path. This involves the creation of new CRISPR effectors or altered *Cas* enzymes, as well as the exploration of different CRISPR variations that have better specificity.[24,58] The enhanced CRISPR-*Cas* systems have the potential to enhance the accuracy and safety of genome editing for the treatment of autoimmune diseases by minimizing unintended effects on non-targeted genes. A significant area for future study involves enhancing delivery methods to efficiently and specifically transport CRISPR components to affected tissues or cells. To enhance the effectiveness of delivering substances, specifically targeting tissues and avoiding immune detection, it is necessary to develop novel delivery systems, such as viral vectors, nanoparticles, and carriers based on cells.[27–29,41] Researchers can enhance delivery tactics to improve the therapeutic efficacy and reduce potential off-target consequences that are related to the systemic dispersion of CRISPR-*Cas* systems. Future research might prioritize the optimization of CRISPR-based techniques to change immune responses and target disease-specific pathways in autoimmune disorders. Novel CRISPR-mediated approaches, such as epigenome editing, RNA targeting, and combinatorial gene editing strategies, are being developed to restore immunological balance and accurately regulate immune function.[43,59] These personalized approaches may provide better therapeutic outcomes with reduced unintended effects by specifically targeting particular populations of immune cells or signaling pathways implicated in autoimmune disease. Functional genomics techniques and advancements in CRISPR screening technology can be used to comprehensively identify novel therapeutic targets and pathways associated with autoimmune diseases. Genome-wide CRISPR deletion and activation screens have the potential to significantly improve the development of targeted medications. These screens can identify key regulators of immune function, disease progression, and therapeutic resistance, hence aiding in the discovery of effective treatments.[48,49] By integrating CRISPR screening data with clinical and molecular profile data, researchers may effectively rank promising targets for future validation and clinical translation. Ultimately, future research should prioritize enhancing the precision, effectiveness, and safety of CRISPR-*Cas* systems in treating autoimmune disorders. This can be accomplished by developing more advanced CRISPR-*Cas* variations, improving delivery systems, fine-tuning tactics to regulate the immune system, and systematically identifying therapeutic targets. To effectively apply CRISPR-based treatments for precision medicine in autoimmune diseases, researchers must overcome these challenges and take advantage of this potential.

16. Conclusion

Recent research on CRISPR-based therapies for autoimmune diseases has demonstrated promising advancements in three key areas: the precision of genetic modification, the techniques of administration, and the strategies for regulating the immune system. There is a strong possibility of improving therapeutic outcomes and minimizing unintended consequences by enhancing the precision and effectiveness of CRISPR-*Cas* systems and developing innovative methods of delivering them to specific tissues. Furthermore, there are emerging prospects in the field of autoimmune sickness therapy customization by the manipulation of immune responses and the targeting of disease-specific pathways. CRISPR-based therapies are on the horizon as a result of these advancements. These therapies have the potential to revolutionize the management of autoimmune diseases by providing precise and customized interventions that address the underlying molecular mechanisms responsible for autoimmunity. This approach aims to minimize side effects while maximizing therapeutic effectiveness. To effectively harness the potential of CRISPR-based therapies for the treatment of autoimmune illnesses, it is essential to have continuous collaboration across multiple disciplines, rigorous preclinical and clinical validation, and meticulous attention to ethical, regulatory, and safety considerations. CRISPR-*Cas* technologies can profoundly transform autoimmune research and clinical practice. These pathways offer promising opportunities to uncover the molecular mechanisms underlying autoimmune diseases and develop state-of-the-art therapies. CRISPR technology, with its ability to precisely modify genomes and regulate immune responses, holds immense potential to revolutionize the treatment of autoimmune illnesses. By elucidating the genetic factors underlying susceptibility and enabling tailored therapies for individual patients, it offers a promising avenue for transforming our approach to these conditions. Autoimmune research has ongoing challenges in identifying disease-associated genes, comprehending disease mechanisms, and developing efficient therapies with minimal off-target effects. CRISPR-based methodologies have the potential to address these concerns. If we can specifically target immune pathways, it is possible to restore immunological balance and achieve long-term remission and personalized treatment for autoimmune diseases. Effective implementation of CRISPR-based therapeutics in clinical settings will require continuous collaboration across several disciplines, rigorous validation in both preclinical and clinical settings, and careful consideration of ethical, regulatory, and safety issues. It is imperative that we promptly advocate for continued study and collaboration to maximize the potential of CRISPR technology in treating autoimmune disorders since we are on the verge of a transformative period in this domain. A multitude of patients worldwide are anticipating that CRISPR-*Cas* systems may initiate a novel era in the management of autoimmune illnesses. However, tofulfil this commitment, it is necessary to secure sustainable funding for interdisciplinary research, innovative methods for technological development, and collaborations across academic, corporate, and healthcare institutions. Collectively, we can overcome the remaining barriers, enhance CRISPR-based therapies for autoimmune diseases, and translate scientific advancements into tangible medical benefits for patients. We should use this chance and intensify our endeavors to explore how CRISPR can transform the fight against autoimmune illnesses.

References

1. Rose NR. Autoimmune diseases: tracing the shared threads. *Hosp Pract.* 1995;32(4):147–154. https://doi.org/10.1080/21548331.1997.11443469 PMID: 9109812.
2. Cooper GS, Bynum ML, Somers EC. Recent insights in the epidemiology of autoimmune diseases: improved prevalence estimates and understanding of clustering of diseases. *J Autoimmun.* 2009;33(3-4):197–207. https://doi.org/10.1016/j.jaut.2009.09.008 Epub 2009 Oct 9. PMID: 19819109; PMCID: PMC2783422.
3. Smolen JS, Aletaha D, McInnes IB. Rheumatoid arthritis. *Lancet.* 2016;388(10055):2023–2038. https://doi.org/10.1016/S0140-6736(16)30173-8 Epub 2016 May 3. Erratum in: Lancet. 2016 Oct 22;388(10055):1984. PMID: 27156434.
4. Compston A, Coles A. Multiple sclerosis. *Lancet.* 2008;372(9648):1502–1517. https://doi.org/10.1016/S0140-6736(08)61620-7 PMID: 18970977.

5. Tsokos GC, Lo MS, Costa Reis P, Sullivan KE. New insights into the immunopathogenesis of systemic lupus erythematosus. *Nat Rev Rheumatol.* 2016;12(12):716–730. https://doi.org/10.1038/nrrheum.2016.186 PMID: 27872476.
6. McInnes IB, Schett G. The pathogenesis of rheumatoid arthritis. *N Engl J Med.* 2011;365(23):2205–2219. https://doi.org/10.1056/NEJMra1004965 PMID: 22150039.
7. Barrangou R, Fremaux C, Deveau H, et al. CRISPR provides acquired resistance against viruses in prokaryotes. *Science.* 2007;315(5819):1709–1712. https://doi.org/10.1126/science.1138140 PMID: 17379808.
8. Doudna JA, Charpentier E. Genome editing. The new frontier of genome engineering with CRISPR-Cas9. *Science.* 2014;346(6213):1258096. https://doi.org/10.1126/science.1258096 PMID: 25430774.
9. Shmakov S, Smargon A, Scott D, et al. Diversity and evolution of class 2 CRISPR-Cas systems. *Nat Rev Microbiol.* 2017;15(3):169–182. https://doi.org/10.1038/nrmicro.2016.184 PMID: 28111461; PMCID: PMC5851899.
10. Jinek M, Chylinski K, Fonfara I, Hauer M, Doudna JA, Charpentier E. A programmable dual-RNA-guided DNA endonuclease in adaptive bacterial immunity. *Science.* 2012;337(6096):816–821. https://doi.org/10.1126/science.1225829 Epub 2012 Jun 28. PMID: 22745249; PMCID: PMC6286148.
11. Ishino Y, Shinagawa H, Makino K, Amemura M, Nakata A. Nucleotide sequence of the iap gene, responsible for alkaline phosphatase isozyme conversion in *Escherichia coli*, and identification of the gene product. *J Bacteriol.* 1987;169(12):5429–5433. https://doi.org/10.1128/jb.169.12.5429-5433.1987 PMID: 3316184; PMCID: PMC213968.
12. Mojica FJ, Díez-Villaseñor C, García-Martínez J, Soria E. Intervening sequences of regularly spaced prokaryotic repeats derive from foreign genetic elements. *J Mol Evol.* 2005;60(2):174–182. https://doi.org/10.1007/s00239-004-0046-3
13. Garneau JE, Dupuis MÈ, Villion M, et al. The CRISPR/Cas bacterial immune system cleaves bacteriophage and plasmid DNA. *Nature.* 2010;468(7320):67–71. https://doi.org/10.1038/nature09523 PMID: 21048762.
14. O'Connell RM, O'Connell RM, Rao DS, et al. Therapeutic potential of targeting microRNAs in the treatment of autoimmune and inflammatory diseases. *Arthritis Res Ther.* 2010;12(1):220. https://doi.org/10.1186/ar2953
15. Lee J, Kim Y, Park D, et al. CRISPR/Cas9-edited T cells for the treatment of autoimmune diseases. *Immune Netw.* 2020;20(3):e23. https://doi.org/10.4110/in.2020.20.e23
16. Wang D, Zhang F, Gao G. CRISPR-based therapeutic genome editing: strategies and *in vivo* delivery by AAV vectors. *Cell.* 2020;181(1):136–150. https://doi.org/10.1016/j.cell.2020.03.023 PMID: 32243786; PMCID: PMC7236621.
17. Wang L, et al. CRISPR/Cas9-mediated immune editing: challenges and opportunities in transplantation. *Front Immunol.* 2020;11:568.
18. Crow YJ. Type I interferonopathies: mendelian type I interferon up-regulation. *Curr Opin Immunol.* 2015;32:7–12. https://doi.org/10.1016/j.coi.2014.10.005 Epub 2014 Oct 30. PMID: 25463593.
19. Mohan C, Putterman C. Genetics and pathogenesis of systemic lupus erythematosus and lupus nephritis. *Nat Rev Nephrol.* 2015;11(6):329–341. https://doi.org/10.1038/nrneph.2015.33 Epub 2015 Mar 31. PMID: 25825084.
20. Najm FJ, Madhavan M, Zaremba A, et al. Drug-based modulation of endogenous stem cells promotes functional remyelination in vivo. *Nature11.* 2015;522(7555):216–220. https://doi.org/10.1038/nature14335 Epub 2015 Apr 20. PMID: 25896324; PMCID: PMC4528969.
21. Xu X, et al. Correction of a pathogenic gene mutation in human embryos. *Nature.* 2019;548(7668):413–419.
22. Chen H, et al. CRISPR/Cas9-mediated gene editing for the development of cell-based therapies in diabetes. *Front Endocrinol.* 2018;9:753.

23. Gupta S, et al. CRISPR/Cas9 in clinical trials for Type 1 diabetes: current status and future perspectives. *Diabetes Res Clin Pract.* 2021;176:108857.
24. Chen JS, Dagdas YS, Kleinstiver BP, et al. Enhanced proofreading governs CRISPR-Cas9 targeting accuracy. *Nature.* 2017;550(7676):407–410. https://doi.org/10.1038/nature24268 Epub 2017 Sep 20. PMID: 28931002; PMCID: PMC5918688.
25. Zhang J, et al. Immune cell engineering with CRISPR/Cas9 for psoriasis therapy. *Curr Gene Ther.* 2019;19(3):160–168.
26. Xiao Y, et al. Precision medicine in psoriasis: a CRISPR/Cas9 approach. *J Investig Dermatol.* 2020;140(12):2307–2315.
27. Li L, Song L, Liu X, et al. Artificial virus delivers CRISPR-Cas9 system for genome editing of cells in mice. *ACS Nano.* 2021;15(6):9253–9264.
28. Li H, et al. CRISPR/Cas9-based gene editing for psoriasis treatment: opportunities and challenges. *Front Immunol.* 2021;12:639893.
29. Li X, et al. Precision medicine in inflammatory bowel disease: CRISPR/Cas9 approaches. *Front Immunol.* 2021;12:663491.
30. Wang J, et al. CRISPR/Cas9-mediated genetic studies in inflammatory bowel disease. *Gut.* 2018;67(2):216–225.
31. Chen Y, et al. Engineering probiotic bacteria using CRISPR/Cas9 for gut microbiome modulation in inflammatory bowel disease. *Nat Commun.* 2019;10(1):577.
32. Zhang Q, et al. CRISPR/Cas9-based gene editing for inflammatory bowel disease therapy. *Curr Opin Gastroenterol.* 2020;36(4):310–316.
33. Wu H, Chen Y, Zhu H, Zhao M, Lu Q. The pathogenic role of dysregulated epigenetic modifications in autoimmune diseases. *Front Immunol.* 2019;10:2305. https://doi.org/10.3389/fimmu.2019.02305 PMID: 31611879; PMCID: PMC6776919.
34. Li L, Song L, Liu X, et al. Artificial virus delivers CRISPR-Cas9 system for genome editing of cells in mice. *ACS Nano.* 2017;11(1):95–111. https://doi.org/10.1021/acsnano.6b04261
35. Ma Y, et al. CRISPR/Cas9-mediated gene editing. *Rheum Arthritis Arthritis Res Ther.* 2018;20(1):94.
36. Wang J, et al. CRISPR/Cas9-based gene therapy for systemic lupus erythematosus. *J Immunol Res.* 2019:6273078.
37. Liu S, et al. Precision genome editing in multiple sclerosis using CRISPR-Cas9. *Front Immunol.* 2020;11:595.
38. Zhang H, et al. CRISPR/Cas9 applications in inflammatory bowel disease. *Front Cell Dev Biol.* 2021;9:661869.
39. Abudayyeh OO, et al. CRISPR/Cas13a targeting of microRNA in autoimmune diseases. *RNA.* 2017;23(5):675–686.
40. Xu J, Zhu W, Liang C, et al. CRISPR/Cas9-mediated IL-17 receptor A (IL-17RA) gene knockout suppresses experimental autoimmune encephalomyelitis, a mouse model of multiple sclerosis. *Biochem Biophys Res Commun.* 2020;521(4):872–879.
41. Wang X, Wang Y, Wu X, et al. CRISPR/Cas9-mediated IL-17RA gene knockout alleviates psoriasis-like skin inflammation in mice. *Biomed Res Int.* 2021:5512460.
42. Lei WT, Shen LJ, Wang Y, et al. The safety and efficacy of CRISPR-Cas9 in preclinical models of autoimmune diseases: a systematic review and meta-analysis. *Front Immunol.* 2020;11:619850.
43. Mandegar MA, Huebsch N, Frolov EB, et al. CRISPR interference efficiently induces specific and reversible gene silencing in human iPSCs. *Cell Stem Cell.* 2016;18(4):541–553. https://doi.org/10.1016/j.stem.2016.01.022 Epub 2016 Mar 10. PMID: 26971820; PMCID: PMC4830697.
44. Sanjana NE, Shalem O, Zhang F. Improved vectors and genome-wide libraries for CRISPR screening. *Nat Methods.* 2014;11(8):783–784. https://doi.org/10.1038/nmeth.3047 PMID: 25075903; PMCID: PMC4486245.

45. Gaudelli NM, Komor AC, Rees HA, et al. Programmable base editing of A*T to G*C in genomic DNA without DNA cleavage. *Nature*. 2020;551(7681):464–471.
46. Anzalone AV, Koblan LW, Liu DR. Genome editing with CRISPR-Cas nucleases, base editors, transposases and prime editors. *Nat Biotechnol*. 2020;38(7):824–844. https://doi.org/10.1038/s41587-020-0561-9 Epub 2020 Jun 22. PMID: 32572269.
47. Miao L, Li L, Huang Y, et al. Delivery of mRNA vaccines with heterocyclic lipids increases anti-tumor efficacy by STING-mediated immune cell activation. *Nat Biotechnol*. 2019;37(10):1174–1185. https://doi.org/10.1038/s41587-019-0247-3 Epub 2019 Sep 30. PMID: 31570898.
48. Shi X, Shi Y, Zhou H, et al. CRISPR-based functional genomic screening in the development of monoclonal antibody-based therapy for immune-mediated diseases. *Front Immunol*. 2021;12:681924.
49. Tzelepis K, Koike-Yusa H, De Braekeleer E, et al. A CRISPR dropout screen identifies genetic vulnerabilities and therapeutic targets in acute myeloid leukemia. *Cell Rep*. 2016;17(4):1193–1205. https://doi.org/10.1016/j.celrep.2016.09.079 PMID: 27760321; PMCID: PMC5081405.
50. Tsai SQ, Joung JK. Defining and improving the genome-wide specificities of CRISPR-Cas9 nucleases. *Nat Rev Genet*. 2016;17(5):300–312. https://doi.org/10.1038/nrg.2016.28 PMID: 27087594; PMCID: PMC7225572.
51. Lander ES, Baylis F, Zhang F, et al. Adopt a moratorium on heritable genome editing. *Nature*. 2019;567(7747):165–168. https://doi.org/10.1038/d41586-019-00726-5 PMID: 30867611.
52. National Academies of Sciences, Engineering, and Medicine (NASEM). *Human Genome Editing: Science, Ethics, and Governance*. Washington, DC: The National Academies Press; 2017.
53. Barrangou R, Doudna JA. Applications of CRISPR technologies in research and beyond. *Nat Biotechnol*. 2016;34(9):933–941.
54. Adamson B, Norman TM, Jost M, et al. A multiplexed single-cell CRISPR screening platform enables systematic dissection of the unfolded protein response. *Cell*. 2016;167(7):1867–1882.e21. https://doi.org/10.1016/j.cell.2016.11.048 PMID: 27984733; PMCID: PMC5315571.
55. Liu XS, Wu H, Ji X, et al. Editing DNA methylation in the mammalian genome. *Cell*. 2016;167(1):233–247.e17. https://doi.org/10.1016/j.cell.2016.08.056 PMID: 27662091; PMCID: PMC5062609.
56. Jaitin DA, Weiner A, Yofe I, et al. Dissecting immune circuits by linking CRISPR-pooled screens with single-cell RNA-Seq. *Cell*. 2016;167(7):1883–1896.e15. https://doi.org/10.1016/j.cell.2016.11.039 PMID: 27984734.
57. Mali P, Yang L, Esvelt KM, et al. RNA-guided human genome engineering via Cas9. *Science*. 2013;339(6121):823–826. https://doi.org/10.1126/science.1232033 Epub 2013 Jan 3. PMID: 23287722; PMCID: PMC3712628.
58. Komor AC, Badran AH, Liu DR. CRISPR-based technologies for the manipulation of eukaryotic genomes. *Cell*. 2017;168(1-2):20–36. https://doi.org/10.1016/j.cell.2016.10.044
59. Gilbert LA, Horlbeck MA, Adamson B, et al. Genome-scale CRISPR-mediated control of gene repression and activation. *Cell*. 2014;159(3):647–661. https://doi.org/10.1016/j.cell.2014.09.029 Epub 2014 Oct 9. PMID: 25307932; PMCID: PMC4253859.

CHAPTER ELEVEN

Advances in CRISPR-Cas systems for blood cancer

Bernice Monchusi[a,1], Phumuzile Dube[a,1], Mutsa Monica Takundwa[a], Vanelle Larissa Kenmogne[a,b], and Deepak Balaji Thimiri Govinda Raj[a,*]

[a]Synthetic Nanobiotechnology and Biomachines, Synthetic Biology and Precision Medicine Centre, Future production Chemicals Cluster, Council for Scientific and Industrial Research, Pretoria, South Africa
[b]Department of Surgery, University of the Witwatersrand, Johannesburg, South Africa
*Corresponding author. e-mail address: dgovindaraj@csir.co.za

Contents

Abstract

CRISPR-Cas systems have revolutionised precision medicine by enabling personalised treatments tailored to an individual's genetic profile. Various CRISPR technologies have been developed to target specific disease-causing genes in blood cancers, and some have advanced to clinical trials. Although some studies have explored the *in vivo* applications of CRISPR-Cas systems, several challenges continue to impede their widespread use. Furthermore, CRISPR-Cas technology has shown promise in improving the response of immunotherapies to blood cancers. The emergence of CAR-T cell therapy has shown considerable success in the targeting and correcting of disease-causing genes in blood cancers. Despite the promising potential of CRISPR-Cas in the treatment of blood cancers, issues related to safety, ethics, and regulatory approval remain significant hurdles. This comprehensive review highlights the transformative potential of CRISPR-Cas technology to revolutionise blood cancer therapy.

[1] Equally contributed.

Progress in Molecular Biology and Translational Science, Volume 208
ISSN 1877-1173, https://doi.org/10.1016/bs.pmbts.2024.07.004

Abbreviations

ABL	Abelson murine leukaemia
AML	Acute myeloid leukaemia
ASXL1	Additional Sex Combs Like 1
BCR	Breakpoint cluster region
CAR-T	Chimeric antigen receptor T
CSR	Chimeric switch receptor
CLL	Chronic lymphocytic leukaemia
CML	Chronic myeloid leukaemia
CRISPR-Cas	Clustered Regularly Interspaced Short Palindromic Repeats CRISPR-associated protein
CRS	Cytokine release syndrome
DLBCL	Diffuse large B-cell lymphomas
FLT-3	FMA-like Tyrosine Kinase 3
FL	Follicular lymphomas
HSCs	Haematopoietic stem cells
ITD	Internal Tandem Duplication
MM	Multiple myeloma
NHL	Non-Hodgkin's lymphoma
PD-1	Programmed cell death protein 1
sgRNAs	Single guide RNAs
TKD	Tyrosine kinase domain
TCR	T cell receptor

1. Genetic landscape of blood cancers

Cytogenetics and next-generation sequencing have revealed the existence of diversely complex mutational profiles associated with hematologic cancers. Although some genes are biased toward certain types of blood cancers, these approaches have shown a limited number of common mutated genes in the spectra. By correcting these disease-causing mutations with genetic editing techniques such as CRISPR-Cas systems, the prognosis can be improved, thus increasing the survival rate of individual patients. Not only that, but also understanding the molecular and functional role played by these different driver mutations could positively impact the discovery of novel therapies.

Gene mutations that modify amino acid sequences or shorten or elongate the affected protein often inactivate protein function. The Chromosomal translocations and point mutations frequently affect the structure of the encoded protein.[1] The chromosomal translocation t(9;22)(q34;q11) results in the Philadelphia (Ph) chromosome that contains a hybrid gene consisting of the Abelson murine leukaemia (*ABL*) gene and the breakpoint cluster region (*BCR*) gene (also referred to as the *BCR::ABL1* chimeric gene) known to cause myeloproliferative neoplasm and chronic myeloid leukaemia (CML).[2,3] Alterations in the *ASXL1* gene that encodes the protein involved in chromatin remodelling have been widely observed in various types of myeloid malignant diseases, such as myelodysplastic syndrome,[4–6] myeloproliferative neoplasms,[5,7] acute myeloid leukaemia (AML)[5,8,9] as well as in patients with CML.[5,10]

AML is characterised by the proliferation of genetically modified haematopoietic stem cells, as well as haematopoietic progenitor cells.[11] Kelly and Gilliland hypothesised that AML is the result of the cooperative effect of at least two mutations classes; namely class I, class II and the recently identified group without classification.[12,13] Class I mutations, described as those that promote proliferation and survival, consist of mutations in the *FLT3, KIT, RAS, PNPN11, JAK2,* and *CBL* genes, *FLT3* being the most common. Mutations in the activation of the FMA-like tyrosine kinase 3 (*FLT3*) gene located on chromosome 13q12[14] have been observed in 25–45% of all patients with AML.[11] This type of mutation includes internal tandem duplication (*ITD*) and the tyrosine kinase domain (*TKD*); with *ITD* is the most identified *FLM3* mutation.[11,13] *PML-RARA, RUNX1-RUNXIT1, CBFB-MYH11, MLL, CEBPA, IDH1, IDH2,* and *NPM1* are class II mutations known to alter cellular differentiation and apoptosis.[13,15] Of these mutations, the nucleophosmin 1 (*NPM1*) mutation is the most identified in patients with AML (30–50% frequency) and involves dysregulation of ribosome biogenesis, DNA regulation, and repair. The co-occurrence of *NPM1* and *FLT3-ITD* mutations was reported in about 20% of AML patients.[11,16]

Chronic lymphocytic leukaemia (CLL) is the most common form of leukaemia that affects adults in the developed world.[17] Mutations in neurogenic locus notch homolog protein 1 (*Notch1*) and splicing factor 3B subunit 1 (*SF3B1*) are the most detected in patients with CLL with a frequency of 5%–20%.[18] These novel gene mutations are associated with the activation of oncogenic pathways and with the influence of the spliceosome, respectively. In addition to these mutations, deletions at 13q, tri12, 11q, and 17p are the most common karyotypic anomalies associated with mature CLL cells and are assumed to contribute to the initiation or progression of disease.[17] Mutations in genes that encode the tumour protein *p53* (*TP53* gene) and the SAM domain and protein 1 containing the HD domain (*SAMHD1* gene) were identified to dominantly drive CLL relapse.[19] Detection of the *TP53* mutation and del17p has been used to independently predict rapid disease progression, as these mutations have been strongly linked to a poor prognosis.[18,20] *TP53* mutations have also been identified with varying frequency in non-Hodgkin lymphoma entities such as Burkitt lymphoma, diffuse large B-cell lymphomas (DLBCL) and follicular lymphomas (FL),[21] as well as in multiple myeloma (MM).[22] The consensus is that this genetic mutation has a negative impact on disease prediction.

Although the most common non-Hodgkin and Hodgkin entities have shown genetic heterogeneity, there have been some similarities in altered pathways and oncogenesis processes. Mature B-cell neoplasms that comprise more than 80% of non-Hodgkin lymphoma cases have been shown to have altered immunoglobulin (Ig) genes.[23] These malignant B-cell diseases have been classified according to morphological, phenotypical, genetic and clinical variations such as DLBCLs, Burkitt lymphoma, high-grade B-cell lymphomas, and FL, with DLBCLs and FL being the most prevalent types.[23–25] Both FL and DLBCL harbour the chromosomal aberration t (14;18)(q32;q21) that favours the unregulated expression of the *BCL2* oncoprotein, as well as mutations in the *MLL2* and *MEF2B* genes, both of which disrupt chromatin biology and function.[25] Other major genetic alterations specific for DLBCL include mutations in the *c-REL* gene and the miR-17–92 microRNA cluster.[25,26] In Hodgkin lymphoma, tumour cells (referred to as Hodgkin and Reed-Sternberg (HRS) cells) originate from B-cells that have lost the typical B-cell phenotypic characteristics.[27] Genetic abnormalities such as mutations in the *NFKBIA, NFKBIE, CYLD, TNFAIP3, REL* and *SOCS1* genes that affect the NF-KB and JAK / STAT signalling pathways in HRS cells have been shown to promote angiogenesis, cell proliferation, and survival.[27,28]

Another hematologic malignancy strongly marked by heterogeneity is MM. Sequencing approaches have been used to identify a variety of mutations related to the pathology of MM. Recurring mutations in certain driver genes, such as *KRAS, NRAS, DIS3, FAM46C,* and *BRAF,* may be crucial in determining the prognosis of the disease.[22,29,30] Various cytogenetic abnormalities have also been detected, including t(4,14), t(1,16), del(17p) considered high risk, and t (1,20) which confer a poor prognosis.[24,29–31] Mutations in the *PRDM1* gene, which acts as a tumour suppressor, have been detected in DLBCL. Although also identified in MM, the exact role of this mutation in MM is currently unknown and still under investigation.[24] Karyotypic anomalies associated with chromosome 13 have been reported to have significant clinical implications in MM patients resulting in a poorer prognosis and shorter survival rates.[32]

The genetic landscape of blood cancers reveals numerous specific mutations and alterations driving the disease, providing a precise target for therapeutic intervention. The emergence of CRISPR technology allows direct and accurate editing of these genetic abnormalities, representing a revolutionary advance in precision medicine. By pinpointing and targeting unique genetic mutations in individual patients, CRISPR facilitates the creation of highly personalised treatments. Therefore, this review will explore the advances of CRISPR-Cas in precision medicine, with particular interest in blood cancers.

2. Exploring CRISPR-Cas techniques

Several gene editing technologies have been designed to rectify genetic abnormalities in complex conditions such as cancer. Choosing programmable nucleases is a crucial factor in the effectiveness of genetic editing. The most well-known options include meganucleases, transcription-activator-like effector nucleases (TALENs), zinc finger nucleases (ZFNs), and CRISPR-Cas9.[33–36] CRISPR technology is notable for its simplicity, improved specificity, and efficiency, and recent advances facilitate a more sophisticated iterative approach to genetic editing.

Multiplexing has become a powerful tool that allows CRISPR to target multiple genetic sites, allowing the modulation of genes at various DNA sites simultaneously.[37] Another significant development is base editing, where instead of cleaving the target gene like traditional nucleases, a CRISPR-related editing enzyme directly converts one base to another. The dCas9 variant has recently been shown to reverse gene editing.[38]

CRISPR technology has transformative potential to treat a wide range of diseases by allowing precise genome modifications. Its application covers rectifying disease-associated mutations for cancer therapy, targeting viral infections, and addressing complex genetic diseases. Several methods can be employed in CRISPR-Cas technology, including gene-targeted modifications, multiplexing, and interventions carried out *ex vivo* and *in vivo* (Fig. 1). The CRISPR gene targeting method involves creating specific single guide RNAs (sgRNAs) to guide the Cas9 enzyme to exact sites within the genome. These components are introduced into cells, where the inherent DNA repair systems of the cell are used to make the intended genetic changes.[39–41] In multiplexing, numerous genes can be targeted

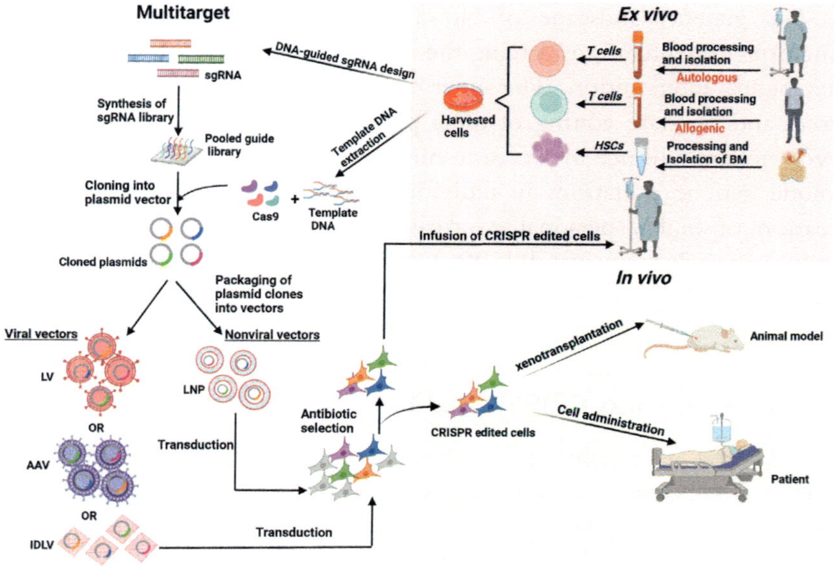

Fig. 1 *Ex vivo* and *in vivo* CRISPR gene therapy. In *ex vivo* CRISPR gene therapy, T lymphocytes (T cells) are extracted from the patient's blood (autologous) or donor's blood (allogenic) or haematopoietic stem cells (HSCs) from the bone marrow (BM). These cells are then edited using CRISPR components to target specific DNA sequences. Electroporation introduces the sgRNAs, Cas9, and donor template into a plasmid vector. CRISPR components are packaged into viral vectors such as lentiviruses (LVs), adeno-associated viruses (AAVs), and integrase-deficient lentiviral vectors (IDLVs), or nonviral vectors such as lipid nanoparticles (LNPs). The edited cells are subsequently reintroduced into the patient or animal model, with or without selection pressure. In *in vivo*, CRISPR gene therapy involves designing sgRNAs without the use of a template DNA to target specific DNA sequences and delivering these sgRNAs into cells using viral or non-viral vectors. The CRISPR components are administered directly *via* injection into the patient or animal model. *Created with Biorender.com.*

simultaneously by Cas enzymes, or multiple sgRNAs can target a single genetic site to improve editing or transcriptional regulation efficiency.[42–49] There has been a growing preference for Cas12a systems in recent years for multiplexed genome editing, as they have inherent RNase activity, allowing efficient processing of the poly-crRNA transcript.[39] *Ex vivo* CRISPR editing allows the modification of genetic traits in cells (such as T lymphocytes and haematopoietic stem cells (HSCs)) derived from patients outside the body, followed by their reintroduction into the body for cancer therapy. *In vivo* CRISPR editing involves the direct injection of genetic material into human subjects or animal models.

The precision offered by CRISPR editing is especially crucial in blood cancers, where it can rectify harmful mutations, improve the efficacy of chimeric antigen receptor T (CAR-T) therapy, and address resistance and relapse mechanisms. These CRISPR methods represent promising strategies for treating blood cancers, as elaborated in the next section.

3. CRISPR-Cas applications in the treatment of haematological cancers

3.1 Targeted therapies using CRISPR-Cas

CRISPR-Cas9 can target specific genetic profiles and pathways involved in hematologic malignancies. One way in which it can be used is by eliminating or inactivating genes (also called gene knockout). A study by Vuelta et al. demonstrated the effective interruption of the *BCR/ABL1* oncogene in murine (Boff-p210) and human (K562) cell lines using CRISPR technology.[50] This was achieved using two sgRNAs that induced a short deletion in the TK domain of *ABL1*, preventing crucial oncogenic effects for the genesis and progression of CML. It was previously demonstrated that the CRISPR-Cas9 system could stop tumour progression by targeting and disrupting the *BCR/ABL1* fusion sequence or individual key genes.[51,52] Garca-Tuón et al. also injected CRISPR-Cas9 edited Boff-p210 cell lines into immunosuppressed mice, subcutaneously.[53] The system was used to edit the *BCR-ABL* oncogene by nullifying the expression of the p201 oncoprotein in cells. These edited cells showed increased apoptosis, while mice developed smaller tumours than those mice injected with the unedited Boff-p210 parental cells of Boff-p210. Mutations in *IDH1* or *IDH2* have been implicated in the leukemogenesis of AML.[15] They can affect AML blast metabolism by blocking the function of cytochrome c oxidase in the mitochondrial respiratory chain[54] or inducing alterations in the pattern of histone modifications that result in atypical DNA methylation.[55] A study demonstrated the successful incorporation of the *IDH2 R140Q* point mutation using the CRISPR-Cas system.[56] This was achieved by inducing specific DNA disruptions in the *IDH2* gene of K562 myeloid leukaemia cells. Interestingly, the single base mutation induced by CRISPR-Cas IDH2 R140Q was shown to exert characteristic functional changes of this mutation. CRISPR-Cas12a was also used to reconstruct unique *FLT3 ITD* sequences from two profiles of patients with AML in Ba/F3 cells by generating insertions and point mutations.[57] These two studies applied CRISPR-directed mutagenesis to reconstruct genetic profiles of known AML driver gene mutations that could be used for future studies, such as to evaluate responsiveness or resistance to established AML treatments in relation to these mutations.

When the *BIRC5* gene in human leukemic cell lines (KG-1 and HL-60) was successfully cleaved at specific sites, the resulting mutant had an enhanced apoptotic process.[58] This gene of interest is known to encode the surviving protein, which regulates cell cycle progression and inhibition of apoptosis. Furthermore, when CRISPR-Cas9-guided knockout of the V set pre-B cell surrogate light chain 1 (*VPREB1*) gene was evaluated in primary myeloma cells, a significant reduction in cellular proliferation was observed.[59]

Another approach by which CRISPR-Cas9 can exhibit targeted therapy is through gene editing, either by directly correcting point mutations within key genes or by reactivating silenced tumour suppressor genes. CRISPR-Cas9 technology was shown to restore wild function to the *ASXL-1* protein by base-correcting a point mutation in the *ASXL1* gene in KBM5 leukemic cell lines with the use of three custom-designed synthetic sgRNA.[60] More recently, Sayed et al. demonstrated that co-correction of *KRAS* and *TP53* mutations in tested cell lines was possible with varying levels of success using base editing.[61] Furthermore, they concluded that correction could lead to restored apoptosis as a result of phase arrest of the G0/G1 cell cycle. CRISPR-Cas systems are also capable of being engineered to activate/upregulate the expression of genes that promote differentiation and apoptosis in cancer cells. Furthermore, CRISPR technologies could be designed to target DNA methylation and histone modification to reverse epigenetic silencing of crucial genes. Generally, apoptosis is crucial for regulated cell death and therefore plays a vital role in targeted therapy.[62]

Targeting and correcting mutations within key genes that drive significant pathways such as signal transduction and immune checkpoint pathways could be therapeutically relevant. A gene with great potential is the Janus Kinase 2 (*JAK2*) gene, a key component of the JAK-STAT signalling pathway that is essential for modulating immune function and hemotopoiesis. Mutations in the *JAK2* gene are common in myeloproliferative neoplasms. This occurs when a G > T point mutation occurs in exon 14 of the *JAK2* gene.[63] CRISPR is able to normalise the pathway and reduce unregulated cell proliferation by disrupting mutations. When a Cas9-based technology was developed in human cell lines to induce and reverse the *JAK2* V617F mutation, the mutation offered cells a growth advantage in coculture with wild-type cells.

Knocking out three suppressive checkpoints (*CBLB, NKG2A, TIGIT*) in NK cells improved the cytotoxicity of NK-92 cell lines towards AML cell lines.[64] When approximately 80% of the NKG2A encoded in the killer cell lectin-like receptor C1 locus (*KLRC1*) of primary NK cells was removed through CRISPR-Cas9 editing, the modified NK cells exhibited increased cytotoxic effects against patient-derived MM cells.[65]

3.2 Ex vivo and *in vivo* applications for blood cancers

CRISPR editing in *ex vivo* and *in vivo* has been used for CRISPR-Cas-based disease therapeutics. One of the main applications of *ex vivo CRISPR* modification is cancer immunotherapy. T-lymphocytes of the haematopoietic system are ideal for *ex vivo* therapy, as they are easily isolated from the blood, expanded, and introduced into the body without inducing an immune response.[66] Recent research reported that the cytogenetic t(8;21)(q22;q22) abnormality in AML patients indicates unique clinical and biological characteristics. Successful chromosomal translocation was achieved by fusion of the *RUNX1* gene on chromosome 21 with the *ETO* gene on chromosome 8 in human mesenchymal cells and HSCs by electroporation.[67] Delivery of Cas9/IL1RAP sgRNA delivery was carried out through CXCL12α-LOADED-LNP/MSCM-NF scaffolds that reduced leukemic stem cells, potentially improving AML therapy.[68]

For *in vivo* therapy, the main targets are genetic diseases.[66] To simulate human diseases *in vivo*, animal models that closely relate to the physiology, anatomy, and evolution of humans such as rabbits and non-human primates should be used. However, due to cost and availability, most research studies use murine models. With the widespread popularity of CRISPR-Cas systems, there is an increasing need for a rich disease model bank. In clinical cancer research, these resource banks allow the essential *in vivo* evaluation of diseased cells in their microenvironment, which influences their physiological behaviour.[69]

A study that used this approach involved designing a combination of two sgRNAs directed at the ABL-Intron 1 and BCR-Intron 8 regions to induce a 133.9 kb deletion of the *BCR/ABL1* oncogene that causes a frame change throughout the DNA binding domain *ABL1*.[94] With any combination of sgRNA, edited K562 leukemic cells exhibited a notable decrease in cloning capacity (85%) and increased apoptosis. It should be noted that hCD34+ human hemopoietic cells were used to study whether any side effects were caused by CRISPR-Cas9. *In vitro* analysis of targeted CD34+ cells did not show differences in proliferation rates in long-term cultures, indicating the absence of additional cancer-associated genomic alterations. Furthermore, K562-transmitted cells with adenovirus vectors carrying all CRISPR reagents were subcutaneously introduced into immune-deprived mice, leading to an 88% reduction in tumour size relative to control tumours. Editing the *ABL* gene in K562 cells significantly inhibited carcinogenesis.[51] Mice xenografted with CRISPR-Cas-corrected KBM5 cells showed significantly longer survival compared to mice with uncorrected xenografts.[60] Correction of the mutated *ASLX1* gene in the CML cell line resulted in functional restoration. When *ABL* gene-edited K562 cells were introduced into a leukaemia mouse xenograft, the bioluminescence imaging of the mice showed a significant decline in the leukaemia cell population. Immunodeficient NSGS mice developed acute lymphoblastic leukaemia when CD34+ haematopoietic stem and progenitor cells harbouring CRISPR-Cas9 generated t(11;19)/MLL-ENL.[70] To date, only a few *in vivo* studies have been conducted in humans. The first clinical use of genomic editing was performed in an 11-month-old infant with end-stage leukaemia.[71]

These *ex vivo* and *in vivo* studies are evidence of the capacity CRISPR-Cas systems could exhibit at different stages of clinical cancer research. Despite its achievements, CRISPR-Cas9 delivery remains a significant obstacle in *in vivo* applications.

3.3 CRISPR-Cas and immunotherapy

Cancer immunotherapy combats tumour cell growth and spread by enhancing or revitalising the immune system.[72] This approach includes various treatments such as cytokine therapy, cancer vaccines, adoptive cellular immunotherapy (ACT), antibody-drug conjugates (ADC), oncolytic virus therapy (OVT), immune checkpoint blockade, and dendritic therapy.[73] T cells readily extracted from peripheral blood have the innate ability to migrate to tissues, where they initiate immune responses and exert cytotoxic effects. They can undergo genome editing using both viral and nonviral methods and are not easily transformed. Genetically engineered T cells, particularly CAR-T cells, exhibit considerable clinical effectiveness in achieving remissions in cases of ALL, B cell non-Hodgkin lymphoma (NHL), and MM.[74–76] CAR-T cells are T cells that have been genetically engineered to display fragments of antigen-binding antibodies that target specific tumour antigens. The immune response of CAR-T cells is enhanced by fusion of its fragments with costimulatory signalling domains and intracellular T-cell signalling domains.[77–79] In hematologic malignancies, some of the primary targets of CAR-T therapy are *CD33* and *CD19* expressed in normal and malignant cells in AML and in all B cell differentiation. In particular, anti-CD19 CAR has shown significant success in the treatment of liquid tumours. In 2017, the FDA approved the use of anti-CD19 CAR as a treatment for refractory B-cell leukaemia and lymphoma.[80]

An innovative approach involves the chimeric switch receptor (CSR) PD1 / CD28, which transforms the inhibitory signal from programmed cell death protein 1 (*PD-1*) into a stimulating signal using *CD28* in CAR-T cells.[81] When *PD-1* is expressed, its extracellular domain also competes with tumour cell surface receptors to bind to *PD-L1*.[82] This change can potentially reduce the incidence of CAR-T therapy cytokine release syndrome (CRS). Currently, no research has shown the effectiveness of knocking CSR into T cells; this highlights a promising avenue that needs to be further explored. However, it is worth noting that CRISPR application in CAR-T cell therapy is primarily in laboratory investigations and has not reached the stage for phase III clinical trials.

4. Clinical trials and therapeutic outcomes

Since its advent as a groundbreaking tool in genetic targeting and modification, CRISPR-*Cas* has advanced into clinical trials. In clinical research, CRISPR-edited CAR-T cell agents are leading the fight against blood cancers due to their efficacy in the management of haematological malignancies.[83] With the help of CRISPR technology, some potentially beneficial therapies are currently being developed. As of June 2024, the ClinicalTrial.gov database (https://www.clinicaltrials.gov/) listed more than 10 active clinical trials using CRISPR technologies to treat hematologic malignancies. Table 1 below highlights some key CRISPR-derived hematologic cancer therapies in different stages of clinical trials.

Although this technology is recognised for its potential to treat severe patients, rapid progress in these revolutionary therapies requires regulatory surveillance and adherence to regulatory standards to ensure the safety and effectiveness of CRISPR-based drugs.[84] The authors also listed the details of the directives, regulations and guidelines clarifying the regulatory framework for clinical studies to ensure that all clinical stages of the trials are conducted in a safe and effective manner.

Table 1 Ongoing clinical trials in the treatment of various hematologic malignancies using CRISPR-derived drugs.

Indication	Enrolment (estimated)	Clinical trial ID number; Phase; Sponsor; Status	Intervention description	Study type; design	Study Location
Relapsed/Refractory T-Cell Acute Lymphoblastic Leukaemia (T-ALL) or T-Cell Lymphoblastic Lymphoma (T-LL)	102 (currently recruiting)	Phase I/II, NCT05885464; BEAM Therapeutics, Recruiting	BEAM-201, an allogeneic anti-CD7 CAR-T with four knocked out genes using base editing.	Interventional; nonrandomised	United States
Relapsed or refractory B-cell malignancies, autoimmune disease	120 (currently recruiting)	Phase I/II, NCT05643742, CRISPR Therapeutics AG; Recruiting	CTX112, an allogeneic anti-CD19 CAR-T with CRISPR-Cas, was employed to improve efficacy.	Interventional; N/A	United States
Hematologic malignancies, solid tumour	250 (currently recruiting)	Phase I/II; NCT05795595; CRISPR Therapeutics AG; Recruiting	CTX131 is an allogeneic anti-CD70 CAR-T with multiple CRISPR-Cas9 modifications to improve cell potency.	Interventional; N/A	United States
Relapsed and refractory acute myeloid leukaemia (AML)	70 (currently recruiting)	Phase I; NCT06128044; Caribou Biosciences; Ongoing	CB-012 is an allogeneic CAR-T cell therapy.	Interventional; non-randomised	United States

With Relapsed or Refractory Multiple Myeloma	26 (Active, not recruiting)	Phase I; NCT04244656; CRISPR Therapeutics AG; Active	CTX120 are Anti-BCMA (B-cell maturation antigen) Allogeneic CRISPR-Cas9 modified T Cells	Interventional; N/A	Australia, Canada, Spain, United States
Non-Hodgkin Lymphoma; B-cell Lymphoma	227 (not recruiting)	Phase I/II; NCT04035434; CRISPR Therapeutics AG; Active (not recruiting)	CTX110, a CD19-directed CAR-T cell immunotherapy	Interventional; N/A	Australia, Canada, France, Germany, Spain, United States
T-Cell Lymphoma	45 (not recruiting)	Phase I; NCT04502446; CRISPR Therapeutics AG; Active (not recruiting)	CTX130 is an anti-CD70 Allogeneic CRISPR-Cas9-Engineered T Cell	Interventional; N/A	Australia, Canada, United States
T-Cell Lymphoma Stage IV; Stage IV Adult Hodgkin; Lymphoma Stage IV Diffuse Large B-Cell Lymphoma	20	Phase I/II; NCT03044743; Yang Yang; Unknown status	CRISPR-Cas9 mediated PD-1 knockout-T cells from autologous origin	Interventional; N/A	China

(continued)

Table 1 Ongoing clinical trials in the treatment of various hematologic malignancies using CRISPR-derived drugs. *(cont'd)*

Indication	Enrolment (estimated)	Clinical trial ID number; Phase; Sponsor; Status	Intervention description	Study type; design	Study Location
Relapsed/Refractory Acute Myeloid Leukaemia (AML)	12 (not yet recruiting)	Phase I; NCT05662904; German Cancer Research Center; Active (not yet recruiting)	Donor-derived CD34+ haematopoietic stem cells with CRISPR-Cas9-mediated CD33 deletion	Interventional; N/A	Germany
Relapsed or Refractory B Cell Leukaemia	80	Phase I/II; NCT03398967; Chinese PLA General Hospital; Status Unknown	CRISPR-Cas9 Gene-Editing CAR-T Cells Targeting CD19 and CD20 or CD22	Interventional; N/A	China
T-cell Acute Lymphoblastic Lymphoma; T-non-Hodgkin Lymphoma	21 (currently recruiting)	Phase I; NCT03690011; Baylor College of Medicine; Recruiting	Peripheral blood T lymphocytes (ATLs) genetically modified to express CAR-T targeting the CD7 molecule (CD7. CAR)	Interventional; N/A	United States

The U.S. clinical trials database (ClinicalTrials.gov) was accessed on June 7, 2024.
CAR-T, chimeric antigen receptor-T; *CRISPR-Cas*, clustered regularly interspaced clustered palindromic repeats – CRISPR-associated protein; *N/A*, not applicable; *PD1*, programmed cell death protein 1.

5. Overcoming challenges in CRISPR-Cas editing

Although CRISPR-mediated gene editing has shown notable advantages in precision medicine, several critical issues require urgent improvement. Minimising off-target effects, enhancing delivery method efficiency, and addressing regulatory challenges (regulatory considerations will be addressed in the subsequent section) challenges remain significant obstacles in the advancement of CRISPR technology. Although the cautious design of sgRNAs could reduce off-target effects, unexpected editing remains a major clinical concern.[85] A study investigating over seven thousand whole genome sequences evaluated how genetic differences influence the targeting accuracy of roughly 3000 gRNAs at 30 therapeutically significant loci.[85,86] Variations were found to affect both the specificity of the target and the nontarget, as single nucleotide polymorphism (SNP) and indels can change the target location and create new nontarget locations, potentially causing side effects and treatment failures. To improve the precision of the editing, high-precision Cas variants such as Sniper2L and SpCas9-HF1 were developed.[87–89] Additionally, computational tools are available to identify sgRNAs with enhanced on-target activity and reduced off-target activity, improving the precision of editing.[90] Despite these advances, editing inaccuracies due to effector domains like transcriptional regulators, reverse transcriptase, and deaminases may still occur.[91,92]

Within an intact immune system, long-term expression and sustained efficacy can be achieved through direct *in vivo* editing using a viral vector. However, this method is not without risks. Cas9 is present in certain bacteria, including Staphylococcus aureus and Streptococcus pyogenes, and prior exposure to these bacteria can prime T cells and produce anti-Cas9 antibodies.[93,94] When a patient receiving CRISPR-Cas9 therapy has preexisting memory T cells, these T cells can target treated cells, remove edited cells, and induce adverse tissue damage.[95] Due to their lower delivery efficiency, physical or synthetic methods of *ex vivo* editing are not scalable for clinical use. Viral vectors could trigger a harmful immune response superseding their delivery and packaging capacity. On the other hand, LNPs present a feasible method of delivering viruses, providing benefits like accessibility, affordability, and excellent compatibility.[96] Recent research has shown that LNPs are highly efficient in precisely targeting and encapsulating sgRNA and Cas9 mRNA for delivery to the mouse liver.[97] Furthermore, the intracellular introduction of Cas9/sgRNA

RNP complexes using LNPs such as nanofibrils has been reported to effectively treat leukaemia.[68] The identification of new RNA-directed endonucleases, such as obligate mobile element-guided activity (OMEGA), offers great promise in overcoming these constraints.[98] OMEGA is a viable candidate for *in vivo* delivery because it is smaller than Cas9 and is a probable member of the IS200/IS605 transposon family.[99]

Another important question to consider is the effectiveness of the enzyme reaction itself. The current efficiency of *in vivo* editing is very low and often insufficient for complete recovery.[100] Several issues, including immunosuppressive resistance and CRS associated with CAR-T therapy, restrict the application of immunotherapy primarily to the experimental stage. Orthogonal receptors and synthetic biology have been highlighted as innovative strategies to address challenges such as antigen escape and manipulation of the tumour microenvironment.[101] Additionally, analytical methods must be optimised to reduce inaccuracies, whether false positives or negatives, during extensive screening processes.[102] An analysis of computational models that eliminates these "probably relevant" genes was created called the Model Based Analysis of Genome-wide CRISPR-Cas9 Knockout (MAGeCK) (http://bitbucket.org/liulab/mageckvispr).[103] Microhomology-mediated end joining (MMEJ) and nonhomologous end joining (NHEJ) commonly result in insertions or deletions in genes, which can lead to functional knockouts.

However, these mechanisms can also produce detrimental consequences, such as cancer-causing chromosomal translocations, significant chromatin deletions, and the insertion of foreign vectors into human cells. Structural variations (SV) present an added risk to genomic stability when performing CRISPR editing.[104] Understanding the significance of SVs for human diseases has been made possible by the development of techniques such as SuperQ, HTGTS, and PEM-seq that can distinguish between different CRISPR-Cas9-induced DNA repair products.[105,106]

6. Ethical and regulatory considerations

CRISPR-*Cas* has attracted a great deal of interest since its inception in 2012 as a promising genome-editing approach. Surprisingly, only 11 years later, it received its first approval as a drug for the treatment of sickle cell disease and beta-thalassemia. This is due to the new regulatory challenges that arise as a result of the new ethical, moral, safety, and technical concerns associated with preclinical and clinical applications. Currently, there are numerous guidelines, directives, and regulations in place to ensure the safety and effectiveness of CRISPR-based medicines based on CRISPR technology.[84]

Despite the more than a decade history of use, limited knowledge on the possible prolonged outcomes of CRISPR-based therapies remains a great concern. A clinical trial (NCT05309733; sponsored by Vor Biopharma) is currently investigating the extended safety and effectiveness of the CRISPR-Cas9 edited HSCs and hematopoietic stem and progenitor cells (HSPCs) therapeutic product VOR33 from CRISPR-Cas9. The observational cohort is all AML patients who have received part of or all of the VOR33 therapeutic. To fully address this issue, more long-term observational studies are highly recommended.

Although one theoretical advantage of genome editing techniques is their potential to edit each gene, genome editing of germlines has caused significant ethical and safety issues such that genetic editing of reproductive cells remains forbidden in some countries, including China.[107] The ethical conversation was further heated by the announcement of the birth of the first babies whose genomes have been edited, Lulu and Nana, whose *CCR5* gene was modified to make them resistant to HIV infection.[108] To address these issues, some researchers have set out to conduct clinical trials to evaluate the safety of CRISPR-Cas systems. In patients with advanced metastatic liposarcoma and MM-resistant therapy, the safety and practicality of multiplexing CRISPR-Cas9 gene editing in T cells were tested.[109] To reduce the lack of T-cell receptors (TCRs) and increase the expression of genetically modified cancer-specific TCR transgens (NY-ESO-1), the genes of the TCR chain encoding TCR (TRAB) and TCR (TRAC) were eliminated. This trial was conducted in conjunction with another phase I clinical trial (ClinicalTrials. gov NCT02793856) of modified CRISPR-Cas9 *PD1* T cells in non-small cell lung cancer to address, to some extent, CRISPR-Cas9 safety risks.[110] Although clinical trials such as these have been performed, the use of CRISPR-Cas technologies remains a restricted therapeutic application due to intrinsic technical and ethical constraints. In addition to these, issues associated with *in vivo* performance and specificity need to be resolved to broaden the use of CRISPR-Cas systems.[111]

7. Future perspectives and conclusion

CRISPR-Cas9 and the Cas family of proteins have been the primary drivers of genome editing in the past decade. Considering that every living organism possesses genetic information, we can project the future of this technology as limitless. Research on hematologic malignancies is about to make significant progress with the potential application of CRISPR-Cas9, opening new avenues for treatment discoveries. To achieve this, the exploration of new RNA-guided endonucleases such as OMEGA for specific applications in blood cancers is crucial. Improving delivery methods for both laboratory-based and *in vivo* applications is necessary to ensure higher efficiency and specificity. Safety and off-target issues can be addressed by introducing novel nonviral delivery methods and developing more precise and efficient CRISPR variants. Integrating immunotherapies with CRISPR-Cas technology could improve targeting and efficacy against blood cancers. To ensure successful clinical translation, it is important to develop documentation and monitoring systems to capture long-term outcomes of patients treated with CRISPR, assessing their durability and possible side effects. The development of global regulatory and ethical guidelines and protocols is essential to address ethical concerns related to gene editing in humans. In addition, collaboration among all relevant stakeholders, including regulatory bodies, clinicians, and researchers, is necessary to guide the approval process of CRISPR-based therapies. Additionally, investing in CRISPR-based research to explore new therapeutics and mechanisms in the biology of blood cancer is crucial to the success of this technology. To ensure the overall success of CRISPR-based treatments in the future, training programmes for health-care professionals and raising awareness and understanding among patients and the general public are essential.

In conclusion, advances in RNA-guided endonucleases, delivery methods, and integration with immunotherapies such as CAR-T cells are crucial to improving efficiency and specificity. Addressing safety and off-target concerns, developing robust monitoring systems, and establishing global regulatory and ethical frameworks are essential for successful clinical translation. Continued research, stakeholder collaboration, and public awareness will be vital in harnessing the full potential of CRISPR to revolutionise blood cancer therapy.

Acknowledgements

BM and PD are funded by NRF Postdoctoral development program. KVL is funded by OWSD PhD fellowship. DBTGR was funded by NRF Competitive Grant. MMT is funded by NRF Thuthuka grant. Authors acknowledge support from DSI emerging research area funding for synthetic biology program and CSIR Strategic funding for synthetic biology and precision medicine centre.

CRediT authorship contribution statement

BM: Writing, review & editing of article draft. **PD:** Writing, review & editing of article draft. **DBTGR:** Review & editing of article draft. **MMT:** Review & editing of article draft. **KVL:** Review & editing of final draft.

References

1. Kontomanolis EN, Koutras A, Syllaios A, et al. Role of oncogenes and tumor-suppressor genes in carcinogenesis: a review. *Anticancer research*. 2020;40(11):6009–6015.
2. Ochi Y. Genetic landscape of chronic myeloid leukemia. *International Journal of Hematology*. 2023;117(1):30–36.
3. Jabbour E, Kantarjian H. Chronic myeloid leukemia: 2018 update on diagnosis, therapy and monitoring. *American journal of hematology*. 2018;93(3):442–459.
4. Palomo L, Acha P, Solé F. Genetic aspects of myelodysplastic/myeloproliferative neoplasms. *Cancers*. 2021;13(9):2120.
5. Zhang A, Wang S, Ren Q, Wang Y, Jiang Z. Prognostic value of ASXL1 mutations in patients with myelodysplastic syndromes and acute myeloid leukemia: a meta-analysis. *Asia-Pacific Journal of Clinical Oncology*. 2023;19(5):e183–e194.
6. Ganguly BB, Kadam NN. Mutations of myelodysplastic syndromes (MDS): An update. *Mutation Research/Reviews in Mutation Research*. 2016;769:47–62.
7. Palumbo GA, Stella S, Pennisi MS, et al. The role of new technologies in myeloproliferative neoplasms. *Frontiers in oncology*. 2019;9:321.
8. Kakosaiou K, Panitsas F, Daraki A, et al. ASXL1 mutations in AML are associated with specific clinical and cytogenetic characteristics. *Leukemia & lymphoma*. 2018;59(10):2439–2446.
9. Fan Y, Liao L, Liu Y, et al. Risk factors affect accurate prognosis in ASXL1-mutated acute myeloid leukemia. *Cancer Cell International*. 2021;21:1–9.
10. Yang FC, Agosto-Peña J. Epigenetic regulation by ASXL1 in myeloid malignancies. *International journal of hematology*. 2023;117(6):791–806.
11. Padmakumar D, Chandraprabha VR, Gopinath P, et al. A concise review on the molecular genetics of acute myeloid leukemia. *Leukemia Research*. 2021;111:106727.
12. Kelly LM, Gilliland DG. Genetics of myeloid leukemias. *Annual review of genomics and human genetics*. 2002;3(1):179–198.
13. Lagunas-Rangel FA, Chávez-Valencia V, Gómez-Guijosa MÁ, Cortes-Penagos C. Acute myeloid leukemia—genetic alterations and their clinical prognosis. *International journal of hematology-oncology and stem cell research*. 2017;11(4):328.
14. Renneville A, Roumier C, Biggio V, et al. Cooperating gene mutations in acute myeloid leukemia: a review of the literature. *leukemia*. 2008;22(5):915–931.
15. Cerchione C, Romano A, Daver N, et al. IDH1/IDH2 inhibition in acute myeloid leukemia. *Frontiers in oncology*. 2021;11:639387.
16. Wang RQ, Chen CJ, Jing Y, et al. Characteristics and prognostic significance of genetic mutations in acute myeloid leukemia based on a targeted next-generation sequencing technique. *Cancer medicine*. 2020;9(22):8457–8467.
17. Chiorazzi N, Chen SS, Rai KR. Chronic lymphocytic leukemia. *Cold Spring Harbor perspectives in medicine*. 2021;11(2):a035220.
18. Stilgenbauer S, Schnaiter A, Paschka P, et al. Gene mutations and treatment outcome in chronic lymphocytic leukemia: results from the CLL8 trial. *Blood, The Journal of the American Society of Hematology*. 2014;123(21):3247–3254.
19. Amin NA, Seymour E, Saiya-Cork K, Parkin B, Shedden K, Malek SN. A quantitative analysis of subclonal and clonal gene mutations before and after therapy in chronic lymphocytic leukemia. *Clinical Cancer Research*. 2016;22(17):4525–4535.

20. Zenz T, Mertens D, Döhner H, Stilgenbauer S. Importance of genetics in chronic lymphocytic leukemia. *Blood reviews*. 2011;25(3):131–137.
21. Rosenquist R, Rosenwald A, Du MQ, et al. Clinical impact of recurrently mutated genes on lymphoma diagnostics: state-of-the-art and beyond. *haematologica*. 2016;101(9):1002.
22. Hu Y, Chen W, Wang J. Progress in the identification of gene mutations involved in multiple myeloma. *OncoTargets and therapy*. 2019:4075–4080.
23. de Leval, Jaffe ES. Lymphoma classification. *The Cancer Journal*. 2020;26(3):176–185.
24. Li S, Young KH, Medeiros LJ. Diffuse large B-cell lymphoma. *Pathology*. 2018;50(1):74–87.
25. Morin RD, Mendez-Lago M, Mungall AJ, et al. Frequent mutation of histone-modifying genes in non-Hodgkin lymphoma. *Nature*. 2011;476(7360):298–303.
26. Lenz G, Wright GW, Emre NT, et al. Molecular subtypes of diffuse large B-cell lymphoma arise by distinct genetic pathways. *Proceedings of the National Academy of Sciences*. 2008;105(36):13520–13525.
27. Weniger MA, Küppers R. Molecular biology of Hodgkin lymphoma. *Leukemia*. 2021;35(4):968–981.
28. Küppers R, Engert A, Hansmann ML. Hodgkin lymphoma. *The Journal of clinical investigation*. 2012;122(10):3439–3447.
29. Robiou du Pont S, Cleynen A, Fontan C, et al. Genomics of multiple myeloma. *Journal of Clinical Oncology*. 2017;35(9):963–967.
30. Corre J, Munshi N, Avet-Loiseau H. Genetics of multiple myeloma: another heterogeneity level? *Blood, The Journal of the American Society of Hematology*. 2015;125(12):1870–1876.
31. Hoang PH, Cornish AJ, Dobbins SE, Kaiser M, Houlston RS. Mutational processes contributing to the development of multiple myeloma. *Blood cancer journal*. 2019;9(8):60.
32. Higgins MJ, Fonseca R. Genetics of multiple myeloma. *Best Practice & Research Clinical Haematology*. 2005;18(4):525–536.
33. Urnov FD, Rebar EJ, Holmes MC, Zhang HS, Gregory PD. Genome editing with engineered zinc finger nucleases. *Nature Reviews Genetics*. 2010;11(9):636–646.
34. Silva G, Poirot L, Galetto R, et al. Meganucleases and other tools for targeted genome engineering: perspectives and challenges for gene therapy. *Current gene therapy*. 2011;11(1):11–27.
35. Joung JK, Sander JD. TALENs: a widely applicable technology for targeted genome editing. *Nature reviews Molecular cell biology*. 2013;14(1):49–55.
36. Doudna JA, Charpentier E. The new frontier of genome engineering with CRISPR-Cas9. *Science*. 2014;346(6213):1258096.
37. Hong A. CRISPR in personalized medicine: Industry perspectives in gene editing. *Semin Perinatol*. 2018;42(8):501–507.
38. Hu JH, Miller SM, Geurts MH, et al. Evolved Cas9 variants with broad PAM compatibility and high DNA specificity. *Nature*. 2018;556(7699):57–63.
39. Das S, Bano S, Kapse P, Kundu GC. CRISPR based therapeutics: a new paradigm in cancer precision medicine. *Molecular Cancer*. 2022;21(1):85.
40. Liu Y, Cao Z, Wang Y, et al. Genome-wide screening for functional long noncoding RNAs in human cells by Cas9 targeting of splice sites. *Nature Biotechnology*. 2018;36(12):1203–1210.
41. Miles LA, Garippa RJ, Poirier JT. Design, execution, and analysis of pooled in vitro CRISPR/Cas9 screens. *The FEBS journal*. 2016;283(17):3170–3180.
42. Chavez A, Scheiman J, Vora S, Pruitt BW, Tuttle M, PR Iyer E, Lin S, Kiani S, Guzman CD, Wiegand, DJ, Ter-Ovanesyan D. Highly efficient Cas9-mediated transcriptional programming. *Nature methods*, 2015;12(4), pp. 326-328.

43. Zalatan JG, Lee ME, Almeida R, et al. Engineering complex synthetic transcriptional programs with CRISPR RNA scaffolds. *Cell*. 2015;160(1):339–350.
44. Brown A, Winter J, Gapinske M, Tague N, Woods WS, Perez-Pinera P. Multiplexed and tunable transcriptional activation by promoter insertion using nuclease-assisted vector integration. *Nucleic acids research*. 2019;47(12):e67.
45. Gilbert LA, Larson MH, Morsut L, et al. CRISPR-mediated modular RNA-guided regulation of transcription in eukaryotes. *Cell*. 2013;154(2):442–451.
46. Tak YE, Kleinstiver BP, Nuñez JK, et al. Inducible and multiplex gene regulation using CRISPR–Cpf1-based transcription factors. *Nature methods*. 2017;14(12):1163–1166.
47. Kocak DD, Josephs EA, Bhandarkar V, Adkar SS, Kwon JB, Gersbach CA. Increasing the specificity of CRISPR systems with engineered RNA secondary structures. *Nature biotechnology*. 2019;37(6):657–666.
48. Campa CC, Weisbach NR, Santinha AJ, Incarnato D, Platt RJ. Multiplexed genome engineering by Cas12a and CRISPR arrays encoded on single transcripts. *Nature Methods*. 2019;16(9):887–893.
49. McCarty NS, Graham AE, Studená L, Ledesma-Amaro R. Multiplexed CRISPR technologies for gene editing and transcriptional regulation. *Nature communications*. 2020;11(1):1281.
50. Vuelta E, Ordoñez JL, Alonso-Pérez V, et al. CRISPR/Cas9 technology abolishes the BCR/ABL1 oncogene in chronic myeloid leukemia and restores normal hematopoiesis. *bioRxiv*. 2020 2020-08.
51. Chen SH, Hsieh YY, Tzeng HE, et al. ABL genomic editing sufficiently abolishes oncogenesis of human chronic myeloid leukemia cells in vitro and in vivo. *Cancers*. 2020;12(6):1399.
52. Martinez-Lage M, Torres-Ruiz R, Puig-Serra P, et al. In vivo CRISPR/Cas9 targeting of fusion oncogenes for selective elimination of cancer cells. *Nature communications*. 2020;11(1):5060.
53. García-Tuñón I, Hernández-Sánchez M, Ordoñez JL, et al. The CRISPR/Cas9 system efficiently reverts the tumorigenic ability of BCR/ABL in vitro and in a xenograft model of chronic myeloid leukemia. *Oncotarget*. 2017;8(16):26027.
54. Grassian AR, Parker SJ, Davidson SM, et al. IDH1 mutations alter citric acid cycle metabolism and increase dependence on oxidative mitochondrial metabolism. *Cancer research*. 2014;74(12):3317–3331.
55. Figueroa ME, Abdel-Wahab O, Lu C, et al. Leukemic IDH1 and IDH2 mutations result in a hypermethylation phenotype, disrupt TET2 function, and impair hematopoietic differentiation. *Cancer cell*. 2010;18(6):553–567.
56. Brabetz O, Alla V, Angenendt L, et al. RNA-guided CRISPR-Cas9 system-mediated engineering of acute myeloid leukemia mutations. *Molecular Therapy-Nucleic Acids*. 2017;6:243–248.
57. Rivera-Torres N, Banas K, Kmiec EB. Modeling pediatric AML FLT3 mutations using CRISPR/Cas12a-mediated gene editing. *Leukemia & lymphoma*. 2020;61(13):3078–3088.
58. Narimani M, Sharifi M, Hakhamaneshi MS, et al. BIRC5 gene disruption via CRISPR/Cas9n platform suppress acute myelocytic leukemia progression. *Iranian biomedical journal*. 2019;23(6):369.
59. Khaled M, Moustafa AS, El-Khazragy N, et al. CRISPR/Cas9 mediated knockout of VPREB1 gene induces a cytotoxic effect in myeloma cells. *PloS one*. 2021;16(1):e0245349.
60. Valletta S, Dolatshad H, Bartenstein M, et al. ASXL1 mutation correction by CRISPR/Cas9 restores gene function in leukemia cells and increases survival in mouse xenografts. *Oncotarget*. 2015;6(42):44061.
61. Sayed S, Sidorova OA, Hennig A, et al. Efficient correction of oncogenic KRAS and TP53 mutations through CRISPR base editing. *Cancer research*. 2022;82(17):3002–3015.

62. Peng F, Liao M, Qin R, et al. Regulated cell death (RCD) in cancer: key pathways and targeted therapies. *Signal transduction and targeted. therapy*. 2022;7(1):286.
63. Baik R, Wyman SK, Kabir S, Corn JE. Genome editing to model and reverse a prevalent mutation associated with myeloproliferative neoplasms. *Plos one*. 2021;16(3):e0247858.
64. Ureña-Bailén G, Dobrowolski JM, Hou Y, et al. Preclinical evaluation of CRISPR-edited CAR-NK-92 cells for off-the-shelf treatment of AML and B-ALL. *International Journal of Molecular Sciences*. 2022;23(21):12828.
65. Bexte T, Alzubi J, Reindl LM, et al. CRISPR-Cas9 based gene editing of the immune checkpoint NKG2A enhances NK cell mediated cytotoxicity against multiple myeloma. *Oncoimmunology*. 2022;11(1):2081415.
66. Song M. The CRISPR/Cas9 system: Their delivery, in vivo and ex vivo applications and clinical development by startups. *Biotechnology progress*. 2017;33(4):1035–1045.
67. Torres RAUL, Martin MC, Garcia A, Cigudosa JC, Ramirez JC, Rodriguez-Perales S. Engineering human tumour-associated chromosomal translocations with the RNA-guided CRISPR–Cas9 system. *Nature communications*. 2014;5(1):3964.
68. Ho TC, Kim HS, Chen Y, et al. Scaffold-mediated CRISPR-Cas9 delivery system for acute myeloid leukemia therapy. *Science Advances*. 2021;7(21):eabg3217.
69. Xu Y, Li Z. CRISPR-Cas systems: Overview, innovations and applications in human disease research and gene therapy. *Computational and structural biotechnology journal*. 2020;18:2401–2415.
70. Reimer J, Knöß S, Labuhn M, et al. CRISPR-Cas9-induced t (11; 19)/MLL-ENL translocations initiate leukemia in human hematopoietic progenitor cells in vivo. *Haematologica*. 2017;102(9):1558.
71. Couzin-Frankel J. Baby's leukemia recedes after novel cell therapy. *Science*. 2015;350(6262):731.
72. Yang Y. Cancer immunotherapy: harnessing the immune system to battle cancer. *The Journal of clinical investigation*. 2015;125(9):3335–3337.
73. Turtle CJ, Hanafi LA, Berger C, et al. CD19 CAR–T cells of defined CD4+: CD8+ composition in adult B cell ALL patients. *The Journal of clinical investigation*. 2016;126(6):2123–2138.
74. Raje N, Berdeja J, Lin YI, et al. Anti-BCMA CAR T-cell therapy bb2121 in relapsed or refractory multiple myeloma. *New England Journal of Medicine*. 2019;380(18):1726–1737.
75. Nastoupil LJ, Jain MD, Feng L, et al. Standard-of-care axicabtagene ciloleucel for relapsed or refractory large B-cell lymphoma: results from the US Lymphoma CAR T Consortium. *Journal of Clinical Oncology*. 2020;38(27):3119–3128.
76. Brentjens RJ, Rivière I, Park JH, et al. Safety and persistence of adoptively transferred autologous CD19-targeted T cells in patients with relapsed or chemotherapy refractory B-cell leukemias. *Blood, The Journal of the American Society of Hematology*. 2011;118(18):4817–4828.
77. Milone MC, Fish JD, Carpenito C, et al. Chimeric receptors containing CD137 signal transduction domains mediate enhanced survival of T cells and increased antileukemic efficacy in vivo. *Molecular therapy*. 2009;17(8):1453–1464.
78. Brentjens RJ, Santos E, Nikhamin Y, et al. Genetically targeted T cells eradicate systemic acute lymphoblastic leukemia xenografts. *Clinical cancer research*. 2007;13(18):5426–5435.
79. Raje NS, Shah N, Jagannath S, et al. Updated clinical and correlative results from the phase I CRB-402 study of the BCMA-targeted CAR T cell therapy bb21217 in patients with relapsed and refractory multiple myeloma. *Blood*. 2021;138:548.
80. Neelapu SS, Locke FL, Bartlett NL, et al. Axicabtagene ciloleucel CAR T-cell therapy in refractory large B-cell lymphoma. *New England Journal of Medicine*. 2017;377(26):2531–2544.

81. Prosser ME, Brown CE, Shami AF, Forman SJ, Jensen MC. Tumor PD-L1 co-stimulates primary human CD8+ cytotoxic T cells modified to express a PD1: CD28 chimeric receptor. *Molecular immunology*. 2012;51(3–4):263–272.
82. Cherkassky L, Morello A, Villena-Vargas J, et al. Human CAR T cells with cell-intrinsic PD-1 checkpoint blockade resist tumor-mediated inhibition. *The Journal of clinical investigation*. 2016;126(8):3130–3144.
83. Rafii S, Tashkandi E, Bukhari N, Al-Shamsi HO. Current status of CRISPR/Cas9 application in clinical cancer research: opportunities and challenges. *Cancers*. 2022;14(4):947.
84. Anliker B, Childs L, Rau J, et al. Regulatory considerations for clinical trial applications with CRISPR-based medicinal products. *The CRISPR Journal*. 2022;5(3):364–376.
85. Lessard S, Francioli L, Alfoldi J, et al. Human genetic variation alters CRISPR-Cas9 on-and off-targeting specificity at therapeutically implicated loci. *Proceedings of the National Academy of Sciences*. 2017;114(52):E11257–E11266.
86. Scott DA, Zhang F. Implications of human genetic variation in CRISPR-based therapeutic genome editing. *Nature medicine*. 2017;23(9):1095–1101.
87. Huang TP, Heins ZJ, Miller SM, et al. High-throughput continuous evolution of compact Cas9 variants targeting single-nucleotide-pyrimidine PAMs. *Nature biotechnology*. 2023;41(1):96–107.
88. Kim YH, Kim N, Okafor I, et al. Sniper2L is a high-fidelity Cas9 variant with high activity. *Nature chemical biology*. 2023;19(8):972–980.
89. Kleinstiver BP, Pattanayak V, Prew MS, et al. High-fidelity CRISPR–Cas9 nucleases with no detectable genome-wide off-target effects. *Nature*. 2016;529(7587):490–495.
90. Liu G, Zhang Y, Zhang T. Computational approaches for effective CRISPR guide RNA design and evaluation. *Computational and structural biotechnology journal*. 2020;18:35–44.
91. Grünewald J, Zhou R, Garcia SP, et al. Transcriptome-wide off-target RNA editing induced by CRISPR-guided DNA base editors. *Nature*. 2019;569(7756):433–437.
92. Zhou C, Sun Y, Yan R, et al. Off-target RNA mutation induced by DNA base editing and its elimination by mutagenesis. *Nature*. 2019;571(7764):275–278.
93. Wagner DL, Amini L, Wendering DJ, et al. High prevalence of Streptococcus pyogenes Cas9-reactive T cells within the adult human population. *Nature medicine*. 2019;25(2):242–248.
94. Simhadri VL, McGill J, McMahon S, Wang J, Jiang H, Sauna ZE. Prevalence of pre-existing antibodies to CRISPR-associated nuclease Cas9 in the USA population. *Molecular therapy. Methods & clinical development*. 2018;10:105–112.
95. Crudele JM, Chamberlain JS. Cas9 immunity creates challenges for CRISPR gene editing therapies. *Nature communications*. 2018;9(1):3497.
96. Mehnert W, Mäder K. Solid lipid nanoparticles: production, characterization and applications. *Advanced drug delivery reviews*. 2012;64:83–101.
97. Miller JB, Zhang S, Kos P, et al. Non-viral CRISPR/Cas gene editing in vitro and in vivo enabled by synthetic nanoparticle co-delivery of Cas9 mRNA and sgRNA. *Angewandte Chemie*. 2017;129(4):1079–1083.
98. Hirano S, Kappel K, Altae-Tran H, et al. Structure of the OMEGA nickase IsrB in complex with ωRNA and target DNA. *Nature*. 2022;610(7932):575–581.
99. Altae-Tran H, Kannan S, Demircioglu FE, et al. The widespread IS200/IS605 transposon family encodes diverse programmable RNA-guided endonucleases. *Science*. 2021;374(6563):57–65.
100. Razzouk S. CRISPR-Cas9: A cornerstone for the evolution of precision medicine. *Annals of Human Genetics*. 2018;82(6):331–357.

101. Young RM, Engel NW, Uslu U, Wellhausen N, June CH. Next-generation CAR T-cell therapies. *Cancer discovery*. 2022;12(7):1625–1633.
102. Sanson KR, Hanna RE, Hegde M, et al. Optimized libraries for CRISPR-Cas9 genetic screens with multiple modalities. *Nature communications*. 2018;9(1):5416.
103. Li W, Xu H, Xiao T, et al. MAGeCK enables robust identification of essential genes from genome-scale CRISPR/Cas9 knockout screens. *Genome biology*. 2014;15:1–12.
104. Zhao B, Rothenberg E, Ramsden DA, Lieber MR. The molecular basis and disease relevance of non-homologous DNA end joining. *Nature Reviews Molecular Cell Biology*. 2020;21(12):765–781.
105. Liu M, Zhang W, Xin C, et al. Global detection of DNA repair outcomes induced by CRISPR–Cas9. *Nucleic acids research*. 2021;49(15):8732–8742.
106. Yin J, Liu M, Liu Y, Hu J. Improved HTGTS for CRISPR/Cas9 off-target detection. *Bio-protocol*. 2019;9(9):e3229.
107. Gabel I, Moreno J. Genome editing, ethics, and politics. *AMA journal of ethics*. 2019;21(12):E1105.
108. Ormond KE, Bombard Y, Bonham VL, et al. The clinical application of gene editing: ethical and social issues. *Personalized medicine*. 2019;16(4):337–350.
109. Stadtmauer EA, Fraietta JA, Davis MM, et al. CRISPR-engineered T cells in patients with refractory cancer. *Science*. 2020;367(6481):eaba7365.
110. Lu Y, Xue J, Deng T, et al. Safety and feasibility of CRISPR-edited T cells in patients with refractory non-small-cell lung cancer. *Nature medicine*. 2020;26(5):732–740.
111. Albitar A, Rohani B, Will B, Yan A, Gallicano GI. The application of CRISPR/Cas technology to efficiently model complex cancer genomes in stem cells. *Journal of Cellular Biochemistry*. 2018;119(1):134–140.

Index

Note: Page numbers followed by "*f*" indicate figures and "*t*" indicate tables.